Strategic Applications of Named Reactions in Organic Chemistry

Strategic Applications of Named Reactions in Organic Chemistry

Contributors

Mathieu Wagner, Yohan Contie et al.

AURIS
Reference

www.aurisreference.com

Strategic Applications of Named Reactions in Organic Chemistry

Contributors: Mathieu Wagner, Yohan Contie et al.

Published by Auris Reference Limited

www.aurisreference.com

United Kingdom

Strategic Applications of Named Reactions in Organic Chemistry

ISBN: 978-1-78154-884-4

British Library Cataloguing in Publication Data
A CIP record for this book is available from the British Library

Printed in the United Kingdom

Exclusively distributed by CBS Publishers & Distributors Pvt. Ltd.

Sales & Distribution Rights only for India, Pakistan, Bangladesh, Sri Lanka, Nepal and Bhutan. This book is not to be sold outside these territories.

Contents

List of Abbreviations .... *vii*

List of Contributors... *ix*

Preface.. ....*xiii*

Chapter 1 **Enantioselective Aldol Reactions and Michael Additions Using Proline Derivatives as Organocatalysts** 1

Mathieu Wagner, Yohan Contie, Clotilde Ferroud, Gilbert Revial

Chapter 2 **The Piancatelli Rearrangement: New Applications for an Intriguing Reaction** 21

Claudia Piutti, and Francesca Quartieri

Chapter 3 **Oxetane Synthesis through the Paternò-Büchi Reaction** 49

Maurizio D'Auria, and Rocco Racioppi

Chapter 4 **Mannich-Type Reactions of Aldimines and Hetero Diels-Alder Reactions of Aldehydes Catalyzed by Anion-Type Lewis Bases Derived from a Single Molecule**... 109

Kaori Ishimaru, Daiki Maeda, Kaori Ono, Yuya Tanimura

Chapter 5 **Italian Chemists' Contributions to Named Reactions in Organic Synthesis: An Historical Perspective**... 123

Gianluca Papeo and Maurizio Pulici

Chapter 6 **Synthesis and Reactions of Acenaphthenequinones-Part-2. The Reactions of Acenaphthenequinones** 163

El Sayed H. El Ashry, Hamida Abdel Hamid, Ahmed A. Kassem and Mahmoud Shoukry

Chapter 7 **Organocatalysis: Key Trends in Green Synthetic Chemistry, Challenges, Scope towards Heterogenization, and Importance from Research and Industrial Point of View** 201

Isak Rajjak Shaikh

Chapter 8 **Environmentally Benign Mortar- Pestle-Induced Acylation and O-Alkylation of Aromatic and Heteroaromatic Compounds under Solvent-Free Micellar Conditions and Computation of Their Drug Likeliness Properties** 279

Kancharla Rajendar Reddy, Kamatala Chinna Rajanna, Kusampally Uppalaiah, Mukka Satish Kumar, and Marri Venkateshwarlu

Citations **299**

Index.. **301**

List of Abbreviations

CTAB	Cetyl trimethyl ammonium bromide
CTAC	Cetyl trimethyl ammonium chloride
DCE	Dichloroethane
DET	Diethyl tartrate
D-A	Donor-acceptor
FCC	Fluidized catalytic cracking
IDCP	Iodonium di-sym-collidine perchlorate
LDH	Layered double hydroxides
NHC	N-heterocyclic carbene
NMR	Nuclear magnetic resonance
OAPEC	Organization of the Arab Petroleum Exporting Countries
PTC	Phase transfer catalysis
PMA	Phosphomolybdic acid
PSA	Polar surface area
SEM	Scanning electron microscopy
SDS	Sodium dodecyl sulphate
SOC	Spin-orbit coupling
TEM	Transmission electron microscopy
XRD	X-ray diffraction

List of Contributors

Mathieu Wagner
Conservatoire National des Arts et Métiers, Laboratoire de Transformations Chimiques et Pharmaceutiques, Paris, France

Yohan Contie
Conservatoire National des Arts et Métiers, Laboratoire de Transformations Chimiques et Pharmaceutiques, Paris, France

Clotilde Ferroud
Conservatoire National des Arts et Métiers, Laboratoire de Transformations Chimiques et Pharmaceutiques, Paris, France

Gilbert Revial
Conservatoire National des Arts et Métiers, Laboratoire de Transformations Chimiques et Pharmaceutiques, Paris, France

Claudia Piutti
Department of Medicinal Chemistry, Nerviano Medical Sciences srl, Business Unit Oncology, Viale Pasteur 10, 20014 Nerviano, MI, Italy

Francesca Quartieri
Department of Chemical Core Technologies, Nerviano Medical Sciences srl, Business Unit Oncology, Viale Pasteur 10, 20014 Nerviano, MI, Italy

Maurizio D'Auria
Dipatimento di Scienze, Università della Basilicata, Viale dell'Ateneo Lucano 19, 85100 Potenza, Italy

Rocco Racioppi
Dipatimento di Scienze, Università della Basilicata, Viale dell'Ateneo Lucano 19, 85100 Potenza, Italy

Kaori Ishimaru
Department of Chemistry, National Defense Academy, Yokosuka, Japan

Daiki Maeda
Department of Chemistry, National Defense Academy, Yokosuka, Japan

Kaori Ono
Department of Chemistry, National Defense Academy, Yokosuka, Japan

Yuya Tanimura
Department of Chemistry, National Defense Academy, Yokosuka, Japan

Gianluca Papeo
Department of Medicinal Chemistry, Nerviano Medical Sciences srl, Business Unit Oncology, Viale Pasteur 10, Nerviano 20014, MI, Italy

Maurizio Pulici
Department of Chemical Core Technologies, Nerviano Medical Sciences srl, Business Unit Oncology, Viale Pasteur 10, Nerviano 20014, MI, Italy

El Sayed H. El Ashry
Department of Chemistry, Faculty of Science, Alexandria University, Alexandria, Egypt

Hamida Abdel Hamid
Department of Chemistry, Faculty of Science, Alexandria University, Alexandria, Egypt

Ahmed A. Kassem
Department of Chemistry, Faculty of Science, Alexandria University, Alexandria, Egypt

Mahmoud Shoukry
Department of Chemistry, Faculty of Science, Alexandria University, Alexandria, Egypt

Isak Rajjak Shaikh
Department of Chemistry, Shri Jagdishprasad Jhabarmal Tibrewala University, Vidyanagari, Jhunjhunu-Churu Road, Chudela, Jhunjhunu District, Rajasthan 333001, India
Razak Institution of Skills, Education and Research (RISER), Shrinagar, Near Rafaiya Masjid and Hanuman Mandir, Nanded, Maharashtra State 431 605, India
Post Graduate and Research Centre, Department of Chemistry, Poona College of Arts, Science and Commerce, Camp Area, Pune 411 001, Maharashtra State, India

Kancharla Rajendar Reddy
Department of Chemistry, Osmania University, Hyderabad 500 007, India

Kamatala Chinna Rajanna
Department of Chemistry, Osmania University, Hyderabad 500 007, India

Kusampally Uppalaiah
Department of Chemistry, Osmania University, Hyderabad 500 007, India

Mukka Satish Kumar
Department of Chemistry, Osmania University, Hyderabad 500 007, India

Marri Venkateshwarlu
Department of Chemistry, Osmania University, Hyderabad 500 007, India

Preface

Organic chemistry is a chemistry sub discipline involving the scientific study of the structure, properties, and reactions of organic compounds and organic materials, i.e., matter in its various forms that contain carbon atoms. The text *Strategic Applications of Named Reactions in Organic Chemistry* deals with organic reactions and their strategic use in the synthesis of complex natural and unnatural products. First chapter focuses on enantioselective aldol reactions and Michael additions using proline derivatives as organocatalysts. Second chapter deals with the general aspects of the reaction along with its more recent applications. The aim of third chapter is to furnish an overview of the latest results obtained by using the Paternò-Büchi reaction. Mannich-type reactions of aldimines and hetero diels-alder reactions of aldehydes catalyzed by anion-type Lewis bases derived from a single molecule have been presented in fourth chapter. In fifth chapter, a brief survey of known (and some lesser known) named organic reactions discovered by Italian chemists, along with their historical contextualization, is presented. The reactions of acenaphthenequinone and its derivatives with different nucleophiles, organic and inorganic reagents are reviewed in sixth chapter. Seventh chapter purports to review catalysis, particularly the organocatalysis and its origin, key trends, challenges, examples, scope, and importance. Last chapter aims at developing greener synthetic protocols for etherification and acylation reactions under solvent-free conditions.

Chapter 1

ENANTIOSELECTIVE ALDOL REACTIONS AND MICHAEL ADDITIONS USING PROLINE DERIVATIVES AS ORGANOCATALYSTS

Mathieu Wagner, Yohan Contie, Clotilde Ferroud, Gilbert Revial

Conservatoire National des Arts et Métiers, Laboratoire de Transformations Chimiques et Pharmaceutiques, Paris, France

ABSTRACT

Six compounds including five proline derivatives have been prepared and tested as chiral organocatalysts for enantioselective aldol reactions and Michael additions. The enantiomeric excesses, which are highly dependent on the molecular structure of catalysts as well as experimental conditions, have reached over 98%.

INTRODUCTION

The aldol reaction and the Michael addition are very precious tools of the synthetic organic chemist. Indeed, these reactions are widely applied to generate carbon-carbon bonds allowing to link building blocks to generate large and complex molecules. The steric control of chiral centers created during these reactions was initially attained using chiral auxiliaries as is the case, for instance, of chiral imines in Michael additions [1] [2] . For over a decade, many studies have been undertaken worldwide to extend the enantioselectivity in catalytic manner using small organic chiral molecules. These metal-free catalysts, highly efficient, more environmentally friendly, and often stable under both aerobic and aqueous reaction conditions, are always extensively investigated [3] -[8] .

The synthesis of azapyridinomacrocycles N-oxides and their use as organocatalysts for enantioselective allylations of 4-nitrobenzaldehyde with allyltrichlorosilane were reported from our laboratory [9] . The described asymmetric inductions were interesting and promising but too low to afford synthetic applications. Therefore we thought to experiment some of these chiral intermediates themselves as organocatalysts in aldol reactions and Michael additions. This kind of reactions has been extensively described as model to establish the efficiency of small molecules regarding the enantioselectivity. Herein we wish to report the results using some new derivatives 1 - 6 (figure 1) as chiral catalysts in both enantioselective aldol and Michael reactions [10] -[17] .

RESULTS AND DISCUSSION

The preparation of new proline derivatives 2, 3, 5 and 6 is displayed in Scheme 1. Compound 2 was synthesized starting from amine 1 [9] with N-Boc-(S)-proline to give 7 followed by elimination of the Boc protecting group. In the case of 3, the amidation was carried out through the same pathway starting with the corresponding epimer of 1 derived from (1S,2S)-1,2-diaminocyclohexane. The proline derivative 9 was produced by condensation of commercially available 2-nitrobenzenesulfonamide with N-Boc-(S)-proline while 10 and 11 [18] resulted from the reaction of N-Boc-prolinamine [19] with respectively 2- and 4-nitrobenzenesulfonyl chloride. After acid deprotection, catalysts 4, 5 and 6 were obtained in very good yields. It should be noted that the 4-nitrophenyl regioisomer of 4 was already described [20] .

Aldol Reaction

With the new derivatives 1 - 6 in hand, we began to evaluate their catalytic behavior in the classical aldol reaction between acetone and 4-nitrobenzaldehyde [21] , and the results are displayed in Table 1. As can be seen, the reaction proceeded smoothly at room temperature under solvent free conditions and was almost total only beyond several dozen hours, except for catalyst 2 (entry 2). If the chemical yields were acceptable (catalysts 2 and 3), the enantiomeric excesses were very low except for catalyst 2. Indeed in that case, a very interesting enantioselectivity of the aldol product was obtained. Regarding the results, we observed the following facts: firstly, the catalyst 2 having proline moieties exhibited higher catalytic efficiencies compared to 1 (entries 1, 2); then, we could notice a strong mismatch effect related to the absolute stereochemistry of the cyclohexane moiety (entries 2, 3). After these initial investigations, catalyst 2, giving the best result, was selected to carry out the optimization of the reaction conditions. We first investigated the solvent effect with or without some additives and the results are displayed in Table 1. While no reaction occurred in aprotic polar solvents (entries 4, 5), apolar solvents (entries 6, 7) induced a significant decrease in catalytic efficiency. Moreover, a catalytic amount of TfOH (entry 9) induced a decrease both rate and yield without significantly affecting the enantiomeric excess. On the other hand, catalytic quantities of AcOH (entry 8) induced an increase of the yield, but involved a slightly decreasing of the enantiomeric excess. With water as additive (entries 10, 11) no significant variations were observed in the reaction rate and enantioselectivity. The reactions, carried out at lower temperatures, have shown as expected a positive effect on the catalytic efficiency and thus the best results were obtained at −20°C but requiring prolonged reaction time (entries 12, 15). Regarding the loading of the catalyst, the reactions carried out with only 10 mol% of 2 (entries 13 - 15) provided the aldol product with almost the same enantioselectivity but with an expected decreasing of the reaction rate. In order to exclude a quite possible retro-aldol process that could lead to partial racemization, the enantiomeric excess of the reaction was measured several times throughout the reaction, showing no significant variation in value.

To explain the experimental results and especially the different behavior between catalysts 2 and 3, we proposed transition states displayed in Figure 2 for the aldol reaction. Thus, the enamine resulting from the condensation of the catalyst (S,R,R)-2 with acetone adding to the aromatic aldehyde forms a pseudo-cycle stabilized by strong binding interactions between the carbonyl oxygen and two acidic hydrogen atoms. In this cycle, the largest substituent "Ar" is located at the equatorial position. This compact and rigid transition state can explain the relative high enantiomeric excess (78%) promoting the R-isomer of the aldol product. With the isomer (S,S,S)-3 this binding interaction can arise with only one hydrogen, the second one being pushed to the rear of the figure induced by the reverse stereochemistry of the cyclohexane moiety. These very different behaviors between 2 and 3 allowed us to highlight one main characteristic required for catalysts, namely the ability to form a compact cyclic transition state induced by the presence of acidic hydrogen atoms [22] . Following these observations, we thought design proline derivatives 4, 5 and 6 having acidic hydrogen of a sulfonamide function. In addition, one could be expected a significant difference between the catalysts, given the presence in the catalyst 4 of an additional carbonyl group increasing the hydrogen acidity [23] . The experimental results of aldol reactions are displayed in Table 1. The low solubility of catalyst 4 in apolar solvents has limited the range of solvents.

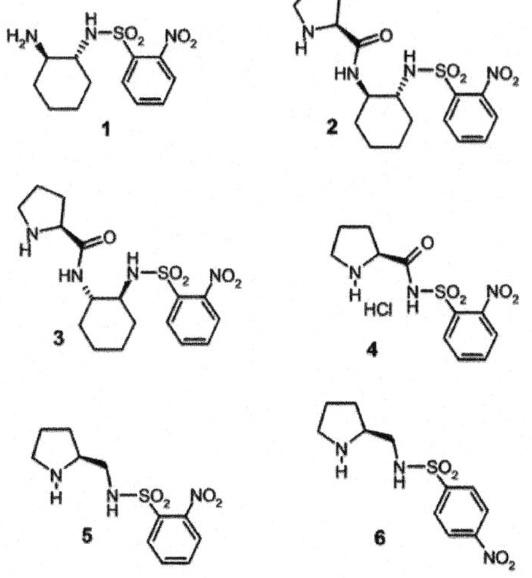

Figure 1: Organocatalysts derived from (R,R)- or (S,S)-1, 2-diaminocyclohexane and/ or (S)-proline.

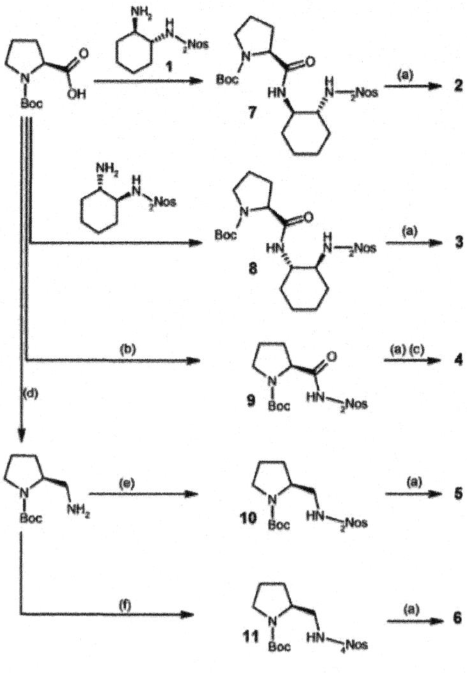

Scheme 1: Reaction conditions: (a) CF_3CO_2H, CH_2Cl_2, rt, 2 h; (b) $_2Nos-NH_2$, DMAP, EDC.HCl, THF, rt, 8 h; (c) HCl/ MeOH, Et_2O; (d) Reference [18]; (e) $_2NosCl$, NEt_3, CH_2Cl_2, rt, 4 h; (f) $_4NosCl$, NEt_3, CH_2Cl_2, rt, 3 h.

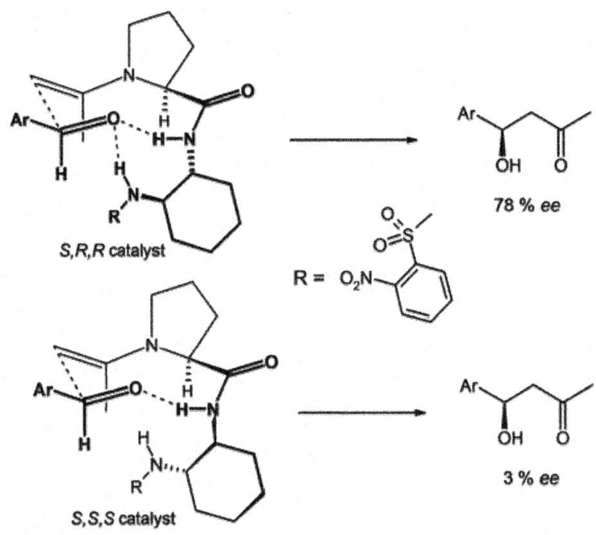

Figure 2: Proposed transition state for the aldol reaction.

Table 1: Catalyst screening and aldol reaction optimization[a].

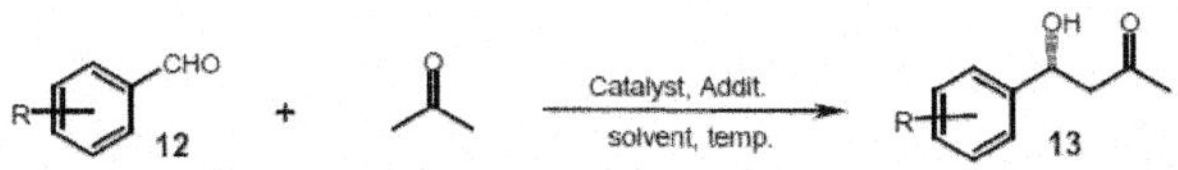

Entry	ArCHO (R)	Catalyst	Solvent	Additives	Temp (°C)	Time (h)	Product (Y%)[b]	ee (%)[c]
1	12a (4-NO₂)	1	–	–	25	48	13a (29)	20
2	12a (4-NO₂)	2	–	–	25	12	13a (54)	78
3	12a (4-NO₂)	3	–	–	25	36	13a (47)	3
4	12a (4-NO₂)	2	DMF	–	25	24	–	–
5	12a (4-NO₂)	2	DMSO	–	25	24	–	–
6	12a (4-NO₂)	2	toluene	–	25	12	13a (46)	44
7	12a (4-NO₂)	2	CHCl₃	–	25	12	13a (49)	44
8	12a (4-NO₂)	2	–	AcOH	25	5	13a (80)	65
9	12a (4-NO₂)	2	–	TfOH	25	48	13a (30)	60
10	12a (4-NO₂)	2	–	H₂O	25	15	13a (87)	70
11	12a (4-NO₂)	2	–	H₂O	5	24	13a (87)	81
12	12a (4-NO₂)	2	–	–	-20	48	13a (85)	87
13	12a (4-NO₂)	2[d]	–	–	25	20	13a (51)	75
14	12a (4-NO₂)	2[d]	–	–	5	24	13a (88)	77
15	12a (4-NO₂)	2[d]	–	–	-20	52	13a (50)	88
16	12a (4-NO₂)	4	–	–	25	35	13a (72)	95
17	12a (4-NO₂)	4	–	–	5	48	13a (46)	95
18	12a (4-NO₂)	4[d]	–	–	25	48	13a (51)	96
19	12a (4-NO₂)	4[d]	–	–	5	51	13a (38)	95
20	12b (2-NO₂)	4	–	–	25	30	13b (58)	95
21	12c (4-Br)	4	–	–	25	30	13c (70)	88
22	12d (4-Cl)	4	–	–	25	30	13d (61)	85
23	12e (H)	4	–	–	25	30	13e (61)	90
24	12f (naphthyl)	4	–	–	25	30	13f (68)	81
25	12g (5-Cl,2-NO₂)	4	–	–	25	30	13g (72)	92
26	12h (2,4-2MeO)	4	–	–	25	30	13h (29)	72
27	12a (4-NO₂)	5	–	–	25	48	13a (68)	31

[a]Reaction conditions: aldehyde (0.33 mmol), acetone (850 µL, 11.5 mmol) using 20 mol% of catalyst and additives (if any); [b]Yield of the isolated product

after chromatography on silica gel;[c]Measured by chiral HPLC, absolute stereochemistry of the reaction products were assigned according to the literature [21] ; [d]10 mol%.

Indeed, while no reaction took place in aprotic solvents, excellent results in solvent-free medium were observed (entries 16 - 19, ee up to 96%). In contrast with catalyst 5 in the same conditions, the reaction provided aldol products in only 31% ee (entry 27). Again, with a lower catalyst loading (10%), the reaction occurred (entries 18, 19) and provided the corresponding aldol product with the same high enantioselectivity (96% ee). Finally, the scope and limitation of this process with various aromatic aldehydes under optimized conditions were explored; the results are displayed in Table 1. In all cases, the enantiomeric excesses were of the same order of magnitude as those previously observed. However, some significant differences between aldehydes bearing withdrawing groups and those bearing donor groups can be noted. Indeed, the aldol reaction rate with compound 12h (entry 26, two donor groups) is lower than that observed with other aldehydes; these two donor groups reducing the electrophilicity of the aldehyde function. To conclude this chapter regarding aldol reactions, promising results were obtained showing, once again, that proline derivatives gave the best results.

Michael Addition

Following these encouraging results, we proposed to test some of these organocatalysts in Michael additions. Organocatalyst activation of this type of reaction has already been intensively studied but still remains a great challenge for the synthetic chemists [13] -[17] . The tests were performed with 2, 4, 5, and 6 to establish their catalytic activities by achieving the reaction model: cyclohexanone/acetone-trans-β-nitrostyrene [24] [25] . The experimental results are displayed inTable 2. As can be seen, the reactions took place smoothly with an extensive reaction time. A surprising result was first obtained with catalyst 2 (entry 1): the enantiomeric excess is near zero while this catalyst was very efficient for the aldol reaction (Table 1, entry 2). With catalyst 4 in acetonitrile, the enantiomeric excess was low at 25°C (10% ee) with a good diastereomeric ratio (entry 2). However, catalyst 5 (entry 5) proved to be very promising since the enantiomeric excess reached almost 80% in the first experiment. Therefore we choose this catalyst as well as its regioisomer 6 to undertake optimization experiments in order to increase the enantiomeric excess.

The influence of solvent and temperature as well as presence or absence of some additives on the steric course of the reaction was examined and the results are shown in Table 2. Before looking to the results in detail, we can observe that the enantiomeric excesses were generally good to excellent (entries 5 - 19). In acetonitrile solution, addition of a catalytic amount of water or acetic acid (entries 7, 8) significantly increased the reaction rate and, more interestingly, the enantiomeric excesses without altering the diastereomeric ratios. On the other hand, addition of a strong acid (entry 9) induced an important decrease of the enantiomeric excess and reaction rate. We have also tested protic and aprotic solvents, iPrOH with acetic acid at room temperature giving almost the same results (entries 10, 11) and toluene with or without additives yielding better enantiomeric excesses (entries 12 - 15). In the absence of solvent other than the ketone itself, the reaction provided adducts with enantiomeric excesses slightly higher than in acetonitrile (entries 5, 16 and 7, 17). Again a decrease in temperature led to better enantiomeric excesses, but slowed down the reaction rates (entries 11, 15, 18 vs. 10, 14, 17). It should be noted that a spectacular result was observed with catalyst 4 (entries 2 - 4) with which a lowering of the temperature (from 25°C to - 15°C) caused an important increase of the enantiomeric excess (from 10% to 60% ee). Similarly to the case of aldol reaction, using a lower catalyst loading (10 mol%, entry 6), the Michael adducts were obtained with a comparable enantiomeric excess but a large reduction of the reaction rate (220 h vs. 120 h). Some improvements of the enantiomeric excesses were observed with catalyst 6 involving the nitro group at the para-position. Enantiomeric excesses were occasionally higher than those obtained with catalyst 5 (entries 28, 15 and 25, 7). We observed a very excellent result (entry 28: 98% ee, 96:4 dr) with a catalytic amount of water in the toluene/cyclohexanone solution at 5°C.

Opposite results were achieved with acetone and trans-β-nitrostyrene without solvent in presence of catalytic amounts of water (entries 29, 30). The enantiomeric excess (25% at 25°C and 29% at 5°C) were the worst results of all examples of Table 2 [26] . One reason that could explain this lower efficiency may be related to the lower size of acetone compared to cyclohexanone which leads to greater degrees of freedom in the transition state. Figure 3 displays the probable catalytic circle [27] of the Michael addition: the reactive enamine from the condensation between chiral secondary amine and cyclohexanone reacts with nitroolefin through the transition state stabilized by some hydrogen bonds including some water molecules [28] [29] . This transition state also explains the increasing of enantioselectivity by the presence of catalytic amounts of water.

Table 2: Catalyst screening and Michael addition optimization[a]

Entry	Ketone	Catalyst	Solvent	Additives	Temp (°C)	Time (h)	Yield(%)[b]	dr [syn/anti][c]	ee (%)[d]
1	n = 3	2	CH₃CN	–	25	48	55	93:7	3
2	n = 3	4	CH₃CN	–	25	48	40	95:5	10
3	n = 3	4	CH₃CN	–	5	90	35	95:5	15
4	n = 3	4	CH₃CN	–	-15	144	29	96:4	60
5	n = 3	5	CH₃CN	–	25	120	86	91:9	79
6	n = 3	5*	CH₃CN	–	25	220	76	90:10	77
7	n = 3	5	CH₃CN	H₂O	25	48	61	90:10	86
8	n = 3	5	CH₃CN	AcOH	25	48	85	89:11	88
9	n = 3	5	CH₃CN	CF₃CO₂H	25	104	53	90:10	65
10	n = 3	5	iPrOH	AcOH	25	120	63	96:4	79
11	n = 3	5	iPrOH	AcOH	5	480	58	93:7	85
12	n = 3	5	toluene	–	25	65	72	90:10	87
13	n = 3	5	toluene	AcOH	25	72	35	89:11	89
14	n = 3	5	toluene	H₂O	25	48	63	91:9	89
15	n = 3	5	toluene	H₂O	5	72	87	91:9	94
16	n = 3	5	neat	–	25	65	53	95:5	85
17	n = 3	5	neat	H₂O	25	65	84	92:8	89
18	n = 3	5	neat	H₂O	5	72	84	92:8	93
19	n = 3	5	neat	AcOH	25	72	35	89:11	89
20	n = 3	6	neat	–	25	41	78	92:8	84
21	n = 3	6	neat	H₂O	25	105	93	93:7	95
22	n = 3	6	neat	AcOH	25	105	89	93:7	95
23	n = 3	6	toluene	–	25	72	62	91:9	87
24	n = 3	6	toluene	AcOH	25	72	97	93:7	86
25	n = 3	6	CH₃CN	H₂O	25	48	93	95:5	94
26	n = 3	6	CH₃CN	H₂O	5	60	84	96:4	94
27	n = 3	6	toluene	H₂O	25	68	86	94:6	75
28	n = 3	6	toluene	H₂O	5	68	81	96:4	98
29	n = 0	6	neat	H₂O	25	48	74	-	25
30	n = 0	6	neat	H₂O	5	72	68	-	29

[a]Reaction conditions: trans-β-nitrostyrene (0.60 mmol), cyclohexanone (300 μL, 2.0 mmol) in 900 μL of solvent using 20 mol% of catalyst and additive (if any); [b]Yield of the isolated product;[c]Measured by GC/MS analysis; [d]Measured by chiral HPLC, absolute stereochemistry of the reaction products were assigned according to the literature [13]; [e]10 mol%.

CONCLUSION

Among the catalysts we have developed in this study, two of them have proved to be very efficient; catalyst 4 for the aldol reaction and catalyst 6 for the Michael addition. It should be noted that, once again, these two catalysts 4 and 6 both derived from a proline structure [30] . Additional studies are currently in progress in our laboratory to investigate further applications of these new catalysts.

EXPERIMENTAL

General

The reactions were monitored by thin-layer chromatography (TLC) on aluminum plates (0.20 mm, 60 F_{254}) us-

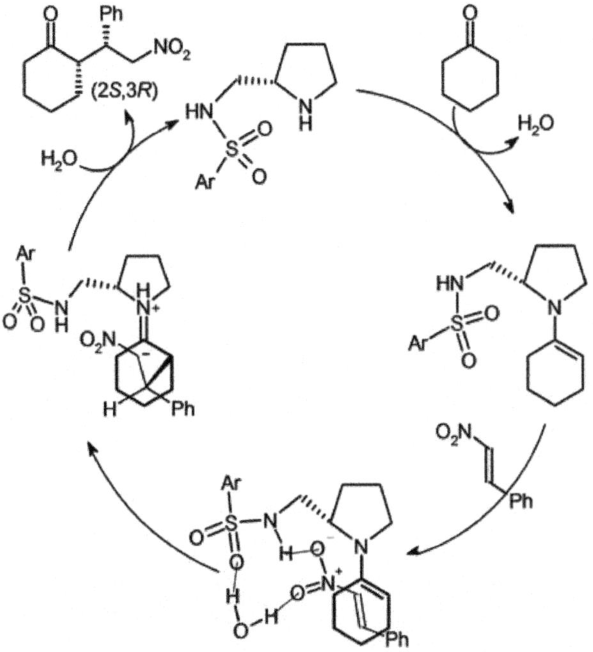

Figure 3: Proposed catalytic cycle of the Michael addition.

ing EtOAc/cyclohexane mixture as eluent. The reaction compounds were first visualized under UV light and then by treatment with iodine vapor. Silica gel (40 - 63 μm) was used for flash chromatographies with EtOAc/ cyclohexane mixtures as eluent. Gas chromatography-mass spectroscopy (GC-MS) was performed with an Agilent 6890N (equipped with a 12 m × 0.20 mm dimethylpolysiloxane capillary column) linked to a Model 5973N (ionization energy: 70 eV). Enantiomeric ratios were determined by chiral HPLC analysis on Daicel Chiralcel AS-H and ADH. Melting points were achieved on a Kofler block. Optical rotations were measured on a Perkin-Elmer model 241. ^{1}H and ^{13}C NMR spectra were recorded in CDCl$_3$ (except specified) respectively at 400 MHz and 100 MHz (Bruker 400). Reagents and materials were obtained from commercial suppliers and were used without further purification.

Preparation of Organocatalysts

(S)-tert-Butyl 2-[(1R,2R)-2-(2-nitrophenylsulfonamido) cy-clohexylcarbamoyl]pyrrolidine-1-carboxylate (7)

To 1.00 g (3.34 mmol) of N-[(1R,2R)-2-aminocyclohexyl]-2-nitrophenylsulfonamide (1) [9] stirred in 33 mL of THF, 740 mg (3.44 mmol) of N-Boc-(S)-proline, and 420 mg (3.44 mmol) of DMAP were successively added. At 0˚C, 660 mg (3.44 mmol) of EDC·HCl was then added. After 8 h at room temperature the reaction medium was diluted with 70 mL of CH$_2$Cl$_2$ and successively washed with 1N HCl, saturated NaHCO$_3$ and dried over MgSO$_4$. After filtration and concentration of the organic layer under reduced pressure, a flash chromatography of the residue (40% EtOAc in cyclohexane) gave 1.47 g (89% yield) of 7 as a white solid.

7: R$_f$ = 0.5 (40% EtOAc in cyclohexane); mp 160˚C; MS (ES+) m/z 497.20 [M+H]$^+$, 519.17 [M+Na]$^+$, 535.15 [M + K]$^+$, 993.38 [2M + H]$^+$; ^{1}H NMR δ 1.13 - 1.40 (m, 4H), 1.48 (s, 9H), 1.60 - 1.77 (m, 3H), 1.78 - 1.88 (m, 1H), 1.89 - 2.13 (m, 3H), 2.15 - 2.33 (m, 1H), 3.07 - 3.64 (m, 3H), 3.67 - 3.89 (m, 1H), 4.18 - 4.40 (m, 1H), 5.57 - 6.03 (m, 1H), 7.72 - 7.77 (m, 2H), 7.81 - 7.84 (m, 1H), 8.12 (d, J = 7.0 Hz, 1H); ^{13}C NMR δ 24.3 (CH$_2$), 24.7 (CH$_2$), 28.4 (3CH$_3$), 32.9 (CH$_2$), 42.1 (CH), 47.1 (CH$_2$), 52.3 (CH), 80.3 (C), 125.1 (CH), 130.1 (CH), 132.7 (CH), 133.3 (CH), 147.7 (C), 174.2 (C).

(S)-N-[(1R,2R)-2-(2-Nitrophenylsulfonamido)-cyclohexyl]pyrroli-dine-2-carboxamide (2)

728 mg (1.47 mmol) of 7 in 15 mL of CF$_3$CO$_2$H/CH$_2$Cl$_2$ (1/4) was stirred 1 h at room temperature. 5 mL of 1N HCl was then added to the reaction medium until pH 1. At 0˚C, the aqueous layer, basified by an excess of NaOH pellets, was extracted with CH$_2$Cl$_2$ and the combined organic layers were dried (MgSO$_4$), filtered and evaporated to give 580 mg of 2 (quantitative yield) as yellow solid.

To 100 mg (0.25 mmol) of 2, dissolved in minimum of HCl/MeOH solution, 25 mL of Et$_2$O was added. After filtration the precipitate was washed with ether to give 103 mg (95% yield) of 2-HCl as yellow solid.

2,HCl: R$_f$ = 0.69 (Al$_2$O$_3$, 5% MeOH in CH$_2$Cl$_2$); mp 94˚C; MS (ES+) m/z 396.95 [M + H]$^+$, 419.09 [M + Na]$^+$; ^{1}H NMR (D$_2$O) δ 0.99 - 1.30 (m, 4H), 1.46 - 1.49 (m, 2H), 1.57 - 1.60 (m, 1H), 1.79 - 1.82 (m, 1H), 1.89 - 2.05 (m, 3H), 2.23 - 2.32 (m, 1H), 3.11 - 3.17 (m, 1H), 3.25 - 3.39 (m, 2H), 3.52 - 3.68 (m, 1H), 4.17 - 4.29 (m, 1H), 7.70 - 7.78 (m, 2H), 7.80 - 7.91 (m, 1H), 7.98 - 8.06 (m, 1H); ^{13}C NMR (D$_2$O) δ 23.6 (CH$_2$), 23.9 (CH$_2$), 24.1 (CH$_2$), 29.8 (CH$_2$),

31.5 (CH_2), 31.9 (CH_2), 46.4 (CH_2), 52.8 (CH), 57.6 (CH), 59.9 (CH), 125.2 (CH), 130.0 (CH), 133.3 (CH), 133.4 (C), 134.5 (CH), 146.9 (C), 169.0 (C).

(S)-tert-Butyl 2-[(1S,2S)-2-(2-nitrophenylsulfonamido)-cyclohexylcarbamoyl]- pyrrolidine-1-carboxylate (8)

Starting from the trans-(S,S)-diaminocyclohexane and using the same protocol as above, 8 was obtained in 91% yields as a white solids.

8: R_f = 0.52 (40% EtOAc in cyclohexane); mp 184°C; MS (ES+) m/z 497.20 [M + H]$^+$, 519.16 [M + Na]$^+$, 535.15 [M + K]$^+$; ^{1}H NMR δ 1.13 - 1.40 (m, 4H), 1.48 (s, 9H), 1.60 - 1.77 (m, 3H), 1.78 - 1.88 (m, 1H), 1.89 - 2.13 (m, 3H), 2.15 - 2.33 (m, 1H), 3.07 - 3.64 (m, 3H), 3.67 - 3.86 (m, 1H), 4.18 - 4.40 (m, 1H), 5.57 - 6.03 (m, 1H), 7.72 - 7.77 (m, 2H), 7.81 - 7.84 (m, 1H), 8.12 (d, J = 7.0 Hz, 1H); ^{13}C NMR δ 24.3 (CH_2), 24.6 (CH_2), 27.4 (CH_2), 28.4 (3CH_3), 31.9 (CH_2), 33.6 (CH_2), 47.0 (CH_2), 51.8 (CH), 59.3 (CH), 59.4 (CH), 80.3 (C), 125.0 (CH), 130.4 (CH), 132.7 (CH), 133.2 (CH), 134.5 (C), 147.7 (C), 174.2 (C).

(S)-N-[(1S,2S)-2-(2-Nitrophenylsulfonamido)-cyclohexyl]pyrrolidine-2-carboxamide (3)

1.00 g (2.0 mmol) of 8 in 20 mL of CF_3CO_2H/CH_2Cl_2 (1/4) was stirred 1 h at room temperature. 5 mL of 1N HCl was then added until pH 1. At 0°C, the separated aqueous layer, basified by an excess of solid NaOH to pH > 10, was extracted with CH_2Cl_2 and the combined organic layers were dried ($MgSO_4$) and evaporated to give 790 mg of 3 (quantitative yield) as white solid.

3: R_f = 0.58 (15% MeOH in CH_2Cl_2); mp 103°C - 105°C; MS (ES+) m/z 397.25 [M + H]$^+$, 419.26 [M + Na]$^+$, 793.53 [2M + H]$^+$; ^{1}H NMR δ 0.96 - 1.40 (m, 4H), 1.44 - 1.72 (m, 4H), 1.73 - 1.97 (m, 3H), 2.04 - 2.20 (m, 1H), 2.85 - 2.96 (m, 2H), 3.12 - 3.27 (m, 1H), 3.58 - 3.73 (m, 1H), 3.74 - 3.86 (m, 1H), 4.85 (brs, 2H), 7.58 - 7.75 (m, 3H), 7.75 - 7.78 (m, 1H), 8.05 - 8.07 (m, 1H); ^{13}C NMR δ 24.5 (CH_2), 24.6 (CH_2), 24.1 (CH_2), 30.6 (CH_2), 32.3 (CH_2), 33.2 (CH_2), 47.0 (CH_2), 52.1 (CH), 59.0 (CH), 60.2 (CH), 124.8 (CH), 130.3 (CH), 132.7 (CH), 133.2 (C), 135.4 (C), 147.7 (C), 173.9 (C).

(S)-tert-Butyl 2-(2-Nitrophenylsulfonylcarbamoyl)-1-carboxylate (9)

To 600 mg (2.80 mmol) of N-Boc-proline stirred at room temperature in 30 mL of THF, 354 mg (2.90 mmol) of DMAP and 506 mg (2.48 mmol) of 2-nitrobenzenesulfonamide were added. At 0°C, 556 mg (2.90 mmol) of EDC·HCl was then added. After 8 h at room temperature the reaction medium was diluted with 60 mL of CH_2Cl_2 and successively washed with 1N HCl, saturated NaCl, and dried over $MgSO_4$. After filtration the organic layer was concentrated over reduced pressure to give 904 mg (91% yield) of 9 as white solid.

9: mp 74°C; MS (ES+) m/z 422.10 $[M + Na]^+$, 438.08 $[M + K]^+$, 821.22 $[2M + Na]^+$, 837.19 $[2M + K]^+$; ^1H NMR δ 1.51 (s, 9H), 1.74 - 2.05 (m, 3H), 2.29 - 2.49 (m, 1H), 3.27 - 3.62 (m, 2H), 4.15 - 4.53 (m, 1H), 7.66 - 7.90 (m, 3H), 8.35 - 8.48 (m, 1H); ^{13}C NMR δ 24.3 (CH_2), 28.3 ($3CH_3$), 30.3 (CH_2), 47.3 (CH_2), 60.7 (CH), 82.1 (C), 124.7 (CH), 132.2 (C), 132.3 (CH), 133.5 (CH), 148.2 (CH), 148.2 (C), 156.8 (C), 171.4 (C).

(S)-N-(2-Nitrophenylsulfonyl-pyrrolidinium)-2-carboxamide (4)

618 mg (1.55 mmol) of 9 in 8 mL of CF_3CO_2H/CH_2Cl_2 (1/4) was stirred 1 h at room temperature and then concentrated under reduced pressure. 10 mL of H_2O/CH_2Cl_2 mixture was then added to the residue and the separated aqueous layer, basified by an excess of solid NaOH to pH > 10, was extracted with CH_2Cl_2. The combined organic layers were dried ($MgSO_4$) and evaporated. The residue was taken up in a minimum of 2.5 N HCl/ MeOH solution and then crystallized by addition of small amounts of ether. After filtration and washing with ether, 477 mg of compound 4-HCl was obtained (92% yield). 4-HCl: mp 192 - 194°C; MS (ES+) m/z 299.91 $[M + H]^+$, MS (ES−) m/z 297.87 [M − H]$^+$, ^1H NMR (D_2O) δ 1.80 - 2.13 (m, 3H), 2.26 - 2.45 (m, 1H), 3.23 - 3.67 (m, 3H), 4.22 - 4.26 (dd, J = 9.0, 6.5 Hz, 1H), 7.72 - 7.85 (m, 3H), 8.04 - 8.14 (m, 1H); ^{13}C NMR (D_2O) δ 23.6 (CH_2), 29.3 (CH_2), 46.4 (CH_2), 62.0 (CH), 124.5 (CH), 130.1 (CH), 132.7 (CH), 133.3 (C), 134.2 (CH), 147.3 (C), 174.6 (C).

(S)-tert-Butyl 2-[(2-nitrophenylsulfonamido)-methyl]-pyrrolidine-1-carboxylate (10)

To 80 mg (0.40 mmol) of N-Boc-prolinamine stirred at room temperature in 5 mL of CH_2Cl_2, 70 μL (0.50 mmol) of NEt_3 and 88.6 mg (0.40 mmol) of 2-nitrobenzenesulfonyl chloride were added. After 4 h, the reaction medium, diluted with 5 mL of CH_2Cl_2, was successively washed with 1N HCl, saturated NaCl, and dried over $MgSO_4$. After filtration the concentration of the organic layer over reduced pressure gave 140 mg (91% yield) of 10 as a yellow oil.

10: ^1H NMR δ 1.37 (s, 9H), 1.67 - 1.84 (m, 3H), 1.88 - 1.97 (m, 1H), 2.95 - 3.41 (m, 4H), 3.75 - 3.93 (m, 1H), 6.39 (brs, 1H), 7.53 - 7.84 (m, 3H), 8.02 - 8.04 (m, 1H); ^{13}C NMR δ 23.9 (CH_2), 28.4 ($3CH_3$), 29.3 (CH_2), 46.7 (CH_2), 47.4 (CH_2), 56.6 (CH), 80.2 (C), 125.1 (CH), 130.9 (CH), 132.5 (CH), 133.4 (CH), 133.8 (C), 148.1 (C), 155.1 (C).

(S)-2-Nitro-N-(pyrrolidin-2-ylmethyl)-benzenesulfonamide (5)

120 mg (0.30 mmol) of 10 in 3 mL of CF_3CO_2H/CH_2Cl_2 (1/4) was stirred 5 h at room temperature. 1 mL of 1N HCl was then added to the reaction medium. The aqueous layer, washed by CH_2Cl_2, was then basified to pH > 10 (NaOH pellets) and extracted with CH_2Cl_2. The combined organic layers, dried over $MgSO_4$ were evaporated to give 81 mg (92% yield) of 5 as a yellow oil.

5: MS (ES+) m/z 286.1 [M + H]$^+$, ^{1}H NMR δ 1.29 - 1.47 (m, 1H), 1.61 - 1.88 (m, 3H), 2.81 - 2.95 (m, 3H), 3.08 - 3.12 (m, 1H), 3.33 - 3.40 (m, 1H), 4.59 (brs, 2H), 7.62 - 7.68 (m, 2H), 7.73 - 7.75 (m, 1H), 7.99 - 8.09 (m, 1H); ^{13}C NMR δ 25.5 (CH_2), 28.8 (CH_2), 46.2 (CH_2), 47.4 (CH_2), 57.8 (CH), 125.1 (CH), 130.9 (CH), 132.6 (CH), 133.3 (CH), 133.9 (C), 148.2 (C).

(S)-4-Nitro-N-(pyrrolidin-2-ylmethyl)-benzenesulfonamide (6)

Starting from 4-nitrobenzenesulfonyl chloride and using the same procedure as above, 11 and 6 were obtained in respectively 92% and 74% yields as yellow oils.

11: ^{1}H NMR δ 1.37 (s, 9H), 1.51 - 1.79 (m, 3H), 1.84 - 2.01 (m, 1H), 2.79 - 2.95 (m, 1H), 2.99 - 3.18 (m, 2H), 3.19 - 3.35 (m, 1H), 3.74 - 3.90 (m, 1H), 7.05 (brs, 1H), 7.97 (d, J = 8.7 Hz, 2H), 8.26 (d, J = 8.7 Hz, 2H); ^{13}C NMR δ 23.7 (CH_2), 28.3 (3CH_3), 29.5 (CH_2), 47.3 (CH_2), 48.5 (CH_2), 56.6 (CH), 80.6 (C), 124.3 (2CH), 128.2 (2CH), 146.2 (C), 149.8 (C), 156.2 (C).

6: MS (ES+) m/z 286.1 [M + H]$^+$; ^{1}H NMR δ 1.29 - 1.38 (m, 1H), 1.60 - 1.87 (m, 3H), 2.74 (dd, J = 12.6, 8.5 Hz, 1H), 2.74 - 3.82 (m, 1H), 2.84 - 2.92 (m, 1H), 2.99 (dd, J = 12.6, 4.1 Hz, 1H), 4.02 - 4.07 (m, 1H), 4.44 (brs, 2H), 7.97 (d, J = 8.7 Hz, 2H), 8.28 (d, J = 8.7 Hz, 2H); ^{13}C NMR δ 25.7 (CH_2), 28.9 (CH_2), 46.2 (CH_2), 46.9 (CH_2), 57.7 (CH), 124.4 (2CH), 128.2 (2CH), 146.2 (C), 149.9 (C).

Aldol Reactions

General Procedure

To 0.33 mmol of aldehyde 12 in 850 μL (11.5 mmol) of acetone, 66 μmol of organocatalyst 4 was added and the mixture was stirred for appropriate time and temperature (see Table 1). The reaction medium was concentrated under reduced pressure and the residue was purified through column chromatography (10% EtOAc in cyclohexane) to afford aldol 13.

(R)-4-Hydroxy-4-(4'-nitrophenyl)-butan-2-one (13a)

From 4-nitrobenzaldehyde (12a), (R)-4-hydroxy-4-(4'-nitro-phenyl)-butan-2-one (13a) was obtained in 72% yield as pale yellow crystals.

13a: R_f = 0.4 (40% EtOAc in cyclohexane); mp 58°C - 60°C (EtOAc/cyclohexane); $[\alpha]^D_{20}$ + 62 (c 0.7, CHCl$_3$); ^{1}H NMR δ 2.17 (s, 3H), 2.78 (d, J = 4 Hz, 2H), 3.0 - 3.5 (m, 1H), 5.19 (t, J = 4.0 Hz, 1H), 7.47 (d, J = 8.0 Hz, 2H), 8.12 (d, J = 8.0 Hz, 2H); ^{13}C NMR δ 30.7 (CH$_3$), 51.5 (CH$_2$), 68.9 (CH), 123.8 (2CH), 126.4 (2CH), 147.3 (C), 149.9 (C), 208.5 (C); HPLC (Chiralcel AS-H, n-hexane/iPrOH: 80:20 v/v, 1.0 mL/min, UV 268 nm): t_r(major, R) = 9.65 min, t_r(minor, S) = 11.43 min, 94% ee.

(R)-4-Hydroxy-4-(2'-nitrophenyl)-butan-2-one (13b)

From 2-nitrobenzaldehyde (12b), (R)-4-hydroxy-4-(2'-nitrophenyl)-butan-2-one (13b) was obtained in 58% yield as a brown solid.

13b: R_f = 0.6 (40% EtOAc in cyclohexane); mp 60°C - 62°C (EtOAc/cyclohexane); $[\alpha]^{20}_D$ +95 (c 0.62, CHCl$_3$); ^{1}H NMR δ 2.17 (s, 3H), 2.65 (dd, J = 17.8, 9.3 Hz, 1H), 3.07 (dd, J = 17.8, 2.0 Hz, 1H), 5.61 (dd, J = 9.3, 2.0 Hz, 1H), 7.33 - 7.40 (m, 1H), 7.54 - 7.65 (m, 1H), 7.79 - 7.86 (m, 1H), 7.87 - 7.92 (m, 1H); ^{13}C NMR δ 30.5 (CH$_3$), 51.0 (CH$_2$), 65.7 (CH), 124.5 (CH), 128.2 (CH), 128.3 (C), 133.8 (CH), 133.8 (CH), 138.4 (C), 208.9 (C); HPLC (Chiralcel AS-H, n-hexane/iPrOH: 40:60 v/v, 1.0 mL/min, UV 255 nm): t_r(major, R) = 5.24 min, t_r(minor, S) = 6.47 min, 95% ee.

(R)-4-(4'-Bromophenyl)-4-hydroxy-butan-2-one (13c)

From 4-bromobenzaldehyde (12c), (R)-4-(4'-bromophenyl)-4-hydroxy-butan-2-one (13c) was obtained in 70% yield as a white solid.

13c: R_f = 0.45 (40% EtOAc in cyclohexane); mp 65°C - 67°C (EtOAc/cyclohexane); $[\alpha]^{20}_D$ +50 (c 0.43, CHCl$_3$); ^{1}H NMR δ 2.10 (s, 3H), 2.69 (dd, J = 17.6, 3.8 Hz, 1H), 2.81 (dd, J = 17.8, 8.5 Hz, 1H), 3.43 (brs, 1H), 5.12 (dd, J = 8.5, 3.3 Hz, 1H), 7.13 - 7.15 (m, 2H), 7.37 - 7.39 (m, 2H); ^{13}C NMR δ 30.8 (CH$_3$), 51.8 (CH$_2$), 69.2 (CH), 121.5 (C), 127.4 (2CH), 131.6 (2CH), 141.8 (C), 208.8 (C); HPLC (Chiralcel AS-H, n-hexane/iPrOH: 85:15 v/v, 1.0 mL/min, UV 221 nm): t_r(major, R) = 9.85 min, t_r(minor, S) = 12.67 min, 88% ee.

(R)-4-(4'-Chlorophenyl)-4-hydroxy-butan-2-one (13d)

From 4-chlorobenzaldehyde (12d), (R)-4-(4'-chlorophenyl)-4-hydroxy-butan-2-one (13d) was obtained in 61% yield as a white solid.

13d: R_f = 0.45 (40% EtOAc in cyclohexane); mp 60°C - 62°C (EtOAc/

cyclohexane); $[\alpha]^{20}_D$ +59.9 (c 0.85, CHCl$_3$); ^{1}H NMR δ 2.12 (s, 3H), 2.79 (dd, J = 17.6, 4.0 Hz, 1H), 2.85 (dd, J = 17.8, 8.3 Hz, 1H), 3.25 (brs, 1H), 5.12 (dd, J = 8.3, 4.0 Hz, 1H), 7.25 - 7.35 (m, 4H); ^{13}C NMR 30.7 δ (CH$_3$), 51.8 (CH$_2$), 69.2 (CH), 127.0 (2CH), 128.7 (2CH), 133.4 (C), 141.2 (C), 208.8 (C); HPLC (Chiralcel AS-H, n-hexane/iPrOH: 82:18 v/v, 1.0 mL/min, UV 221 nm): t$_r$(major, R) = 11.80 min, t$_r$(minor, S) = 13.80 min, 85% ee.

(R)-4-Hydroxy-4-phenyl-butan-2-one (13e)

From benzaldehyde (12e), (R)-4-hydroxy-4-phenyl-butan-2-one (13e) was obtained in 61% yield as colorless liquid.

13e: R$_f$ = 0.45 (40% EtOAc in cyclohexane); $[\alpha]^{20}_D$ +68 (c 0.52, CHCl$_3$); ^{1}H NMR δ 2.19 (s, 3H), 2.74 (dd, J = 17.6, 3.5 Hz, 1H), 2.82 (dd, J = 17.6, 9.0 Hz, 1H), 3.15 (d, J = 3.5 Hz, 1H), 5.15 (ddd, J = 9.0, 3.5, 3.5 Hz, 1H), 7.26 - 7.36 (m, 5H); ^{13}C NMR δ 30.8 (CH$_3$), 52.0 (CH$_2$), 69.8 (CH), 125.6 (2CH), 128.3 (CH), 128.5 (2CH), 142.9 (C), 209.0 (C); HPLC (Chiralcel AS-H, n-hexane/iPrOH: 80:20 v/v, 1.0 mL/min, UV 225 nm): t$_r$(major, R) = 9.98 min, t$_r$(minor, S) = 10.98 min, 90% ee.

(R)-4-Hydroxy-4-(naphthalen-2-yl)-butan-2-one (13f)

From naphthalene-2-carbaldehyde (12f), (R)-4-hydroxy-4-(naphthalen-2-yl)-butan-2-one (13f) was obtained in 68% yield as a colorless oil.

13f: R$_f$ = 0.55 (40% EtOAc in cyclohexane); $[\alpha]^{20}_D$ +49 (c 0.72, CHCl$_3$); ^{1}H NMR δ 2.23 (s, 3H), 2.99 (d, J = 5.5 Hz, 1H), 3.0 - 3.5 (m, 1H), 5.96 (t, J = 5.5 Hz, 2H), 7.47 - 7.54 (m, 3H), 7.69 (d, J = 7.3 Hz, 1H), 7.79 (d, J = 8.0 Hz, 1H), 7.87 (dd, J = 7.5, 7.5 Hz, 1H), 8.01 (d, J = 8.0 Hz, 1H); ^{13}C NMR δ 30.8 (CH$_3$), 51.4 (CH$_2$), 66.7 (CH), 122.7 (CH), 123.0 (CH), 125.6 (CH), 125.6 (CH), 126.2 (CH), 128.1 (CH), 129.1 (CH), 129.9 (C), 133.8 (C), 138.2 (C), 209.2 (C); HPLC (Chiralcel AS-H, n-hexane/iPrOH: 78:22 v/v, 1.0 mL/min, UV 217 nm): t$_r$(major, R) = 8.98 min, t$_r$(minor, S) = 9.81 min, 81% ee.

(R)-4-(5-Chloro-2-nitrophenyl)-4-hydroxy-butan-2-one (13g)

From 5-chloro-2-nitrobenzaldehyde (12g), (R)-4-(5-chloro-2-nitrophenyl)-4-hydroxy-butan-2-one (13 g) was obtained in 72% yield as a brown oil.

13g: R$_f$ = 0.65 (40% EtOAc in cyclohexane); ^{1}H NMR δ 2.17 (s, 3H), 2.63 (dd, J = 17.8, 9.5 Hz, 1H), 3.03 (dd, J = 17.8, 2.3 Hz, 1H), 3.5 - 4.0 (m, 1H), 5.63 (dd, J = 9.5, 2.3 Hz, 1H), 7.33 (dd, J = 8.3, 2.5 Hz, 1H), 7.84 (d, J = 2.5 Hz, 1H), 7.88 (d, J = 8.3 Hz, 1H); ^{13}C NMR δ 30.4 (CH$_3$), 50.9 (CH$_2$), 63.4 (CH), 126.1 (CH), 128.4 (CH), 128.6 (CH), 140.7 (C), 140.8 (C), 145.1 (C), 208.5 (C); HPLC (Chiralcel AS-H, n-hexane/iPrOH: 60:40 v/v, 1.0 mL/min,

UV 225 nm): t_r(major, R) = 7.35 min, t_r(minor, S) = 8.26 min, 92% ee.

(R)-4-(2,4-Dimethoxyphenyl)-4-hydroxy-butan-2-one (13h)

From 2,4-dimethoxybenzaldehyde (12h), (R)-4-(2,4-dimethoxyphenyl)-4-hydroxy-butan-2-one (13h) was obtained in 29% yield as yellow oil.

13h: R_f = 0.45 (40% EtOAc in cyclohexane); ^{1}H NMR δ 2.10 (s, 3H), 2.73 (dd, J = 17.1, 8.8 Hz, 1H), 2.81 (dd, J = 17.1, 3.5 Hz, 1H), 3.21 (d, J = 4.5 Hz, 1H), 3.72 (s, 3H), 3.74 (s, 3H), 5.23 - 5.27 (m, 1H), 6.37 (d, J = 2.5 Hz, 1H), 6.42 (dd, J = 8.5, 2.3 Hz, 1H), 7.24 (d, J = 8.3 Hz, 1H); ^{13}C NMR δ 30.6 (CH$_3$), 50.6 (CH$_2$), 55.3 (CH$_3$), 55.4 (CH$_3$), 65.5 (CH), 98.6 (CH), 104.2 (CH), 123.5 (C), 127.2 (CH), 157.0 (C), 160.2 (C), 209.3 (C); HPLC (Chiralcel AS-H, n-hexane/iPrOH: 73:27 v/v, 1.0 mL/min, UV 225 nm): t_r(major, R) = 12.80 min, t_r(minor, S) = 13.90 min, 72% ee.

Michael Additions

(2S)-2-[(1R)-2-Nitro-1-phenylethyl)-cyclohexanone (14a)

From 90 mg (0.60 mmol) of trans-β-nitrostyrene in 200 μL (2.0 mmol) of cyclohexanone and 900 μL of toluene with a catalytic amount of H$_2$O, 34 mg (0.12 mmol) of 6 was added at room temperature. The mixture was stirred 68 h at room temperature. After evaporation under reduced pressure the residue was purified by flash chromatography (10% EtOAc in cyclohexane) to afford 120 mg (81% yield) of 14a as white crystals.

14a: R_f = 0.5 (10% EtOAc in cyclohexane); mp 134°C; $[\alpha]^{20}_D$ –36 (c 0.28, CHCl$_3$); EIMS m/z 200 (M$^+$-47, 61%), 183 (22), 171 (76), 157 (13), 141 (14), 129 (26), 115 (25), 104 (32), 91 (60); ^{1}H NMR δ 1.12 - 1.19 (m, 1H), 1.48 - 1.65 (m, 4H), 1.82 - 2.02 (m, 1H), 2.24 - 2.42 (m, 2H), 2.57 - 2.64 (m, 1H), 3.68 (td, J = 12.0, 12.0, 4.0 Hz, 1H), 4.54 (dd, J = 12.0, 9.6 Hz, 1H), 4.86 (dd, J = 12.0, 4.0 Hz, 1H), 7.08 - 7.25 (m, 5H); ^{13}C NMR δ 25.0 (CH$_2$), 28.5 (CH$_2$), 33.2 (CH$_2$), 42.7 (CH$_2$), 44.0 (CH), 52.5 (CH), 78.93 (CH$_2$), 127.5 (CH), 128.2 (2CH), 128.9 (2CH), 137.8 (C), 212.0 (C); HPLC (Chiralcel AD7 column, n-hexane/iPrOH: 80:20 v/v, flow rate 0.8 mL/min, wavelength = 260 nm): t_R = 8.1 min (2S,3R), t_R = 9.2 (2R,3S), t_R = 10.5 min (major, 2S,3R), 98% ee.

(–)-(R)-5-Nitro-4-phenyl-pentan-2-one (14b)

From 90 mg (0.60 mmol) of trans-β-nitrostyrene and 150 μL (2.0 mmol) of acetone, and 2.0 mg (0.11 mmol) of water, 35 mg (0.12 mmol) of organocatalyst 6 was added at room temperature. After 48 h at 5°C, 56 mg (45% yield) of 14b was obtained as white crystals.

14b: R_f = 0.45 (10% EtOAc in cyclohexane); mp 111°C; $[\alpha]^{20}_D$ levorotatory (CHCl$_3$); EIMS m/z 160 (M$^+$-47, 61%), 145 (46), 115 (21), 104 (44), 91 (18), 77 (12); ^{1}H NMR δ 2.11 (s, 3H), 2.91 (d, J = 7.2 Hz, 2H), 3.95 - 4.05 (m, 1H), 4.59 (dd, J = 12.4, 7.6 Hz, 1H), 4.70 (dd, J = 12.5, 6.8 Hz, 1H), 7.2 - 7.35 (m, 5H); ^{13}C NMR δ 30.4 (CH$_3$), 39.0 (CH$_2$), 46.1 (CH$_2$), 79.4 (CH), 127.3 (2CH), 127.9 (CH), 129.1 (2CH), 138.8 (C), 205.4 (C); HPLC (Chiralpak IC, n-hexane/tert-BuOMe: 35:65 v/v, flow rate = 0.8 mL/min, wavelength 220 nm): t_R = 12.18 min (major, R), t_R = 17.53 min (minor, S), 25% ee.

REFERENCES

1. Pfau, M., Revial, G., Guingant, A. and d'Angelo, J. (1985) Enantioselective Synthesis of Quaternary Carbon Centers through Michael-Type Alkylation of Chiral Imines. Journal of the American Chemical Society, 107, 273-274. http://dx.doi.org/10.1021/ja00287a061

2. Pfau, M., Tomas, A., Lim, S. and Revial, G. (1995) Diastereo Selectivity in the Michael-Type Addition of Imines Reacting as their Secondary Enamine Tautomers. Journal of Organic Chemistry, 60, 1143-1147. http://dx.doi.org/10.1021/jo00110a015

3. List, B. (2006) The Ying and Yang of Asymmetric Amino-Catalysis. Chemical Communications, 2006, 819-824. http://dx.doi.org/10.1039/b514296m

4. Taylor, M.S. and Jacobsen, E.N. (2006) Asymmetric Catalysis by Chiral Hydrogen-Bond Donors. Angewandte Chemie International Edition, 45, 1520-1543. http://dx.doi.org/10.1002/anie.200503132

5. Pellissier, H. (2007) Asymmetric Organocatalysis. Tetrahedron, 63, 9267-9331. http://dx.doi.org/10.1016/j.tet.2007.06.024

6. Enders, D., Grondal, C. and Hüttl, M.R.M. (2007) Asymmetric Organocatalytic Domino Reactions. Angewandte Chemie International Edition, 46, 1570-1581. http://dx.doi.org/10.1002/anie.200603129

7. Chai, Z. and Zhao, G. (2012) Efficient Organocatalysts Derived from Simple Chiral Acyclic Amino Acids in Asymmetric Catalysis. Catalysis Science & Technology, 2, 29-41. http://dx.doi.org/10.1039/c1cy00347j

8. Pellissier, H. (2012) Asymmetric Organocatalytic Cycloadditions. Tetrahedron, 68, 2197-2232. http://dx.doi.org/10.1016/j.tet.2011.10.103

9. Sylla-Iyarreta Veitia, M., Joudat, M., Wagner, M., Falguière, A., Guy, A. and Ferroud, C. (2011) Ready Available Chiral Azapyridinomacrocycles n-Oxides; First Results as Lewis Base Catalysts in Asymmetric Allylation of 4-Nitro- benzaldehyde. Heterocycles, 83, 2011-2030.

10. Geary, L.M. and Hultin, P.G. (2009) The State of the Art in Asymmetric Induction: The Aldol Reaction as a Case Study. Tetrahedron: Asymmetry, 20, 131-173.http://dx.doi.org/10.1016/j.tetasy.2008.12.030

11. Trost, B.M. and Brindle, C.S. (2010) The Direct Catalytic Asymmetric Aldol Reaction. Chemical Society Reviews, 39, 1600-1632. http://dx.doi.org/10.1039/b923537j

12. Liu, Y.X., Yang, L., Ma, Z.W., Wang, C.C. and Tao, J.C. (2011) Research Progress in Supported Proline and Proline Derivatives as Recyclable Organocatalysts for Asymmetric C-C Bond Formation Reactions. Chinese Journal of Catalysis, 32, 1295-1311.

13. Roca-Lopez, D., Sadaba, D., Delso, I., Herrera, R.P., Tejero, T. and Merino, P. (2010) Asymmetric Organocatalytic Synthesis of γ-Nitrocarbonyl Compounds through Michael and Domino Reactions. Tetrahedron: Asymmetry, 21, 2561-2601.

14. Quintard, A., Belot, S., Marchal, E. and Alexakis, A. (2010) Aminal-Pyrrolidine Organocatalysts—Highly Efficient and Modular Catalysts for α-Functionalization of Carbonyl Compounds. European Journal of Organic Chemistry, 2010, 927-936.http://dx.doi.org/10.1002/ejoc.200901283

15. Tsakos, M. and Kokotos, C.G. (2012) Organocatalytic "Difficult" Michael Reaction of Ketones with Nitrodienes Utilizing a Primary Amine-Thiourea Based on Di-tert-Butyl Aspartate. European Journal of Organic Chemistry, 2012, 576-580.http://dx.doi.org/10.1002/ejoc.201101402

16. Sun, Z.W., Peng, F.Z., Li, Z.Q., Zou, L.W., Zhang, S.X., Li, X. and Shao, Z.H. (2012) Enantioselective Conjugate Addition of both Aromatic Ketones and Acetone to Nitroolefins Catalyzed by Chiral Primary Amines Bearing Multiple Hydrogen-Bonding Donors. Journal of Organic Chemistry, 77, 4103-4110. http://dx.doi.org/10.1021/jo300011x

17. Wang, L., Zang, X. and Ma, D. (2012) Organocatalytic Michael Addition of Aldehydes to Trisubstituted Nitroolefins. Tetrahedron, 68, 7675-7679.

18. Dahlin, N., Bøgevig, A. and Adolfsson, H. (2004) N-Arene-Sulfonyl-2-Aminomethylpyrrolidines. Novel Modular Ligands and Organocatalysts for Asymmetric Catalysis. Advanced Synthesis & Catalysis, 346, 1101-1105.http://dx.doi.org/10.1002/adsc.200404098

19. Veverková, E., Štrasserová, J., Šebesta, R. and Toma, Š. (2010) Asymmetric Mannich Reaction Catalyzed by N-Arylsulfonyl-L-proline Amides. Tetrahedron: Asymmetry, 21, 58-61. http://dx.doi.org/10.1016/j.tetasy.2009.12.013

20. List, B., Lerner, R.A. and Barbas III, C.F. (2000) Proline-Catalyzed

Direct Asymmetric Aldol Reactions. Journal of the American Chemical Society, 122, 2395-2396.http://dx.doi.org/10.1021/ja994280y

21. Bisai, V., Bisai, A. and Singh, V.K. (2012) Enantioselective Organocatalytic Aldol Reaction Using Small Organic Molecules. Tetrahedron, 68, 4541-4580.http://dx.doi.org/10.1016/j.tet.2012.03.099

22. Huang, X.-Y., Wang, H.-J. and Shi, J. (2010) Theoretical Study on Acidities of (S)-Proline Amide Derivatives in DMSO and Its Implications for Organocatalysis. The Journal of Physical Chemistry A, 114, 1068-1081. http://dx.doi.org/10.1021/jp909043a

23. Lao, J.-H., Zhang, X.-J., Wang, J.-J., Li, X.-M., Yan, M. and Luo, H.-B. (2009) The Effect of Hydrogen Bond Donors in Asymmetric Organocatalytic Conjugate Additions. Tetrahedron: Asymmetry, 20, 2818-2822. http://dx.doi.org/10.1016/j.tetasy.2009.11.029

24. List, B., Pojarliev, P. and Martin, H.J. (2001) Efficient Proline-Catalyzed Michael Additions of Unmodified Ketones to Nitro-Olefins. Organic Letters, 3, 2423-2425.http://dx.doi.org/10.1021/ol015799d

25. Betancort, J.M. and Barbas III, C.F. (2001) Catalytic Direct Asymmetric Michael Reactions: Taming Naked Aldehyde Donors. Organic Letters, 3, 3737-3740.http://dx.doi.org/10.1021/ol0167006

26. Lu, A., Liu, T., Wu, R., Wang, Y., Wu, G., Zhou, Z., Fang, J. and Tang, C.A. (2011) Recyclable Organocatalyst for Asymmetric Michael Addition of Acetone to Nitroolefins. Journal of Organic Chemistry, 76, 3872-3879.

27. Singh, K.N., Singh, P., Singh, P., Lal, N. and Sharma, S.K. (2012) Pyrrolidine Based Chiral Organocatalyst for Efficient Asymmetric Michael Addition of Cyclic ketones to β-Nitrostyrenes. Bioorganic & Medicinal Chemistry Letters, 22, 4225-4228.

28. Sevin, A., Tortajada, J. and Pfau, M. (1986) Toward a Transition-State Model in the Asymmetric Alkylation of Chiral Ketone Secondary Enamines by Electron-Deficient Alkenes. A Theoretical MO Study. Journal of Organic Chemistry, 51, 2671-2675.http://dx.doi.org/10.1021/jo00364a011

29. Sevin, A., Masure, D., Giessnerprettre, C. and Pfau, M. (1990) A Theoretical Investigation of Enantioselectivity— Michael Reaction of Secondary Enamines with Enones. Helvetica Chimica Acta, 73, 552-573. http://dx.doi.org/10.1002/hlca.19900730303

Chapter 2

THE PIANCATELLI REARRANGEMENT: NEW APPLICATIONS FOR AN INTRIGUING REACTION

Claudia Piutti [1], and Francesca Quartieri [2]

[1]Department of Medicinal Chemistry, Nerviano Medical Sciences srl, Business Unit Oncology, Viale Pasteur 10, 20014 Nerviano, MI, Italy

[2]Department of Chemical Core Technologies, Nerviano Medical Sciences srl, Business Unit Oncology, Viale Pasteur 10, 20014 Nerviano, MI, Italy

ABSTRACT

Nearly forty years ago, at the University of Rome, Giovanni Piancatelli and co-workers discovered the acid-catalyzed water-mediated rearrangement of 2-furylcarbinols into 4-hydroxycyclopentenones. These motifs are core components of several pharmacologically active compounds and precursors of many natural products. The main features of this reaction are the simple experimental conditions, the stereochemical outcome and the generality of the procedure. Consequently, a re-emergence of this reaction has been seen recently, including developments of the Piancatelli rearrangement with some interesting inter- and intramolecular variants. This review will mainly focus on the general aspects of the reaction along with its more recent applications.

INTRODUCTION

In 1976, while studying the reactivity of heterocyclic steroids, Piancatelli and co-workers observed for the first time the rearrangement of a 2-furylcarbinol into a 4-hydroxycyclopent-2-enone in an acidic aqueous system. Following the original report on this transformation [1], the same group continued to systematically investigate the rearrangement [2]. In summary, the heating of 2-furylcarbinols (compounds 1a–c) in an acetone-water solvent system in the presence of strong acids (e.g., formic, polyphosphoric or p-toluenesulfonic acid) led to the formation of 4-hydroxy-5-substituted- cyclopent-2-enones (compounds 2a–c), useful intermediates for the synthesis of natural products (Scheme 1).

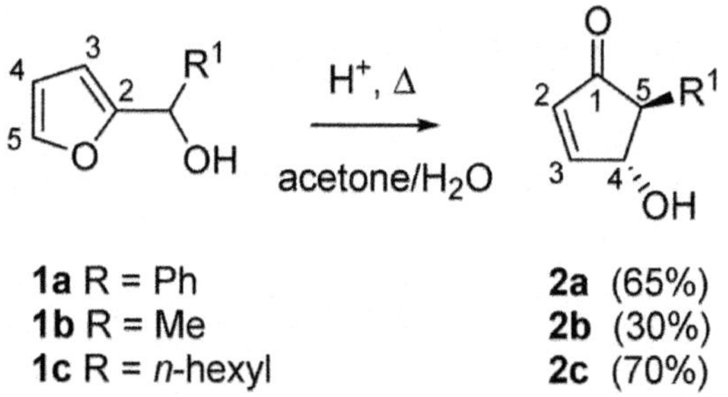

1a R = Ph
1b R = Me
1c R = *n*-hexyl

2a (65%)
2b (30%)
2c (70%)

Scheme 1: The Piancatelli rearrangement.

The high level of stereochemical control inherent in the rearrangement delivered exclusively the *trans* isomer, as demonstrated by the ^{1}H-NMR coupling constant between the two vicinal hydrogens (J_{trans} = 2.5 Hz). The proposed mechanism involves the formation of a carbocation driven by a protonation-dehydration sequence of the 2-furylcarbinol, the nucleophilic attack of a water molecule then generates intermediate **A** which undergoes ring opening (Scheme 2). The resulting 1,4-dihydroxypentadienyl cation **B**, that adopts a conformation in which the two hydroxy groups are *anti*, provides the *trans*-4-hydroxy-5-substituted-cyclopent-2-enone (**2**) as a racemate, through a 4π-conrotatory cyclization [2].

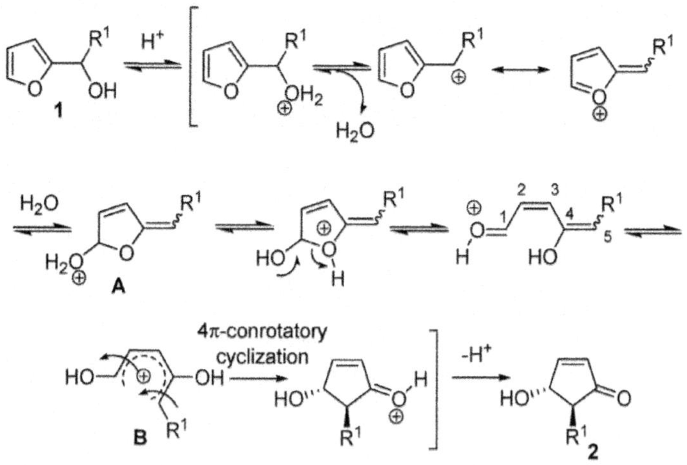

Scheme 2: Proposed mechanism of the rearrangement.

The mechanism shown in Scheme 2 resembles the Nazarov cyclization, in which a divinylketone, activated by Brønsted or Lewis acids, rearranges to a cyclopent-2-enone progressing through a pentadienylic cation (**C**) and a conrotatory ring-closure (Scheme 3) [3,4,5].

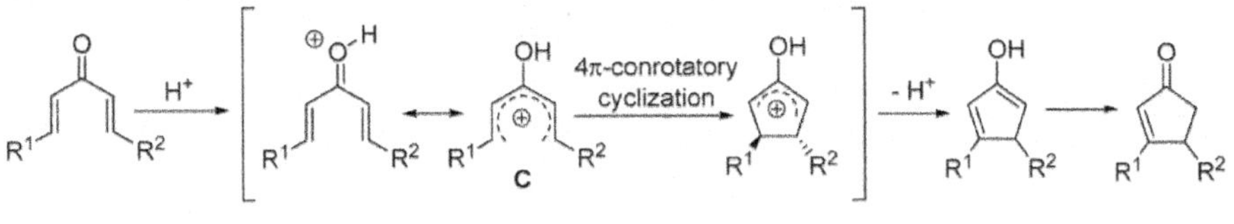

Scheme 3: Nazarov cyclization mechanism.

De Lera and co-workers supported the pericyclic nature of this rearrangement with theoretical calculations of the energy content of the possible isomeric 1,4-dihydroxypentadienyl cations involved in the mechanism [6]. Furthermore, they attributed the high *trans* stereoselectivity of the reaction to a preferred *out,out*-geometry of the cationic intermediate (**B**).

According to the authors' hypothesis, the enolic hydroxy group at C4 can isomerize to the *outwards* orientation, while the one at C1 should derive from a stereoselective furan ring opening that provides the less congested *out,out*-isomeric form, thus disfavoring isomer with the hydroxy group with an *inwards* position [6].

Despite the electrocyclic process is the most widely accepted, two other possible mechanisms have been proposed for the rearrangement. The first one was described only by D'Auria [7], who invoked zwitterionic intermediates in an attempt to rationalize the formation of the *cis* isomer, along with the more abundant *trans* one, when performing the rearrangement on **1c** (R = small alkyl groups) in boiling water without any acid catalysis (Scheme 4).

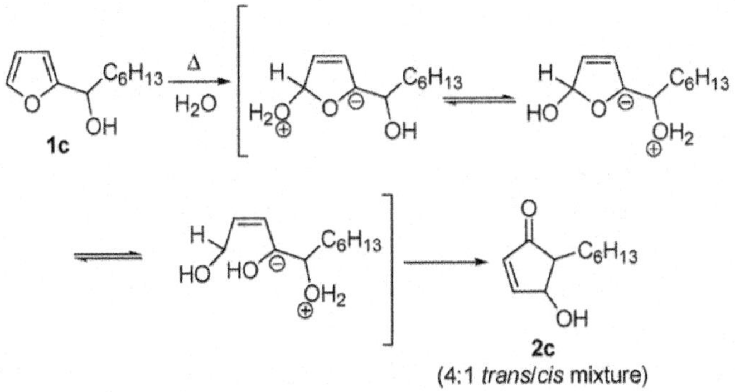

Scheme 4: Proposed zwitterion-mediated mechanism.

The second alternative mechanism (Scheme 5) was proposed by Yin and co-workers, while investigating the rearrangement of 2-furylcarbinols bearing an hydroxyalkyl chain at the 5 position [8]. For example, conversion of 3 into oxabicyclic cyclopentenone 6, progressing through intermediate spiroketal enol ether 4 [9,10], was rationalized by envisioning an aldol-type intramolecular addition. Thus, 4 generates intermediate (D) in the presence of a Brønsted or Lewis acid. Upon addition of water and prototropic shift, D evolves into keto-enol E that finally undergoes an intramolecular aldol reaction, leading to intermediate 4-hydroxy-5-substituted-cyclopent-2-enone (5). This in turn delivers the more stable derivative 6 [8,11].

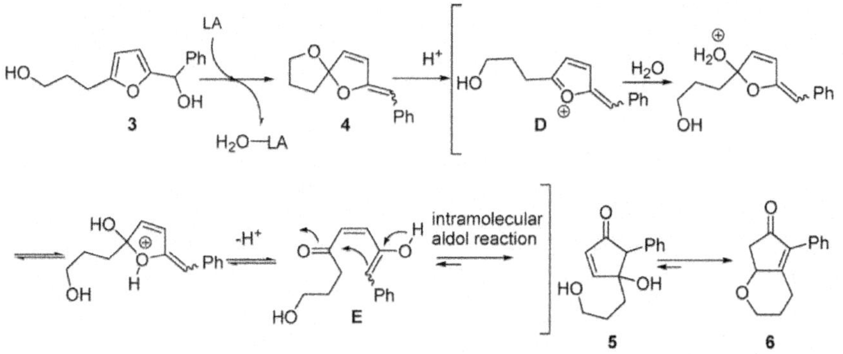

Scheme 5: Proposed aldol-type mechanism.

The scope and limitations of the Piancatelli rearrangement were previously reviewed in 1982 and 1994 [12,13]. In this review the latest and most interesting features of the reaction, including its new applications will be described.

AN OVERVIEW OF THE PIANCATELLI REARRANGEMENT

Piancatelli observed that more reactive substrates like 5-methyl-2-furylcarbinols required milder conditions to rearrange in order to avoid side-products. In such cases, weak Lewis acids as $ZnCl_2$ could drive the reaction to the desired 4-hydroxy-4-methyl-5-substituted-cyclopent-2-enones, although an equimolar ratio of $ZnCl_2$ and substrate was required [14]. It was also noted that alkyl groups on the hydroxy-bearing carbon atom render the starting material more stable and less prone to dehydrate, thus resulting in longer reaction time. In addition the corresponding cationic intermediates are more reactive, consequentially leading to lower yields and the formation of side-products [14]. Among the 5-substituted-2-furylcarbinols, only 5-methyl-2-furylcarbinols [14] and 4-bromo-5-phenyl-2-furylcarbinols undergo rearrangement [15]. Moreover, 5-nitro-2-furyl-derivatives result in no reaction even under harsh conditions, while the 5-methoxy and 5-chloro-2-furylcarbinols lead to the formation of 4-ylidenebutenolides [16,17,18].

The corresponding 3-bromo (**7a**) and 4-bromo-2-furylcarbinols (**7b**) undergo a stereoselective rearrangement, although under more forcing conditions. The *trans* relationship between the substituents in the resulting cyclopenten-2-ones **8a** and**8b** is however maintained (Scheme 6) [15].

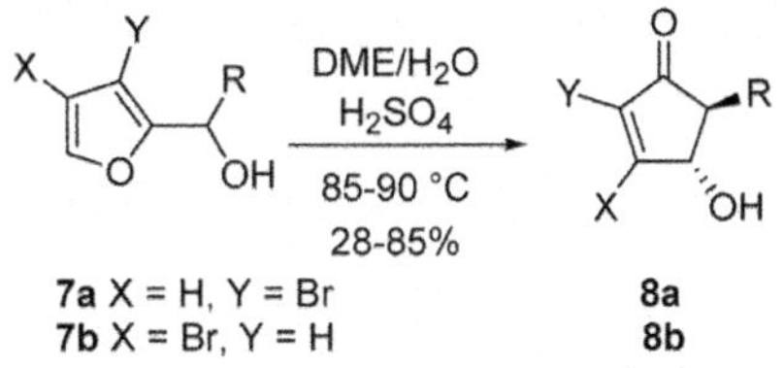
DME/H2O
H2SO4
85-90 °C
28-85%
7a X = H, Y = Br
7b X = Br, Y = H
8a
8b

Scheme 6: Rearrangement on brominated substrates.

The same rearrangement was applied to 2-furyl-alkenyl carbinols **9** leading to 5-alkenyl-derivatives **10**, useful intermediates for the synthesis of prostaglandin analogues as the upper side chain at C5 is suitable for further manipulations. Since these compound structures (**9**) are remarkably reactive, a simple solvolysis in an acetone/water mixture was sufficient for the rearrangement to occur (Scheme 7) [19].

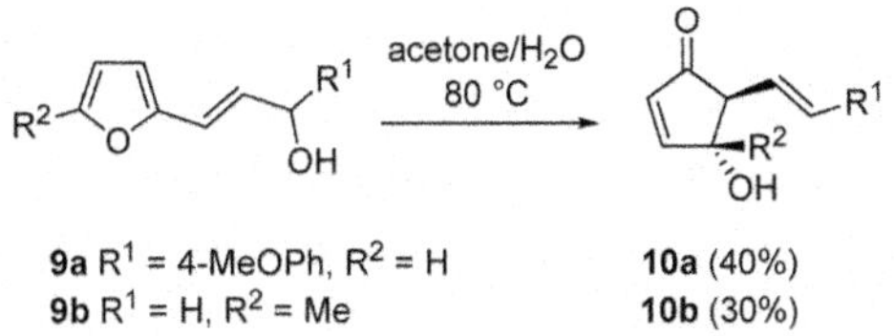
acetone/H2O
80 °C
9a R1 = 4-MeOPh, R2 = H
9b R1 = H, R2 = Me
10a (40%)
10b (30%)

Scheme 7: Rearrangement on 2-furyl-alkenyl carbinols.

Piancatelli's group also investigated the reactivity of 2-furyl-hydroxymethylphosphonates (11). In this case however, an acidic treatment of 11 led to levulinic acid derivatives 12 (Marckwald-type products) or to diethyl 2,5-dioxohex-3-enylphosphonate 13 according to the substitution pattern on the furan ring. However, in order to access compounds 15a–d, the hydroxy group had to be converted into a more reactive leaving moiety (e.g., chlorine, 14a–d, Scheme 8) [20].

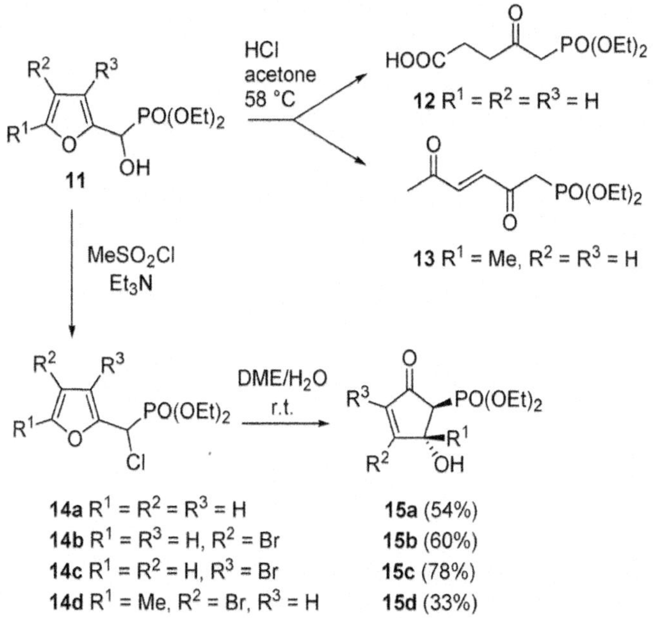

Scheme 8: Rearrangement on 2-furyl-hydroxymethylphosphonates.

During the 80s the rearrangement was widely studied by researchers at Sumitomo Chemical Company, Ltd. (Osaka, Japan), in collaboration with Prof. Piancatelli, resulting in several patent applications [21,22,23,24]. After an extensive experimentation, they found that treatment of 2-furylcarbinols (**1**) in an aqueous medium within a specific pH range (3.5–5.8) could afford 4-hydroxycyclopent-2-enones (**2**) in good yields, which also included the normally less reactive substrates (when R was an alkyl, alkenyl and alkynyl group). Furthermore, these experimental conditions increased the reaction rate and minimized the formation of by-products.

Recently, the rearrangement was also studied in a batch reactor under microwave irradiation (300 W), thus dramatically shortening the reaction time (minutes *vs.* several hours) and improving the yields (up to 95%). The scale-up of the rearrangement was optimized by employing a microreactor that allowed the development of a continuous flow process [25]. A very intriguing conversion involving 4-hydroxy-5-substituted-cyclopent-2-enones (**2**) is their isomerization to 4-hydroxy-2-substituted-analogues (16), excellent intermediates for prostaglandins synthesis (Scheme 9). This transformation had already been described by Stork *et al.*, who treated 2 with chloral in the presence of triethylamine [26]. Piancatelli *et al.*, also observed this migration during the chromatographic purification of 2 on neutral (for 2a) or basic alumina (for 2b–c) [27]. The corresponding isomerized products 16 were isolated in high yield.

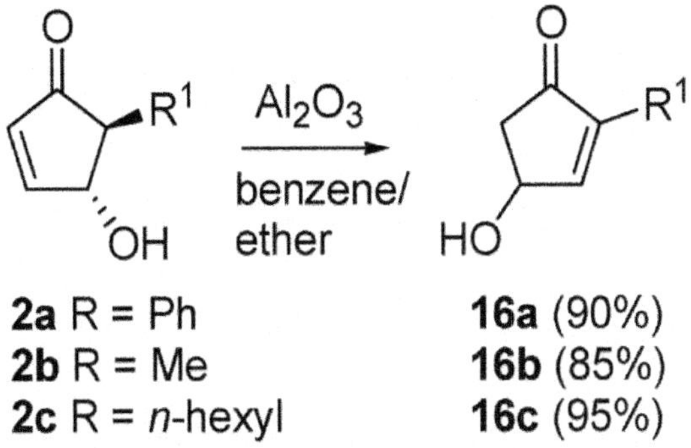
R1
Al_2O_3
benzene/
ether
OH
HO
R1
2a R = Ph
2b R = Me
2c R = n-hexyl
16a (90%)
16b (85%)
16c (95%)

Scheme 9: Isomerization to 2-substituted-4-hydroxycyclopent-2-enones.

By extending this isomerization to a series of derivatives and after a careful examination of the mechanism, it was demonstrated that the reaction occurred *via* an intramolecular shift of the hydroxy group on the intermediate enolate **17**(Scheme 10) [28], thus ruling out a dehydration-hydration sequence involving the nucleophilic attack of an external water molecule, as previously supposed [29].

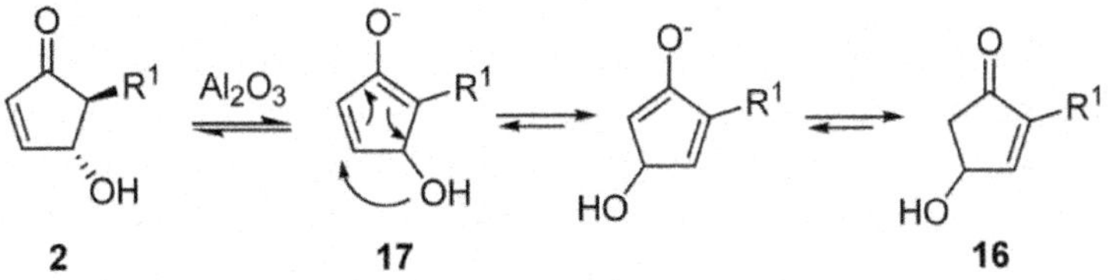
R1
Al_2O_3
OH
2
17
HO
16

Scheme 10: Possible isomerization mechanism.

In fact, when adsorbed on methanol-deactivated alumina under anhydrous conditions, 2 delivered isomer 16quantitatively, with no detectable amounts of the 4-methoxy derivative. On the other hand, 4-acetoxy-5-phenylcyclopent-2-enone isomerizes to 4-acetoxy-2-phenylcyclopent-2-enone after adsorption on neutral alumina and elution with a mixture of benzene and diethyl ether, thus confirming the proposed mechanism. With aromatic R groups, neutral alumina was sufficient for the isomerization to occur, while with aliphatic substituents the employment of basic alumina was necessary to promote the hydroxy shift [28].

4-Hydroxy-2-substituted-derivatives 16 can be synthesized in a one-pot procedure from 2-furylcarbinols (1) simply by switching from an acidic pH, necessary for the rearrangement, to a basic environment, to promote the isomerization. For example, 5-methyl-2-furylallylcarbinol (18) was converted into cyclopentenone 19 in refluxing water over 12 hours at pH 5. Intermediate 19 was not isolated, but it was isomerized to 2-allyl-4-hydroxy-3-methyl-cyclopent-2-enone (20) by adjusting the pH to 7.9 and prolonging the reflux for an additional two hours (Scheme 11) [23,24].

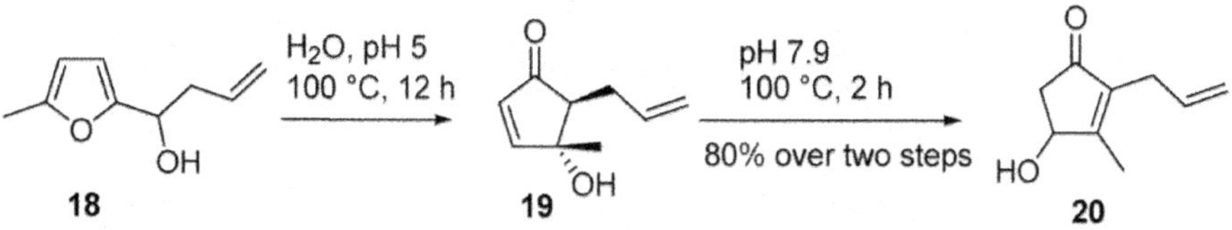

Scheme 11: Example of one-pot isomerization.

The straight transformation of 2-furylcarbinols (1) into 4-hydroxy-2-substituted-cyclopent-2-enones (16) can also be achieved in a buffered aqueous solution in the presence of $MgCl_2$ heated at high temperature (150 °C) in an autoclave, thus allowing to isolate the desired products in high yield (80%–84%) [23,24].

Applications of the Original Piancatelli Rearrangement

One of the most important applications of the Piancatelli rearrangement is in the synthesis of prostaglandins and their derivatives. Demonstration of the versatility of the domino sequence "2-furylcarbinol rearrangement/isomerization" was shown by Piancatelli himself, who synthesized key intermediates for the preparation of the prostanoic acid skeleton starting from 2-furylcarbinols bearing a second functional group in the side chain [29].

Some of the products arising from the application of this domino sequence are: 3E,5Z-misoprostol (21) [30], enisoprost (22) [31], 4-fluoro-enisoprost [32], 2-normisoprostol [33], prostaglandin E$_1$ (23) [34], ent-phytoprostane E$_1$ (24) and 16-epi-phytoprostane E$_1$ [35], bimatoprost (25, Lumigan™) and travoprost (26, Travatan™) [36,37] (Figure 1).]

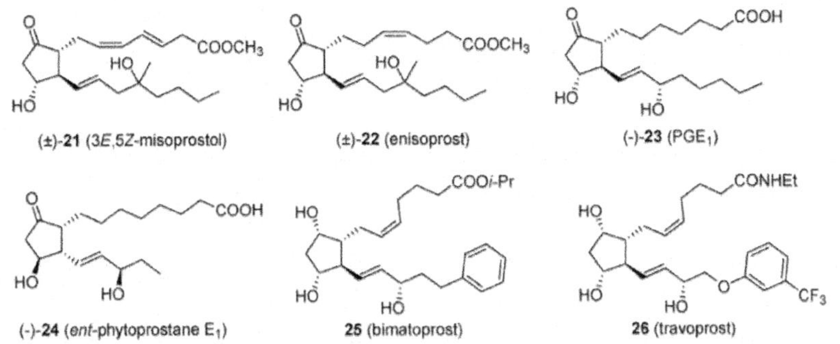

Figure 1: Products obtained *via* the Piancatelli rearrangement.

Bimatoprost (**25**) and travoprost (**26**) can be synthesized from the common intermediate 4-silyloxycyclopentenone **28**, whose production was set-up on a kg-scale by exploiting a "Piancatelli rearrangement/chloral-mediated isomerization" sequence, starting from furfural, and subsequent enzymatic resolution of the resulting 4-hydroxycyclopentenone **27** [38] (Scheme 12).

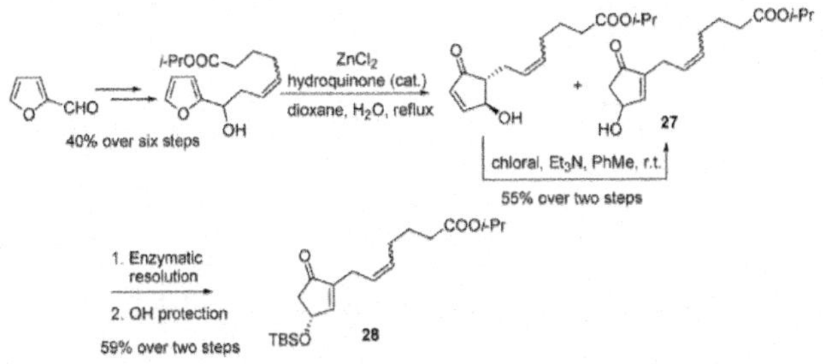

Scheme 12: Large scale synthesis of intermediate **28**.

4-Hydroxycyclopentenones bearing functionalized side chains at the position 2 can also be prepared from *bis*-thioalkylfurans [39]. 5-heptyl-4-hydroxycyclopent-2-enone core structure (**29**) was obtained by Piancatelli rearrangement for the synthesis of selective and potent PPARγ agonists (**30**) (Scheme 13) [40].

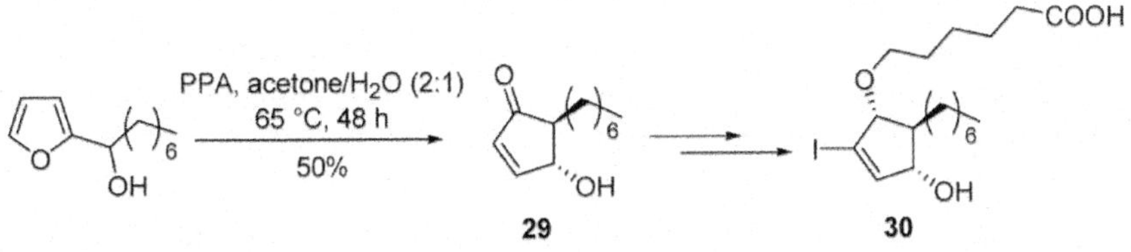

Scheme 13: Synthesis of PPARγ agonists.

The rearrangement of 2-furylcarbinols has also been applied to the total synthesis of natural products [41,42]. Very recently, the core framework of the proposed structure of sargafuran (31) was accomplished *via* the Piancatelli rearrangement on intermediate 32 in the presence of $MgCl_2$ as the key-step [43]. The use of this Lewis acid allowed to isolate the rearranged product 33 in moderate yield (58%), while other acids (e.g., polyphosphoric acid, $ZnCl_2$) led to lower recovery (up to 34%) because of the competitive dehydration of 32 and the formation of unidentified side-products. Subsequent protection of the hydroxy group of intermediate 33, addition of the furan moiety, dehydration and deprotection led to 34, a simplified analogue of sargafuran [43] (Scheme 14).

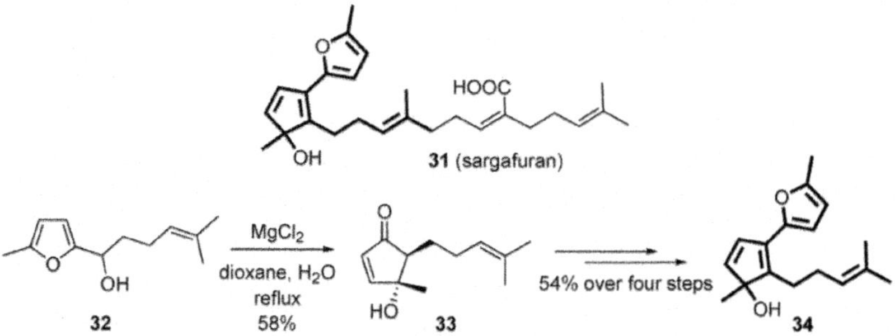

Scheme 14: Synthesis of the racemic core framework (34) of sargafuran.

RECENT VERSIONS OF THE PIANCATELLI REARRANGEMENT

Several new and efficient applications of this rearrangement have recently appeared in the literature utilizing alternative nucleophiles to water, and the synthetic utility of this reaction has been widely developed. Pharmacologically valuable scaffolds, which were usually prepared via multi-step sequences, can be synthesized in very mild conditions, directly and smartly, in only one step.

Intermolecular Aza-Piancatelli Rearrangement

Recently Read de Alaniz's group employed 2-furylcarbinols and a series of anilines to access *trans*-4-amino-5-substituted-cyclopent-2-enones (35), appealing structures for the synthesis of biologically active compounds [44] (Scheme 15).

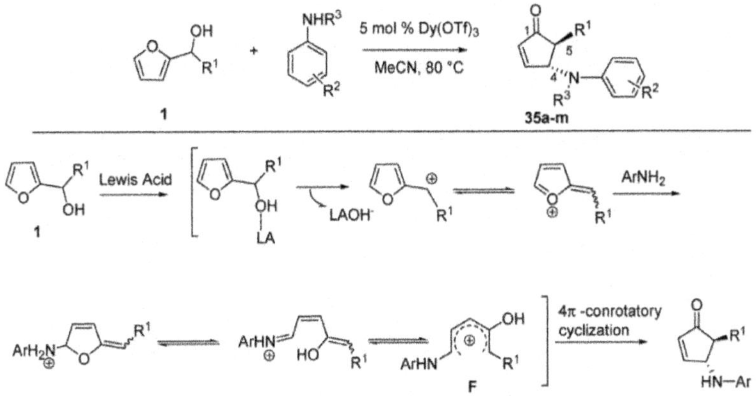

Scheme 15: Aza-Piancatelli rearrangement and a proposed reaction mechanism.

This was allowed thanks to the identification of catalysts such as lanthanoid salts which were able to selectively activate 2-furaldehydes in the presence of an excess of nucleophilic amines [45]. Under optimized conditions, the aza-version of the Piancatelli rearrangement was carried out in acetonitrile at 80 °C, together with a catalytic amount (5 mol %) of Dy(OTf)$_3$[46], preferred over Sc(OTf)$_3$ which gave similar results, but it is more expensive.

The mechanism is proposed to involve the elimination of the hydroxy-group through coordination and activation by the Lewis acid. The resulting furylcation undergoes nucleophilic attack by the aniline at the 5 position of the ring, thus starting the cascade reaction that forms the product (Scheme 15). The 1-amino-4-hydroxy pentadienyl cation F is analogous to B(Scheme 2) and it is supposed to undergo the 4π-conrotatory electrocyclization, thus explaining the high *trans*-diasteroselectivity of the reaction.

The synthesized products (Table 1, 35a–m) were isolated as single diastereomers and they demonstrate that the rearrangement is compatible with several substituents. Yields ranging from 62% to 93% were obtained with different anilines in combination with 2-furylcarbinols (1) bearing R^1 as a simple phenyl (35a–c), a substituted aryl possessing an electron-donating (35d–e) or an electron-withdrawing group (35f–g). Furthermore, the presence of an aliphatic group as R^1(35h–k) was also tolerated and the reaction worked well with both primary and secondary anilines (35l–m).

Table 1: Aza-Piancatelli rearrangement products and yields

Compound	R^1	R^2	R^3	Yield (%) *
35a	Ph	H	H	86
35b	Ph	4-I	H	93
35c	Ph	4-MeO	H	62
35d	4-MeOC$_6$H$_4$	4-I	H	68
35e	4-MeOC$_6$H$_4$	2,4,6-triMe	H	89
35f	4-CF$_3$C$_6$H$_4$	4-I	H	83
35g	4-CF$_3$C$_6$H$_4$	2,4,6-triMe	H	78
35h	Me	4-I	H	68
35i	Me	2,4,6-triMe	H	74
35j	i-Pr	4-I	H	73
35k	i-Pr	2,4,6-triMe	H	89
35l	Ph	H	CH$_3$	74
35m	Ph	4-Br	CH$_3$	88

* Conditions: 5% mol Dy(OTf)$_3$, MeCN, 80 °C.

No isomerization products (4-amino-2-substituted-cyclopent-2-enones) were found using the aza-Piancatelli rearrangement because of the mild reaction conditions [46]. The reaction was successful with sterically hindered 2,4,6-trimethylaniline (35e, 35g, 35i, 35k), while a competitive Friedel-Craft alkylation occurred when 2,6-dimethylaniline was employed delivering 36 and 37 (Scheme 16).

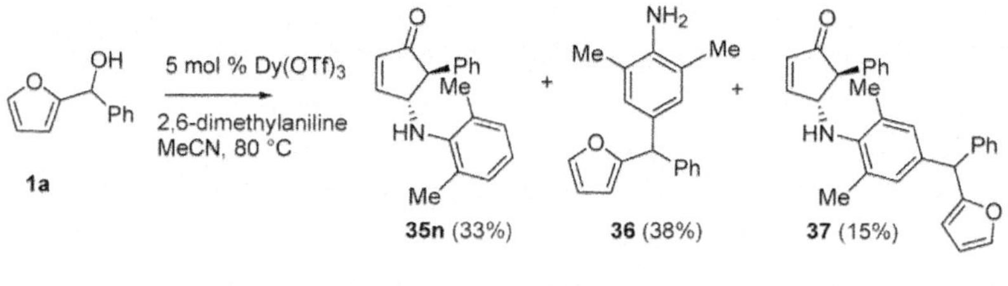

Scheme 16: Rearrangement with 2,6-dimethylaniline.

Compound **35c** (Table 1) was converted into racemic **38** which represents a key intermediate for the preparation of a mimetic of the morpholinic core of the hNK1 antagonist Aprepitant (**39**) (Scheme 17) [47,48].

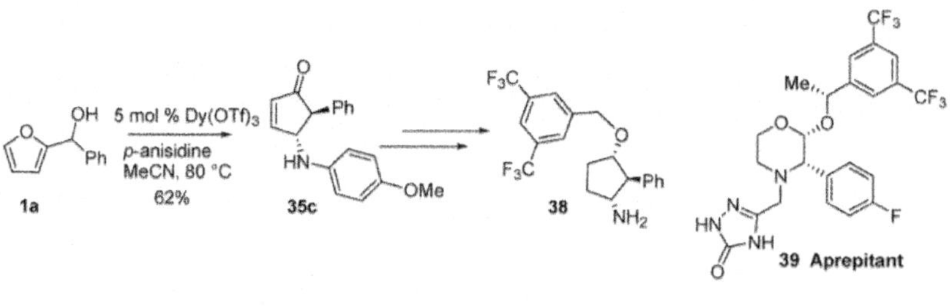

Scheme 17: Application to the synthesis of an hNK1 inhibitor.

Subba Reddy *et al.* [49] performed the aza-Piancatelli rearrangement using phosphomolybdic acid (PMA) as the catalyst (0.03 mol %) in acetonitrile at reflux, isolating *trans*-4-amino-5-substituted cyclopent-2-enones in greater than 80% yield, in about 1 h.

A further advance in the aza-Piancatelli rearrangement has been recently published and allowed the synthesis of cyclopent-2-enones with a quaternary carbon atom at the 5 position in high diasteroselectivity [50]. Taking advantage of the reactivity of polarized donor-acceptor (D-A) cyclopropanes (40) [51,52] due to ring strain [53] and their behavior as carbocation upon Lewis acid activation, the Read de Alaniz group found an alternative method to trigger the rearrangement and obtain highly functionalized cyclopent-2-enones (41) (Scheme 18, A). The protocol obviated the problems that Piancatelli and D'Auria [15] encountered when they tried to prepare cyclopent-2-enones with a quaternary carbon through the Piancatelli rearrangement starting from a furan with a tertiary carbinol side chain (42) and a large amount of dehydrated compound was found together with the rearranged product (Scheme 18, B).

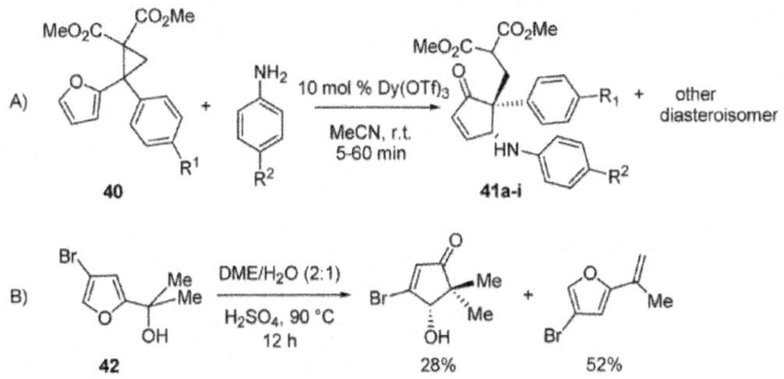

Scheme 18: Rearrangement of differently functionalized furans.

The best results in terms of yields (57%–89%), diasteroselectivity (6:1 up to 60:1) and reaction times (5–60 min) were found with 10 mol % Dy(OTf)$_3$ in acetonitrile at room temperature. The influence of R^1 and R^2 was investigated (Table 2).

Para-substituted anilines with electron-withdrawing groups led to higher diastereoselectivities (41c *vs.* 41a, 41i *vs.* 41h) than the corresponding anilines bearing electron-donating groups (41b *vs.* 41a, 41g *vs.* 41h). But when R^1 was a methoxy moiety, good yields and diasteroselectivities (41d–f) were observed with anilines possessing either an electron-donating or electron-withdrawing group.

Table 2: Rearrangement with (D–A) cyclopropanes.

Compound	R^1	R^2	Yield (%) *	dr
41a	H	H	89	13:1
41b	H	MeO	57	6:1
41c	H	CF$_3$	72	60:1
41d	4-MeO	MeO	87	25:1
41e	4-MeO	H	84	32:1
41f	4-MeO	CF$_3$	63	22:1
41g	4-CN	MeO	58	2:1
41h	4-CN	H	76	1:1
41i	4-CN	CF$_3$	65	5:1

* Conditions: 10 mol % Dy(OTf)$_3$, MeCN, r.t.

When R^1 was an electron-withdrawing group (e.g., CN, CF$_3$), diasteroselectivities were poor (41g–i) and the resulting products were less stable. In fact an intramolecular Michael addition frequently occurred, even during column chromatography, delivering bicyclic compounds that could not be isolated from the reaction mixture. With $R^1 = R^2 = CF_3$, compound 43 was obtained in high yield (83%) by forcing this side-reaction under basic conditions.

The X-ray crystal structure analysis demonstrated that only the major diastereomer deriving from the rearrangement underwent cyclization [50] (Figure 2).

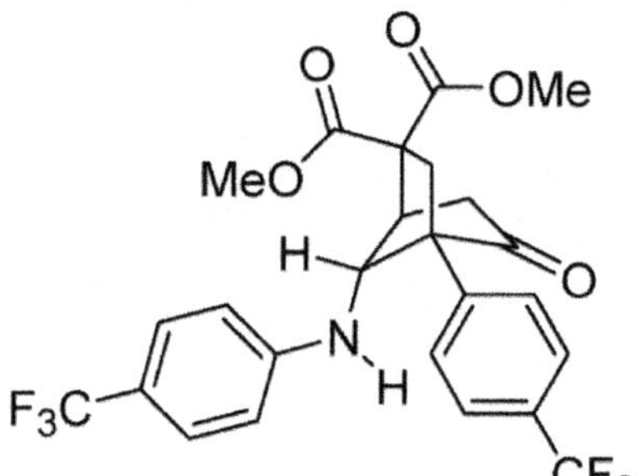

43

Figure 2: Michael addition side-product.

Diastereoselectivity was observed to be dependent on the reaction temperature. In fact, when the rearrangement was carried out at 80 °C, the diastereomeric ratio improved to 30:1 for 41a, to 16:1 for 41b and to 22:1 for 41i.

Intramolecular Aza-Piancatelli Rearrangement

In 2011 Read de Alaniz published the first example of an intramolecular version of the aza-Piancatelli rearrangement [54]. This approach was based upon Piancatelli's observation that suitable 5-substituted-2-furylcarbinols could rearrange under certain conditions [14] and upon the protocol of Yin on the synthesis of oxabicyclic cyclopentenones (6) starting from 2-furylcarbinols (3) bearing a hydroxyalkyl side chain at the 5 position of the furan ring (Scheme 5) [8,11].

The authors worked with 2-furylcarbinols bearing an aminoalkyl chain at the 5 position of the furan ring (44–45) and generated azaspirocyclic scaffolds (46–47, Scheme 19) [54]. This densely functionalized framework was obtained in only one step, as a single diastereomer, with high efficiency and in high yield, differently from the known procedures that require several synthetic steps for the construction of the tertiary carbon center bearing the nitrogen atom, and the formation of the spirocyclic ring [55,56].

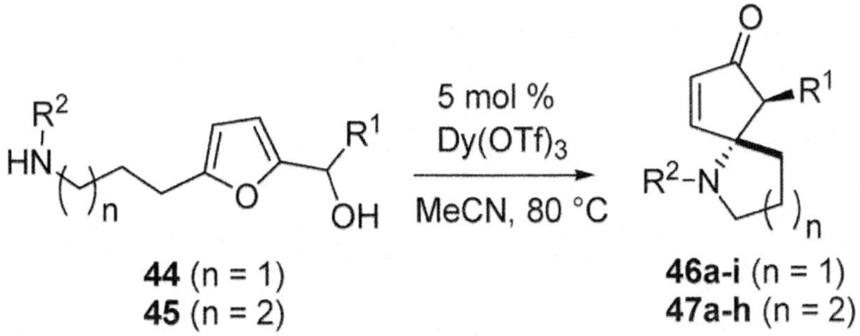

Scheme 19. Intramolecular aza-Piancatelli rearrangemet and synthesis of azaspiro-cycles.

The optimized conditions for the intermolecular rearrangement (Dy(OTf)$_3$ 5 mol % in refluxing acetonitrile) [44] turned out to be suitable also for the intramolecular version, and the mechanism previously hypothesized was also assumed to be in action (Scheme 15). As shown in Table 3 for the formation of 5-azaspirocycles **46** (n = 1), the rearrangement worked equally well for most common substituents, but the nature of R^2 had a significant impact on the outcome of the reaction, both in terms of yield and reaction rate.

Table 3. 5-Azaspirocycles (**46**, $n = 1$) *via* aza-Piancatelli rearrangement

Compound	R^1	R^2	Time	Yield (%) *
46a	Ph	Ph	15 min	96
46b	Ph	4-MeOC$_6$H$_4$	150 min	74
46c	4-NO$_2$C$_6$H$_4$	Ph	5 h	67
46d	4-BrC$_6$H$_4$	Ph	5 h	84
46e	H	Ph	15 h	90
46f	H	4-MeOC$_6$H$_4$	48 h	57 [#]
46g	4-MeOC$_6$H$_4$	4-CF$_3$C$_6$H$_4$	15 min	78
46h	4-MeOC$_6$H$_4$	4-IC$_6$H$_4$	5 min	84
46i	CH$_3$	Ph	15 min	91

* Conditions: 5 mol % Dy(OTf)$_3$, MeCN, 80 °C; [#] 10 mol % cat. Required.

When R^2 was an electron-rich group, as in the case of PMP-cleavable *N*-protecting group, yields were lower and longer reaction times were required (46a *vs.* 46b, 46e *vs.* 46f), while the presence of CF$_3$- or I- on the aryl at R^2 led to faster rearrangement (46g and 46h). NO$_2$- or Br-substituted aryl groups at R^1 increased the reaction times (46c and 46d). Furthermore, a methyl group at R^1 was very well tolerated (46i). A similar effect of the R1 and R2 substitution occurred in the synthesis of 6-azaspirocycles 47 ($n = 2$). Electron-rich anilines required longer reaction times than unsubstituted ones (47a *vs.* 47b; 47d *vs.* 47f, Table 4), and in the case of 47f a 20 mol % catalyst was required to achieve a modest yield. On the contrary, the presence of a halogen had a beneficial effect on the reaction rate (47c, 47e and 47h).

Table 4. 6-Azaspirocycles (**47**, $n = 2$) via intramolecular aza-Piancatelli rearrangement

Compound	R^1	R^2	Time	Yield (%) *
47a	Ph	Ph	1 h	75
47b	Ph	4-MeOC$_6$H$_4$	75 h	74
47c	Ph	4-IC$_6$H$_4$	5 min	69
47d	H	Ph	48 h	70
47e	H	4-IC$_6$H$_4$	8 h	90
47f	H	4-MeOC$_6$H$_4$	72 h	37 [#]
47g	nBu	Ph	15 h	54
47h	nBu	4-IC$_6$H$_4$	2 h	65

* Conditions: 5 mol % Dy(OTf)$_3$, MeCN, 80 °C. [#] 20 mol % cat. Required.

Intramolecular Oxa-Piancatelli Rearrangement

Read de Alaniz also explored the oxa-Piancatelli version of the rearrangement on substrates bearing a hydroxy-alkyl moiety at the 5 position of the furan ring (3, see Scheme 5, and 48) [57] and thereby accessing oxaspirocycles (49 and 50) (Scheme 20) in only one step and in a highly diasteroselective manner, thus avoiding the usually performed multistep procedures [58,59,60].

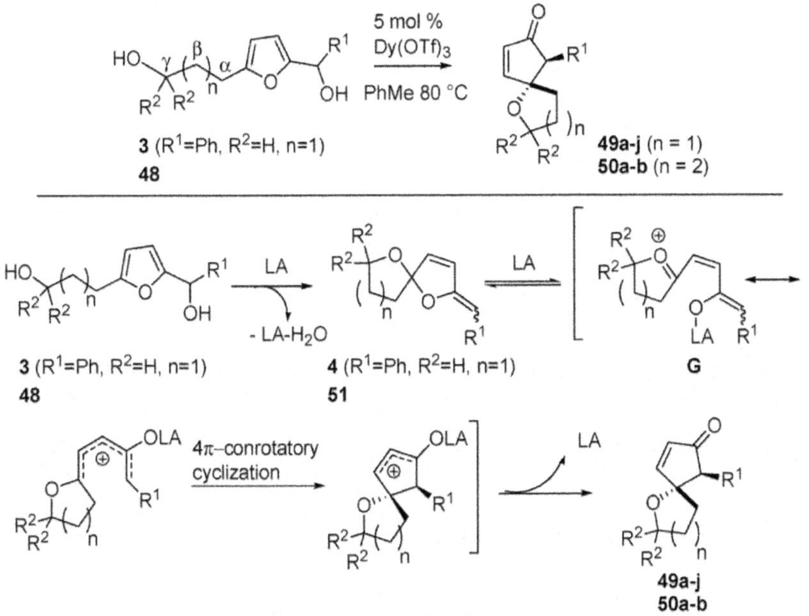

Scheme 20: Intramolecular oxa-Piancatelli rearrangement and proposed mechanism.

According to Scheme 20, 2-furylcarbinols 3 and 48 lead to the formation of spiroketal enol ether 4 and 51 upon acidic catalysis. The coordination of $Dy(OTf)_3$ delivers intermediate G which, according to the canonical 4π-conrotatory cyclization, affords spirocyclic ethers 49 and 50. The optimized conditions required 5 mol % $Dy(OTf)_3$ in toluene at 80 °C. It has to be mentioned that the use of refluxing acetonitrile, the best solvent for the aza-Piancatelli rearrangement, in this case led to decomposition.

In most cases the products were isolated in reasonable to excellent yields and with the expected high *trans*-selectivity (Table 5). The reaction proceeded efficiently (compounds 49a–c), but in the case of strong electron-withdrawing groups as R^1, harsher conditions (higher temperature and longer reaction time) were required in respect to the aza-rearrangement (compare 49d and 46c, Table 3). The presence of a heterocycle was also tolerated (49e).

Table 5: Oxaspirocycles (**43** and **47**) via oxa-Piancatelli rearrangement.

Compound	R^1	R^2	n	Time	Yields (%) *
49a	Ph	H	1	6 h	91
49b	4-MeOC$_6$H$_4$	H	1	20 h	89
49c	4-BrC$_6$H$_4$	H	1	24 h	83
49d	4-NO$_2$C$_6$H$_4$	H	1	7 days	75 [#]
49e	2-thiophene	H	1	4 h	90
49f	Ph	CH$_3$	1	2 h	74
49g	i-Pr	H	1	-	0
49h	i-Pr	CH$_3$	1	1 h	25
49i	n-Bu	H	1	-	0
49j	n-Bu	CH$_3$	1	2 h	98
50a	Ph	H	2	-	0
50b	Ph	CH$_3$	2	16 h	20

*Conditions: 5 mol % Dy(OTf)$_3$, toluene, 80 °C. [#] 16 h, 100 °C.

In contrast to the aza-Piancatelli reaction, aliphatic R^1 groups delivered complex reaction mixtures (49g and 49i), except in the case where a *gem*-dimethyl moiety was present in position γ (Scheme 20). The beneficial effect of this substitution is not clear, but the Authors hypothesize a more favoured formation of intermediate G because the *gem*-dimethyl group prevents Lewis acid from coordinating the vicinal oxygen atom. The *gem*-dimethyl effect is pronounced also in the formation of the 6-oxaspirocycles (50a and 50b).

Further investigations of the effects of substituents along the aliphatic chain (α, β and γ positions, Scheme 20) were carried out revealing that only groups in the γ or β were tolerated, while the presence of a α substitution hampered the rearrangement.

Domino Aza-Piancatelli/Hetero-Michael Reaction

Recently Subba Reddy's group illustrated a further application of the aza-Piancatelli reaction by a new route to 3,4-dihydro-2H-benzo[*b*][1,4]oxazine and thiazine derivatives, core structures shared among several biologically and pharmacologically active compounds [61]. This new procedure is based on a domino aza-Piancatelli/hetero-Michael addition conducted by reacting 2-furylcarbinols with 2-aminophenols or 2-aminothiophenols (Scheme 21).

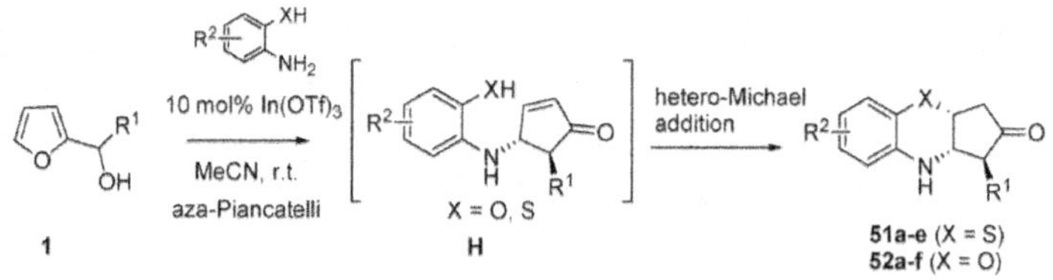

Scheme 21: Domino aza-Piancatelli/hetero-Michael addition products.

Preliminary experiments with 10% phosphomolybdic acid gave products in rather good yields (75%, 2.5 h), but the use of indium salts as catalysts [62,63,64,65] (10 mol% In(OTf)$_3$) in acetonitrile at room temperature allowed the optimization of the synthesis of 3,4-dihydro-2H-benzo[b][1,4] thiazine derivative (51a) up to 86% in terms of yields, with high diastereoselectivity and shorter reaction times (2 h) [61].

The initial steps of the mechanism are the same as proposed for the aza-Piancatelli (see Scheme 15). Then after the 4π-conrotatory cyclization which leads to intermediate H, a thia/oxa-conjugate addition finally gives compounds 51 or 52, that show a *trans* relationship between R^1 and N, resulting from the aza-Piancatelli rearrangement, and a *cis* orientation between N and X (O or S) arising from the conjugate addition (Scheme 21).

In Table 6, some compounds obtained with the domino approach are illustrated. The isolated yields were high both with 2-aminothiophenols and 2-aminophenols, although 2-aminophenols required slightly longer reaction times, being less reactive.

Table 6: Derivatives obtained via domino aza-Piancatelli/hetero-Michael reaction with In(OTf)$_3$.

Compound	R^1	R^2	X	Time (h)	Yields (%) *
51a	Ph	H	S	2.0	86
52a	Ph	5-Me	O	3.5	73
51b	Ph	4-Cl	S	2.5	85
52b	Ph	4-Cl	O	3.5	78
51c	4-MeOC$_6$H$_4$	H	S	2.0	87
52c	4-MeOC$_6$H$_4$	H	O	3.0	75
51d	4-MeOC$_6$H$_4$	4-Cl	S	2.5	88
52d	4-MeOC$_6$H$_4$	4-Cl	O	2.5	82
51e	4-FC$_6$H$_4$	H	S	2.5	85
52e	4-FC$_6$H$_4$	H	O	3.5	81
52f	4-FC$_6$H$_4$	4-Cl	O	3.5	80

* Conditions: 10 mol % In(OTf)$_3$, MeCN, r.t.

Curiously, nearly two months later, another group published [66] the same reaction sequence shown in Scheme 21 using La(OTf)$_3$ (5 mol %) as the catalyst and acetonitrile at reflux as the solvent. For the same substrates, yields were lower if compared with those obtained with In(OTf)$_3$ and reaction times were longer (Table 7). Moreover 2-aminothiophenols failed to react under these conditions.

Table 7: Oxazines obtained via domino aza-Piancatelli/hetero-Michael reaction with La(OTf)$_3$.

Compound	R^1	R^2	Yields (%)[*]
52g	Ph	H	81
52b	Ph	4-Cl	66
52c	4-MeOC$_6$H$_4$	H	46
52d	4-MeOC$_6$H$_4$	4-Cl	51
52e	4-FC$_6$H$_4$	H	46
52f	4-FC$_6$H$_4$	4-Cl	78

* Conditions: 5 mol % La(OTf)$_3$, MeCN, 80 °C, 4 h.

Complex mixtures or no reaction were observed in the attempts of building [1,4]heterocycles different from benzoxazine, starting from building blocks such as o-phenylenediamines, pyrocatechols or 2-aminopyridin-3-ol. The introduction of an electron-withdrawing group (e.g., methanesulfonyl, tosyl) on one of the nitrogens of o-phenylenediamines proved beneficial for the isolation of the desired products (Scheme 22, 53a–d). On the other hand, despite the electron-withdrawing group in R^2, the presence of an electron-poor moiety in R^1 hampered the reaction to occur (53e).

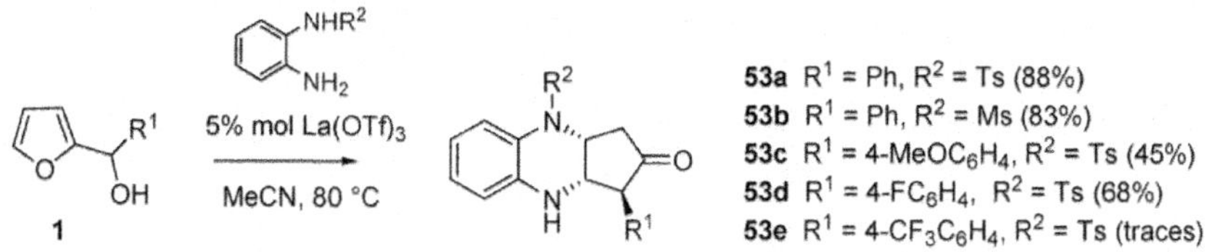

53a R^1 = Ph, R^2 = Ts (88%)
53b R^1 = Ph, R^2 = Ms (83%)
53c R^1 = 4-MeOC$_6$H$_4$, R^2 = Ts (45%)
53d R^1 = 4-FC$_6$H$_4$, R^2 = Ts (68%)
53e R^1 = 4-CF$_3$C$_6$H$_4$, R^2 = Ts (traces)

Scheme 22: Diazines (**53**) via domino aza-Piancatelli/hetero-Michael addition.

Intramolecular C-Piancatelli Rearrangements

A very recent variant of this transformation, which is also the first example of an intramolecular C-Piancatelli rearrangement, has been reported by Yin and co-workers [67]. When treating 2-furylcarbinols functionalized with an electron-rich aromatic tertiary amide (54) in the presence of acidic catalysts, a nucleophilic Friedel-Craft attack from the orthoposition onto the oxa-carbenium intermediate occurs, forming the spirofurooxindole derivative (55) (Scheme 23) [68,69]. Since these compounds are structurally similar to intermediate A (Scheme 2) in the proposed Piancatelli rearrangement mechanism, the authors envisioned the possibility to promote the rearrangement through ring opening and cyclization under suitable reaction conditions, thus accessing the novel core scaffold spirocyclopenten-2-oneoxindoles (56) as a mixture of cis/trans isomers [67]. The use of a catalyst (Pd^{II} complexes or

Lewis acids) led to a mere 40% yield, while a simple solvolysis without the use of a catalyst proved to be more successful. The use of 1,2-dichloroethane (DCE) at 130 °C increased yields over 80%.

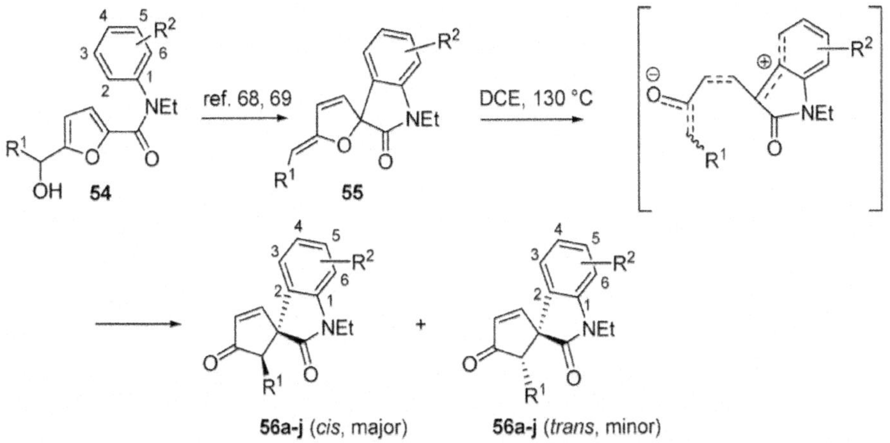

Scheme 23: Synthesis of spirocyclopentenoneoxindoles (**56**) from 2-furylcarbinols (**54**).

The influence of substituents R^1 and R^2 on diasteroselectivity is highlighted in Table 8. The lack of a methoxy group at the 3-position of R^2 leads to a lower diasteroselectivity (56a–b vs. 56c, 56e vs. 56f, 56g vs. 56h), while an increase is seen if an ortho-substituent is present at R^1 (56d vs. 56g). 3,4,5-Trimethoxy derivatives combined with an o-substituent enabled almost exclusively cis-isomer isolation (56e, 56g and 56i).

Table 8: Synthesis of spirocyclopenten-2-oneoxindoles.

Compound	R^1	R^2	cis/trans	Yield (%) *
56a	Ph	3,4,5-tri-MeO	4/1	94
56b	Ph	3,5-di-MeO	5/1	91
56c	Ph	4,5-di-MeO	1.5/1	89
56d	4-ClC$_6$H$_4$	3,4,5-tri-MeO	5.6/1	93
56e	2-FC$_6$H$_4$	3,4,5-tri-MeO	>99/1	87
56f	2-FC$_6$H$_4$	4,5-di-MeO	3/2	85
56g	2-ClC$_6$H$_4$	3,4,5-tri-MeO	>99/1	92
56h	2-ClC$_6$H$_4$	4,5-di-MeO	1.2/1	89
56i	2-MeC$_6$H$_4$	3,4,5-tri-MeO	>99/1	94
56j	2-MeC$_6$H$_4$	3,5-di-MeO	5/1	91

* Conditions: DCE, 130 °C.

CONCLUSIONS

Since its discovery in 1976, the Piancatelli rearrangement has appeared as a versatile reaction for the construction of substituted cyclopent-2-enones convenient for the synthesis of prostaglandin derivatives. Several groups have been investigating new applications of the reaction that culminated in the recent publication of a number of fascinating papers. The use of alternative nucleophiles to water in both inter- and intramolecular reactions has allowed access to several attractive and complex chemotypes, such as azaspirocycles and spirocyclic ethers. Moreover, the combination with a subsequent intramolecular conjugate addition permits the synthesis of thiazines and oxazines in a straightforward manner. These are promising scaffolds for the synthesis of complex natural products and biologically active compounds.

In summary, the Piancatelli rearrangement turned out to be an elegant approach for the stereoselective construction of complex scaffolds that traditional methods would generate *via* multistep processes and with limited diversity on the substituents. Thanks to the use of milder reaction conditions, this recently re-discovered rearrangement holds promises for a number of new applications. An enantioselective version would represent a further progress in the field, for instance in the synthesis of oxaspirocycles with the use of chiral Lewis acids.

ACKNOWLEDGMENTS

Authors would like to thank Alessandra Badari, Francesco Casuscelli, Daniele Donati and Eduard Felder for their valuable comments.

REFERENCES

1. Piancatelli, G.; Scettri, A. Heterocyclic steroids-III: The synthetic utility of a 2-Furyl steroid. *Tetrahedron* 1977, *33*, 69–72.

2. Piancatelli, G.; Scettri, A.; Barbadoro, S. A useful preparation of 4-Substituted-5-hydroxy-3-oxocyclopentene.*Tetrahedron Lett.* 1976, *17*, 3555–3558.

3. Spencer, W.T., III; Vaidya, T.; Frontier, A.J. Beyond the divinyl ketone: Innovations in the generation and nazarov cyclization of pentadienyl cation intermediates. *Eur. J. Org. Chem.* 2013, 3621–3633.

4. Habermas, K.L.; Denmark, S.E.; Jones, T.K. The nazarov cyclization. In *Org. React.*; Paquette, L.A., Ed.; John Wiley & Sons, Inc.: New York, NY, USA, 2004; Volume 45, pp. 1–158.

5. Davis, R.L.; Tantillo, D.J. Theoretical studies on pentadienyl cation

electrocyclizations. *Curr. Org. Chem.* 2010, *14*, 1561–1577.

6. Nieto Faza, O.; Silva López, C.; Álvarez, R.; de Lera, Á.R. Theoretical study of the electrocyclic ring closure of hydroxypentadienyl cations. *Chem. Eur. J.* 2004, *10*, 4324–4333.

7. D'Auria, M. A new simple procedure for the isomerization of 2-Furylcarbinols to cyclopentenones. *Heterocycles* 2000,*52*, 185–194.

8. Yin, B.-L.; Wu, Y.-L.; Lai, J.-Q. Novel conversions of furandiols and spiroacetal enol ethers into cyclopentenones: Implications of the isomerization mechanism of 2-furylcarbinols into cyclopentenones. *Eur. J. Org. Chem.* 2009, 2695–2699.

9. Yin, B.-L.; Yang, Z.-M.; Hu, T.-S.; Wu, Y.-L. Molecular diversity of tonghaosu: synthesis of lactam-containing tonghaosu analogs. *Synthesis* 2003, 1995–2000.

10. Yang Gao, Y.; Wu, W.-L.; Ye, B.; Zhou, R.; Wu, Y.-L. Convenient syntheses of tonghaosu and two thiophene substituted spiroketal enol ether natural products. *Tetrahedron Lett.* 1996, *37*, 893–896.

11. Yin, B.-L.; Wu, Y.; Wu, Y.-L. Acid catalysed rearrangement of a spiroketal enol ether. An easy synthesis of chrycorin. *J. Chem. Soc. Perkin Trans. 1* 2002, 1746–1747.

12. Piancatelli, G. Advances in cyclopentenone synthesis from furans. *Heterocycles* 1982, *19*, 1735–1744.

13. Piancatelli, G.; D'Auria, M.; D'Onofrio, F. Synthesis of 1,4-Dicarbonyl compounds and cyclopentenones from furans.*Synthesis* 1994, 867–889.

14. Piancatelli, G.; Scettri, A.; David, G.; D'Auria, M. A new synthesis of 3-Oxocyclopentenes. *Tetrahedron* 1978, *34*, 2775–2778.

15. D'Auria, M.; D'Onofrio, F.; Piancatelli, G.; Scettri, A. A convenient synthesis of 2-Bromo- and 3-Bromo-4-hydroxy-2-cyclopenten-1-ones. *Gazz. Chim. Ital.* 1986, *116*, 173–175.

16. D'Auria, M.; Piancatelli, G.; Scettri, A. A useful preparation of 5-Nitro-2-furan derivatives. *Tetrahedron* 1980, *36*, 1877–1878.

17. D'Auria, M.; Piancatelli, G.; Scettri, A. Synthesis of 4-Ylidenebutenolides and 4-Oxo-2-enoic acid methyl esters from 5-Methoxy-2-furyl carbinols. *Tetrahedron* 1980, *36*, 3071–3074.

18. Gilman, H.; Franz, R.A.; Hewlett, A.P.; Wright, G.F. Decomposition of 5-Halogeno-2-furylmethyl ethers to benzalcrotonolactones.*J. Am. Chem. Soc.* 1950, *72*, 3–8.

19. Antonioletti, R.; de Mico, A.; Piancatelli, G.; Scettri, A.; Ursini, O. Cyclopentenones derivatives by solvolysis of 2-Furylidene

carbinols. *Gazz. Chim. Ital.* 1986, *116*, 745–746.

20. Castagnino, E.; D'Auria, M.; de Mico, A.; D'Onofrio, F.; Piancatelli, G. Studies of the reactivity of 2-Furylhydroxymethylphosphonates: Synthesis of 1-Oxo-4-hydroxycyclopent-2-en-5-ylphosphonates. *J. Chem. Soc. Chem. Commun.* 1987, 907–908.

21. Saito, K.; Yamachika, H. Process for producing 3-Oxocyclopentenes. U.S. Patent 4356326, 1982.

22. Saito, K.; Yamachika, H. Process for producing 4-Hydroxycyclopentenones. U.S. Patent 4371711, 1983.

23. Saito, K.; Yamachika, H. Process for preparing cyclopentenolones. U.S. Patent 4510329, 1985.

24. Saito, K.; Takisawa, Y.; Yamachika, H. Process for preparing cyclopentenolones. U.S. Patent 4398043, 1983.

25. Ulbrich, K.; Kreitmeier, P.; Reiser, O. Microwave- or microreactor-assisted conversion of furfuryl alcohols into 4-Hydroxy-2-cyclopentenones. *Synlett* 2010, *13*, 2037–2040.

26. Stork, G.; Kowalski, C.; Garcia, G. Route to prostaglandins *via* a general synthesis of 4-Hydroxycyclopentenones. *J. Am. Chem. Soc.* 1975, *97*, 3258–3260.

27. Piancatelli, G.; Scettri, A. A simple conversion of 4-Substituted-5-hydroxy-3-Oxocyclopentenes into the 2-Substituted analogs. *Synthesis* 1977, 116–117.

28. Scettri, A.; Piancatelli, G.; D'Auria, M.; David, G. General route and mechanism of the rearrangement of the 4-Substituted-5-hydroxy-3-oxocyclopentenes into the 2-Substituted analogs. *Tetrahedron* 1979, *35*, 135–138.

29. Piancatelli, G.; Scettri, A. A useful preparation of (±) *t*-Butyl 3-hydroxy-5-oxo-1-cyclopenteneheptanoate and its 3-Deoxy-derivative, important prostaglandin intermediates. *Tetrahedron Lett.* 1977, *18*, 1131–1134.

30. Collins, P.W.; Kramer, S.W.; Gasiecki, A.F.; Weier, R.M.; Jones, P.H.; Gullikson, G.W.; Bianchi, R.G.; Bauer, R.F. Synthesis and gastrointestinal pharmacology of a 3*E*,5*Z* diene analog of misoprostol. *J. Med. Chem.* 1987, *30*, 193–197.

31. Dygos, J.H.; Adamek, J.P.; Babiak, K.A.; Behling, J.R.; Medich, J.R.; Ng, J.S.; Wieczorek, J.J. An efficient synthesis of the antisecretory prostaglandin enisoprost. *J. Org. Chem.* 1991, *56*, 2549–2552.

32. Collins, P.W.; Kramer, S.W.; Gullikson, G.W. Synthesis and gastrointestinal pharmacology of the 4-Fluoro analog of enisoprost. *J.*

Med. Chem. 1987, 30, 1952–1955.

33. Harikrishna, M.; Mohan, H.R.; Dubey, P.K.; Subbaraju, G.V. Synthesis of 2-Normisoprostol, methyl 6-(3-hydroxy-2-((E)-4-hydroxy-4-methyloct-1-enyl)-5-oxocyclopentyl)hexanoate. Synth. Commun. 2009, 39, 2763–2775.

34. Rodríguez, A.; Nomen, M.; Spur, B.W.; Godfroid, J.-J. An efficient asymmetric synthesis of prostaglandin E1. Eur. J. Org. Chem. 1999, 1999, 2655–2662.

35. Rodríguez, A.R.; Spur, B.W. First total synthesis of the E type I phytoprostanes. Tetrahedron Lett. 2003, 44, 7411–7415.

36. Henschke, J.P.; Liu, Y.; Chen, Y.-F.; Meng, D.; Sun, T. Process for the preparation of prostaglandin analogues and intermediates thereof. U.S. Patent 2009/0259058, 2009.

37. Henschke, J.P.; Liu, Y.; Chen, Y.-F.; Meng, D.; Sun, T. Process for the preparation of prostaglandin analogues and intermediates thereof. U.S. Patent 2011.

38. Henschke, J.P.; Liu, Y.; Huang, X.; Chen, Y.; Meng, D.; Xia, L.; Wei, X.; Xie, A.; Li, D.; Huang, Q.; et al. The manufacture of a homochiral 4-silyloxycyclopentenone intermediate for the synthesis of prostaglandin analogues. Org. Process. Res. Dev. 2012, 16, 1905–1916.

39. Peake, S.L. Bis-thiolakylfurans useful as cyclopentenon prostaglandin intermediates. U.S. Patent 4390707, 1983.

40. Paz-Otero, M.; Santín, E.P.; Rodríguez-Barrios, F.; Vaz, B.; de Lera, Á.R. Selective, potent PPAR@c agonists with cyclopentenone core structure. Bioorg. Med. Chem. Lett. 2009, 19, 1883–1886.

41. Li, C.-C.; Wang, C.-H.; Liang, B.; Zhang, X.-H.; Deng, L.-J.; Liang, S.; Chen, J.-H.; Wu, Y.-D.; Yang, Z. Synthetic study of 1,3-Butadiene-based IMDA Approach to Construct a [5-7-6] Tricyclic core and its application to the total synthesis of C8-epi-Guanacastepene O. J. Org. Chem. 2006, 71, 6892–6897.

42. Beingessner, R.L.; Farand, J.A.; Barriault, L. Progress toward the total synthesis of (±)-Havellockate. J. Org. Chem. 2010, 75, 6337–6346.

43. Katsuta, R.; Aoki, K.; Yajima, A.; Nukada, T. Synthesis of the core framework of the proposed structure of sargafuran. Tetrahedron Lett. 2013, 54, 347–350.

44. Veits, G.K.; Wenz, D.R.; Read de Alaniz, J. Versatile method for the synthesis of 4-Aminocyclopentenones: Dysprosium(III) triflate catalyzed aza-piancatelli rearrangement. Angew. Chem. Int. Ed. 2010, 49, 9484–

9487.

45. Li, S.W.; Batey, R.A. Mild lanthanide (III) catalyzed formation of 4,5-Diaminocyclopent-2-enonesfrom 2-Furaldehyde and secondary amines: A domino condensation/ring-opening/electrocyclization process. *Chem. Commun.* 2007, 3759–3761.

46. Veits, G.K.; Read de Alaniz, J. Dysprosium(III) catalysis in organic synthesis. *Tetrahedron* 2012, *68*, 2015–2026.

47. Finke, P.E.; Meurer, L.C.; Levorse, D.A.; Mills, S.G.; MacCoss, M.; Sadowski, S.; Cascieri, M.A.; Tsao, K.-L.; Chicchi, G.G.; Metzger, J.M.; *et al.* Cyclopentane-based human NK1 antagonists. Part 1: Discovery and initial SAR. *Bioorg. Med. Chem. Lett.* 2006, *16*, 4497–4503.

48. Meurer, L.C.; Finke, P.E.; Owens, K.A.; Tsou, N.N.; Ball, R.G.; Mills, S.G.; MacCoss, M.; Sadowski, S.; Cascieri, M.A.; Tsao, K.-L.; *et al.* Cyclopentane-based human NK1 antagonists. Part 2: Development of potent, orally active, water-soluble derivatives. *Bioorg. Med. Chem. Lett.* 2006, *16*, 4504–4511.

49. Subba Reddy, B.V.; Narasimhulu, G.; Subba Lakshumma, P.; Vikram Reddy, Y.; Yadav, J.S.; Sridhar, B.; Purushotham Reddy, P.; Kunwar, A.C. Phosphomolybdic acid: A highly efficient solid acid catalyst for the synthesis of *trans*-4,5-Disubstituted cyclopentenones. *Tetrahedron Lett.* 2012, *53*, 1776–1779.

50. Wenz, D.R.; Read de Alaniz, J. Aza-Piancatelli rearrangement initiated by ring opening of donor-acceptor cyclopropanes. *Org. Lett.* 2013, *15*, 3250–3253.

51. Carson, C.A.; Kerr, M.A. Heterocycles from cyclopropanes: Applications in natural product synthesis. *Chem. Soc. Rev.*2009, *38*, 3051–3060.

52. Yu, M.; Pagenkopf, B.L. Recent advances in donor-acceptor (DA) cyclopropanes. *Tetrahedron* 2005, *61*, 321–347.

53. Wong, H.N.C.; Hon, M.Y.; Tse, C.W.; Yip, Y.C.; Tanko, J.; Hudlicky, T. Use of cyclopropanes and their derivatives in organic synthesis. *Chem. Rev.* 1989, *89*, 165–198.

54. Palmer, L.I.; Read de Alaniz, J. Direct and highly diastereoselective synthesis of azaspirocycles by a Dysprosium(III) triflate catalyzed aza-piancatelli rearrangement. *Angew. Chem. Int. Ed.* 2011, *50*, 7167–7170.

55. Dake, G. Recent approaches to the construction of 1-Azaspiro[4.5] decanes and related 1-Azaspirocycles. *Tetrahedron Lett.* 2006, *62*, 3467–3492.

56. Burkhard, J.A.; Wagner, B.; Fischer, H.; Schuler, F.; Müller, K.;

Carreira, E.M. Synthesis of azaspirocycles and their evaluation in drug discovery. *Angew. Chem. Int. Ed.* 2010, *49*, 3524–3527.

57. Palmer, L.I.; Read de Alaniz, J. Rapid and stereoselective synthesis of spirocyclic ethers via the intramolecular piancatelli rearrangement. *Org. Lett.* 2013, *15*, 476–479.

58. Zhang, Q.-W.; Fan, C.-A.; Zhang, H.-J.; Tu, Y.-Q.; Zhao, Y.-M.; Gu, P.; Chen, Z.-M. Brønsted acid catalyzed enantioselective semipinacol rearrangement for the synthesis of chiral spiroethers. *Angew. Chem. Int. Ed.* 2009, *48*, 8572–8574.

59. Adrien, A.; Gais, H.-J.; Köhler, F.; Runsink, J.; Raabe, G. Modular asymmetric synthesis of functionalized azaspirocycles based on the sulfoximine auxiliary. *Org. Lett.* 2007, *9*, 2155–2158.

60. Noguchi, N.; Nakada, M. Synthetic studies on (+)-Ophiobolin α asymmetric synthesis of the spirocyclic CD-Ring moiety. *Org. Lett.* 2006, *8*, 2039–2042.

61. Subba Reddy, B.V.; Reddy, Y.V.; Lakshumma, P.S.; Narasimhulu, G.; Yadav, J.S.; Sridhar, B.; Reddy, P.P.; Kunwar, A.C. In(OTf)$_3$-Catalyzed tandem aza-piancatelli rearrangement/michael reaction for the synthesis of 3,4-Dihydro-2*H*-benzo[*b*][1,4]thiazine and Oxazine derivatives. *RSC Adv.* 2012, *2*, 10661–10666.

62. Yadav, J.S.; Subba Reddy, B.V.; Hara Gopal, A.V.; Patil, K.S. InBr$_3$-Catalyzed three-component reaction: A facile synthesis of propargyl amines. *Tetrahedron Lett.* 2009, *50*, 3493–3496.

63. Yadav, J.S.; Subba Reddy, B.V.; Maity, T.; Narayana Kumar, G.G.K.S. In(OTf)$_3$-Catalyzed synthesis of 4-Thiocyanotetrahydropyrans via a Three-component reaction. *Tetrahedron Lett.* 2007, *48*, 8874–8877.

64. Yadav, J.S.; Subba Reddy, B.V.; Rao, K.V.; Saritha Raj, K.; Prasad, A.R.; Kiran Kumar, S.; Kunwar, A.C.; Jayaprakash, P.; Jagannath, B. InBr$_3$-Catalyzed cyclization of glycals with aryl amines. *Angew. Chem. Int. Ed.* 2003, *42*, 5198–5201.

65. Subba Reddy, B.V.; Sreelatha, M.; Kishore, C.; Borkar, P.; Yadav, J.S. InCl$_3$-Promoted a Novel prins cyclization for the synthesis of Hexahydro-1*H*-furo[3,4-*c*]pyran derivatives. *Tetrahedron Lett.* 2012, *53*, 2748–2751.

66. Liu, J.; Shen, Q.; Yu, J.; Zhu, M.; Han, J.; Wang, L. A concise domino synthesis of benzo-1,4-heterocycle compounds via a piancatelli/C–N coupling/michael addition process promoted by La(OTf)$_3$. *Eur. J. Org. Chem.* 2012, 6933–6939.

67. Yin, B.; Huang, L.; Wang, X.; Liu, J.; Jiang, H. Metal-Free rearrangement

of spirofurooxindoles into spiropentenoneoxindoles and indoles: Implications for the mechanism and stereochemistry of the piancatelli rearrangement. *Adv. Syn. Catal.* 2013, *355*, 370–376.

68. Yin, B.; Huang, L.; Zhang, X.; Ji, F.; Jiang, H. Cu(II)-Promoted transformations of α-thienylcarbinols into spirothienooxindoles: Regioselective halogenation of dienyl sulfethers containing electron-rich aryl rings. *J. Org. Chem.* 2012, *77*, 6365–6370.

69. Yin, B.-L.; Lai, J.-Q.; Zhang, Z.-R.; Jiang, H.-F. A novel entry to spirofurooxindoles involving tandem dearomatization of furan ring and intramolecular friedel–crafts reaction. *Adv. Synth. Catal.* 2011, *353*, 1961–1965.

Chapter 3

OXETANE SYNTHESIS THROUGH THE PATERNÒ-BÜCHI REACTION

Maurizio D'Auria, and Rocco Racioppi

Dipatimento di Scienze, Università della Basilicata, Viale dell'Ateneo Lucano 19, 85100 Potenza, Italy

ABSTRACT

The Paternò-Büchi reaction is a photochemical reaction between a carbonyl compound and an alkene to give the corresponding oxetane. In this review the mechanism of the reaction is discussed. On this basis the described use in the reaction with electron rich alkenes (enolethers, enol esters, enol silyl ethers, enanines, heterocyclic compounds has been reported. The stereochemical behavior of the reaction is particularly stressed. We pointed out the reported applications of this reaction to the synthesis of naturally occuring compounds.

INTRODUCTION

Oxetanes are important component in the scaffold of compounds with relevant biological properties: an oxetane ring is present in the scaffold of Taxol® (1), an important drug used in the treatment of ovarian cancer [1], merrilactone A (2), a sesquiterpene dilactone with neurotrophic activity [2], and several antiviral oxetanes, such as 3, 4, and 5, have been described in the literature (Figure 1) [3,4,5]. In 1909 Paternò, studying the photochemical reaction of benzophenone with amylene, showed the formation of the corresponding [2+2]cycloadduct (Scheme 1) [6].

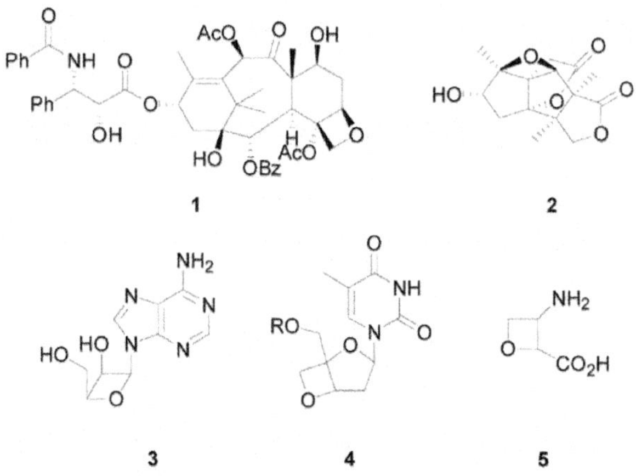

Figure 1: Natural and biologically active compounds containing the oxetane ring.

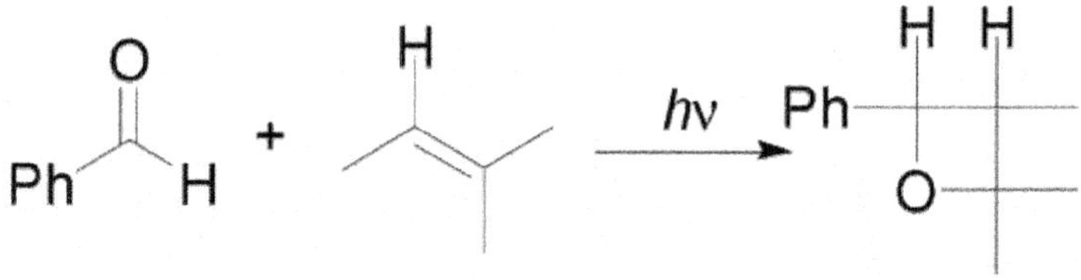

Scheme 1: The reaction performed by Paternò.

Only in 1953 Büchi reported the exact identification of the product, showing that this photochemical reaction could represent an interesting synthetic process [7]. After these pioneering works, several reviews have covered the relevant papers published in this field [8,9,10,11,12,13,14,15,16,17,18,19,20,21,22,23,24].

The reaction is a photocycloaddition reaction of a carbonyl compound in the excited state with an alkene in the ground state. The carbonyl compound must posses a n,π* S_1 or T_1 excited states. The frontier orbitals approach has been used to explain the formation of oxetanes. HSOMO-LUMO interaction is the main process observed where the half-occupied π*orbital of the carbonyl compound interacts with the unoccupied π* molecular orbital of an electron deficient alkene: the results of this interaction is the formation of a C,O-biradical (A in Scheme 2). If the LSOMO-HOMO interaction is prevalent (interaction of the half-occupied n orbital of the carbonyl O atom with the π orbital of an electron-rich alkene), the formation of a C,C-biradical (B in Scheme 2) is the main process [25,26].

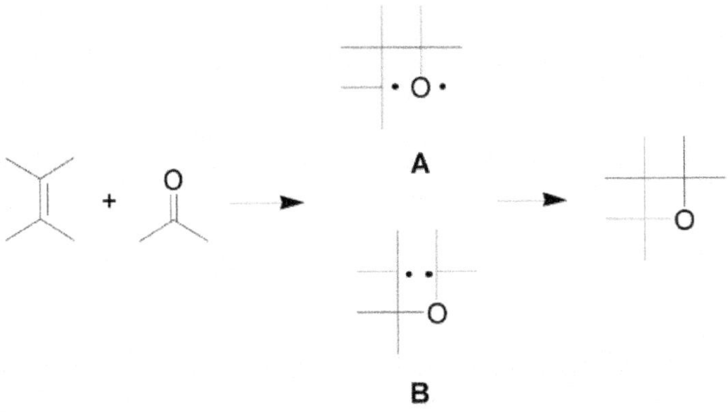

Scheme 2: Possible mechanisms of the Paternò-Büchi reaction.

The majority of Paternò-Büchi reactions occur from the carbonyl triplet state which is accessed by an intersystem crossing. The biradicals are the direct consequence of the precursor spin and their lifetimes are connected with the mode of spin inversion processes and the mechanism, which leads to the formation of closed-shell products. 1,4-Biradicals derived from the addition of a carbonyl compound in its first excited triplet state and an alkene were studied spectroscopically [27,28,29,30] and they were trapped by radical quenchers [31,32,33,34]. The biradical intermediate in the reaction between benzophenone and an electron-rich alkene has been determined by using laser flash photolysis. An absorption with λ_{max} 535 nm has been observed [27,29].

In the reaction between 1,4-dioxene and benzaldehyde, theoretical calculations showed that the only transition able to give the observed transient absorption is that from LSOMO to LUMO (549 nm); the same result was obtained for the reaction between furan and benzaldehyde [21].

The regioselectivity of the reaction can be explained considering the hard-soft acids and bases theory [35], or using an approach where atoms arrange themselves so that the obtained molecule reaches the minimum electrophilicity [36].

A theoretical study of the Paternò-Büchi reaction showed that there are two conical intersection points located near the C-C and C-O bonded biradical regions of the ground state. These two conical intersections support a mechanism where the decay from the excited state is accompanied by a geometric rotation of the terminal group, in the case of C-O attack, and by an orbital rotation at the oxygen center, in the case of C-C attack. Thus, for the triplet, the reaction path can be predicted by the most stable biradical rule [37]. A conformational analysis of the biradicals also appeared [38,39].

A CAS SCF geometry optimization by using TZV basis set of the intermediate biradicals showed that the diradical region corresponding to the C-C attack lies about 10 kcal mol^{-1} lower in energy than the C-O region [40,41]. Unfortunately, this result is not in agreement with reported experimental results. The formation of an exciplex is used to explain the reaction behavior of simple alkene. Evidence of monoelectron transfer processes is reported for electron-rich alkenes [42].

HOW TO PERFORM THE REACTION

The Paternò-Büchi reaction can be performed in solution. The effect of the solvent is relevant and non-polar solvents are preferred. When an aromatic carbonyl compound is used as substrate the irradiation has to be performed at 300 nm through Pyrex, while the use of aliphatic carbonyl compounds needs the use of 254 nm irradiation through quartz or Vycor. The quantum yields of the reaction are not high. The photochemical coupling of benzophenone, or another carbonyl compound, with itself to five the corresponding pinacone derivative is a competitive reaction allowing to give low quantum yields (typically, 10^{-1}–10^{-2}).

REACTIONS WITH ELECTRON RICH UNSATURATED COMPOUNDS

The first oxetane deriving from ethyl vinyl ether and benzophenone was described by Paternò [43]. He described the formation of the oxetane in the reaction of benzophenone with diethyl ether. He suggested the *in situ* formation of ethyl vinyl ether. The reaction between acetone and ethyl vinyl ether in the presence of ultrasound gave a different mixture of regioisomers than that obtained without the use of ultrasound [44].

A high regioselectivity was obtained when ethyl vinyl ether was used as an alkene with benzaldehyde and benzophenone [45]. The reaction of benzophenone with alkenyl sulphide also showed a high regioselectivity (Scheme 3) [46].

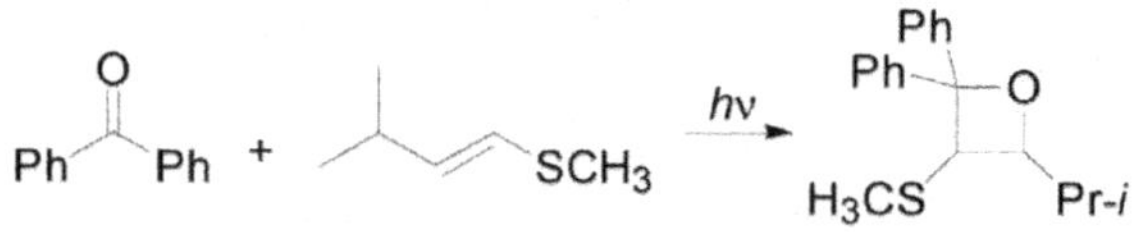

Scheme 3: Reaction of benzophenone with alkenyl sulphide.

Substituted pyruvates reacted with 1,3-dioxole and this reaction allows one to obtain the *endo* isomer [47,48,49]. The analysis of the conformations of the biradical intermediate was used in order to explain the observed stereochemical behavior [50]. Unsubstituted cyclopropyl enol ethers gave, when reacted with a carbonyl compound, the corresponding oxetane rings. When a substituent is introduced on the cyclopropyl ring, the formation of oxepanes was observed deriving from cyclopropyl ring opening at the level of the biradical intermediate [51].

When the enol ether **6** was irradiated in the presence of benzaldehyde, and the reaction mixture was purified after an acid hydrolysis, the oxetane **7** was isolated in good yields with good diastereoselectivity (Scheme 4) [52].

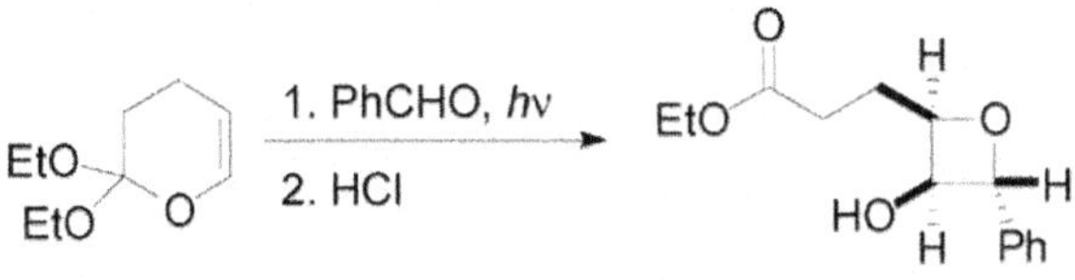

6   **7**

Scheme 4: Reaction of an enol ether.

1,2,3-Indanetrione reacted with 2,3-diphenyl-1,4-dioxene to give the oxetane on the carbonyl in the 2 position [53]. On the contrary, when the less electron rich 2,3-dimethyl-2-butene is used as alkene, a complex mixture of products is obtained [54]. The irradiation of 2,3-dihydrofuran with benzophenone in benzene gave adduct **8** (Scheme 5) [55,56,57].

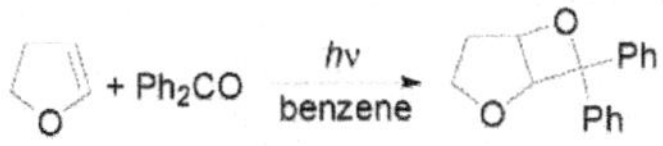

8

Scheme 5: Reaction of 2,3-dihydrofuran.

The selectivity depended on 2,3-dihydrofuran concentration: this behavior was explained with a switch from a triplet mechanism to a singlet mechanism at higher concentration [58,59,60,61,62]. This behavior can be explained [63]. The best interaction between the frontier orbitals is that from the LSOMO of acetaldehyde and the HOMO of 2,3-dihydrofuran. The atomic coefficients on the olefinic carbon atoms in 2,3-dihydrofuran were -0.26 at C-2 and -0.38 at C-3. The atomic coefficient on the oxygen atom in the LSOMO of singlet excited acetaldehyde was 0.48, while the atomic coefficient at the C-1 of acetaldehyde in the HSOMO was 0.49.

The nature of the LUMO of 2,3-dihydrofuran excludes the possibility of a concerted mechanism. The reaction has to provide for the formation of extremely labile singlet biradical. In this case, the oxygen atom of acetaldehyde has to attack the C-3 carbon atom in 2,3-dihydrofuran in order to give the more stable biradical intermediate. The reaction, in this case, allowed the formation of only the *exo* isomer. In the triplet state, the main interaction is that between the LSOMO of the triplet state acetaldehyde and the HOMO of the dihydrofuran. This interaction leads to the formation of the corresponding CC biradical intermediate (Scheme 6).

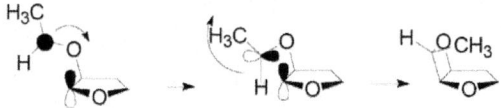

Scheme 6: Ring closure of the oxetane ring in the reaction of 2,3-dihydrofuran with acetaldehyde.

The HSOMO on the biradical intermediate was mainly localized on the aromatic ring and it is extended to the radical carbon. The LSOMO was mainly localized on the dihydrofuran ring. The coupling between the radical carbons in these two orbitals was possible (the atomic orbitals involved can superimpose themselves) only if the *endo* isomer was formed (Scheme 6).

When benzaldehyde was used as carbonyl compound, the reaction showed a good regio- and stereoselectivity [61,64,65,66]. The adducts were obtained with an overall yield of 98% as a > 98:2 regioisomeric mixture; the major isomer is 88:12 *endo/exo* mixture. The stereoselectivity was partially lost when benzaldehyde reacted with a 2,3-dihydrofuran derivative substituted in the 2 position [64,67].

A change of regioselectivity was observed when the reaction between 2,3-dihydrofuran and benzaldehyde was performed in high polar solvent: this result has been interpreted as proof of an electron transfer mechanism [58,59].

The reaction of dihydrofuran with benzaldehyde is the first example where a spin controlled selectivity is observed [60,68]. In singlet photoreactions, stereoselectivity is often controlled by the optimal geometries for radical-radical combinations. On the contrary, in triplet photoreactions the optimal geometries are those able to favor the intersystem crossing from the triplet excited state to the singlet one. The singlet biradicals should be too short-lived to enable rotation about the endocyclic C-O or C-C bonds and therefore, conformation memory effects on the stereochemistry are expected. The geometries in the triplet state can be quite different from the former ones due to differences of the spin-orbit coupling (SOC) values. The lifetimes of many triplet biradical intermediates are definitely high enough to enable bond rotations. Therefore, the formation of the thermodynamically favored product can be expected because the radical-radical combination step should not be influenced by the approach geometry: "memory effects" should be erased due to the relatively long lifetimes. After transition from the triplet to the singlet potential energy surface, immediate product formation is expected. Thus, the ISC proceeds in a concerted fashion with the formation of a new bond or the cleavage of the primarily formed single bond. As a consequence, the stereoselectivity of the Paternò-Büchi reaction is the result of a combination of several rate constants for cyclization versus cleavage reactions.

Benzaldehyde reacts in its triplet state. This way, a triplet biradical intermediate is formed. To obtain the products, intersystem crossing into the singlet manifold is necessary. The most important factor influencing an intersystem crossing for flexibile triplet biradicals is spin-orbit coupling. The angle between p orbitals at the radical centers is approximately 90° for maximun spin orbit coupling. For the pronounced *endo* selectivity in the reaction between aromatic aldehydes and 2,3-dihydrofuran, we can consider the two biradical conformers **9** and **10** to be responsible, with the alkyloxy substitutent localized in a pseudoequatorial position and **9** being more populated because of fewer steric interactions (Scheme 7).

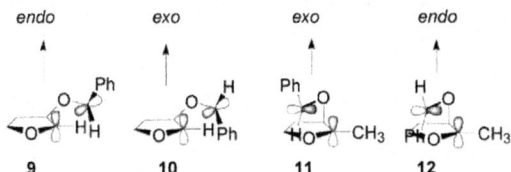

Scheme 7: Possible biradicals in the reactions between 2,3-dihydrofuran and benzaldehyde.

When a methyl group is present, the increasing *gauche* interactions with the β-alkyloxy substituent lead to a certain concentration of 11 and 12, again with 11 being preferred because of fewer steric interactions. Another explanation for the regio- and stereochemistry appeared [69]. In fact, theoretical calculations showed that the biradical 13 is more stable than 14by 1.49 kcal mol^{-1} (Figure 2).

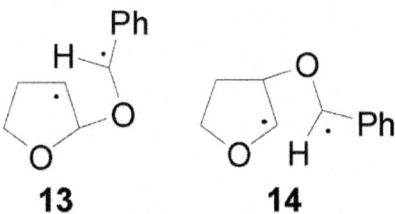

Figure 2. Possible biradical intermediates in the reaction of 2,3-dihydrofuran with benzaldehyde.

Furthermore, the HSOMO was mainly localized on the aromatic ring and it is extended to the radical carbon. The LSOMO was mainly localized on the dihydrofuran ring. The coupling between the radical carbons in these two orbitals was possible (the atomic orbitals involved can superimpose themselves) only if the *endo* isomer was formed (Scheme 8) [69].

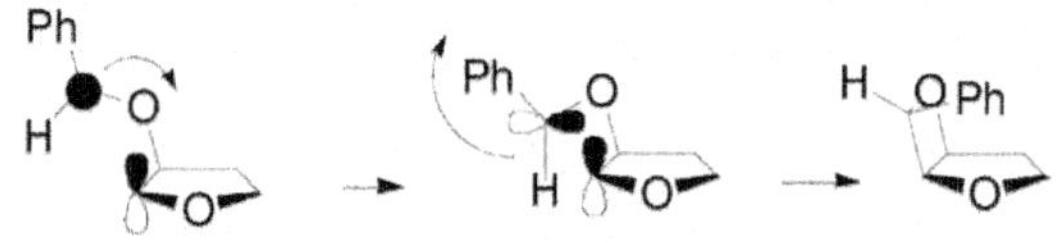

Scheme 8: Ring closure reaction in the formation of the *endo* isomer of the adduct between 2,3-dihydrofuran and benzaldehyde.

α- and β-Naphthaldehydes, on the contrary, gave high *exo* selectivity (Scheme 9). The reaction occurred also in the presence of triplet quenchers, while fluorescence quenching in the presence of dihydrofuran was observed.

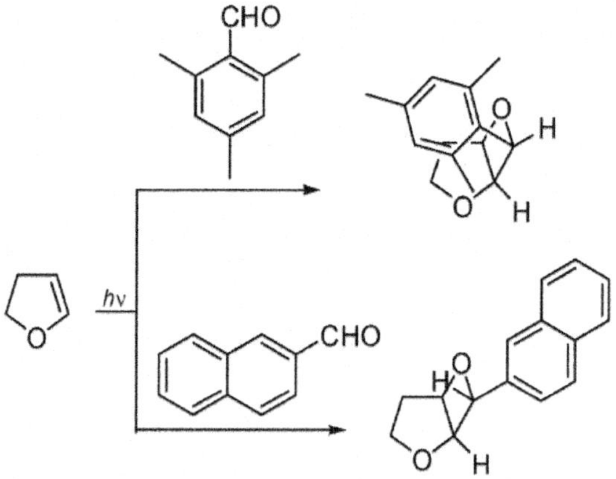

Scheme 9: Different stereochemical behavior in the reactions of 2,3-dihydrofuran.

In this case, the singlet excited state was responsible for the high *exo* selectivity [61,70]. In this case, the coefficients on the HSOMO and LSOMO allowed the coupling of the radical carbons only if the *exo* isomer is obtained [71].

When 2,3-dihydrofuran derivatives react with α,β-unsaturated carbonyl compounds, 2+2 cycloaddition between the olefins occurs [72]. Benzaldehyde reacts with L-ascorbic acid giving a mixture of regioisomeric compounds with *exo*stereoselectivity [73].

Irradiation of the oxetane **15** in the presence of 1-methoxynaphthalene and 2,7-dimethoxynaphthalene as sensitizers gave an efficient carbonyl-alkene metathesis of bicyclic oxetanes through an electron transfer process (Scheme 10) [74].

Scheme 10: Metathesis reaction on an oxetane.

A cycloreversion reaction was supposed also in the photochemical reaction of α,α-diphenyl substituted diazotetrahydrofuranones [75]. High regio- and stereoselectivity (*syn* addition) was observed in the reaction of pentafluorobenzaldehyde with enol acetates [76]. The reaction between a 2,3-thiophenone derivative and vinyl acetate gave the corresponding adduct on a ketonic function in low yields [77]. The reaction between benzaldehyde and a silyl derivative of cinnamyl alcohol gave the corresponding oxetane with high stereoselectivity (Scheme 11) [78].

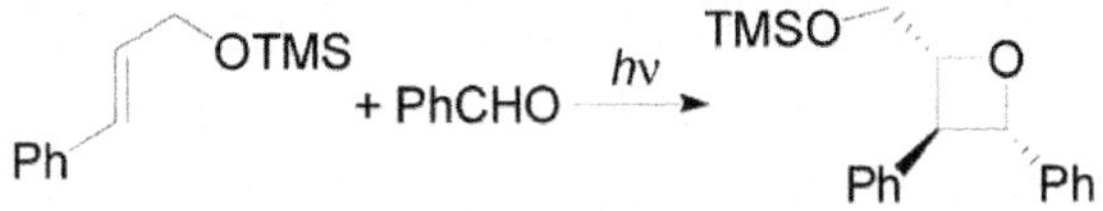

Scheme 11: Reaction of a silyl derivative of cinnamyl alcohol.

3-(Silyloxy)oxetanes **17** were successfully prepared from silyl enol ethers **16** containing carbon-chlorine, carbon-silicon, or carbon-sulfur bonds (Scheme 12) [79,80,81]. Ether and ester groups were compatible with the reaction. The presence of an alkene moiety was also compatible. When a β-alkyl substituted silyl enol ether was used, a *trans* relationship between α and β susbstituents in the oxetanes was observed. This result did not depend on the (*E*)- or (*Z*)- nature of the alkene. The products were obtained with high simple diastereoselectivity (ds 74%–95%) [82].

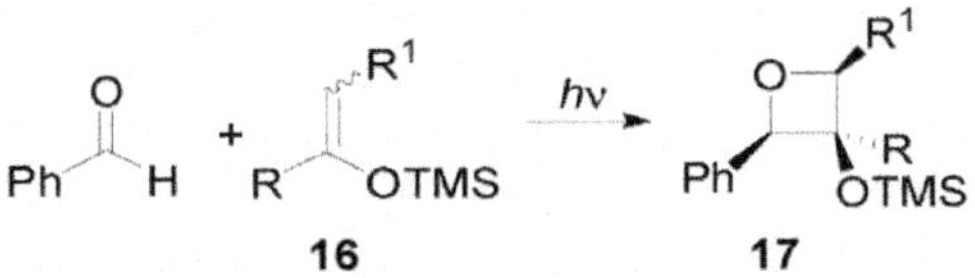

Scheme 12: Reaction of silyl enol ether.

In the triplet biradical, free rotation leads to the highly preferred, sterically least congested conformation. The further reaction pathway of this species includes ISC and an assumed selection step (cleavage vs. ring closure) at the singlet 1,4-diradical level which accounts for the high simple diastereoselectivity at C-2/C-3.

The oxetanes can be converted into the corresponding diols by using hydrogenolysis under Pd catalysis. Acid-sensitive substrates can be hydrogenated using $Pd(OH)_2$ as catalyst [83]. Also $LiAlH_4$ can be used in order to induce the cleavage of the oxetane ring [84].

The presence of a stereogenic center in the β-alkyl group (as in **18**) induced a facial diastereoselectivity. In some cases, high diastereoisomeric ratios were observed (Scheme 13) [85].

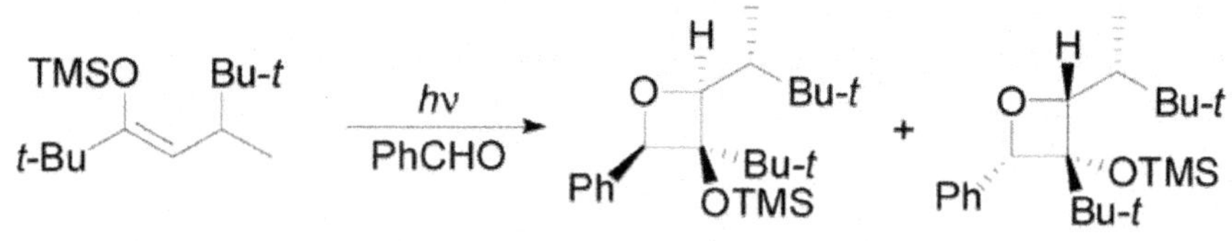

Scheme 13: Diastereoselectivity in the reaction of a silyl enol ether.

The diastereoselectivity was probably due to the presence of a conformational preference represented in the Scheme 14. This conformation allows the *Si* attack [84].

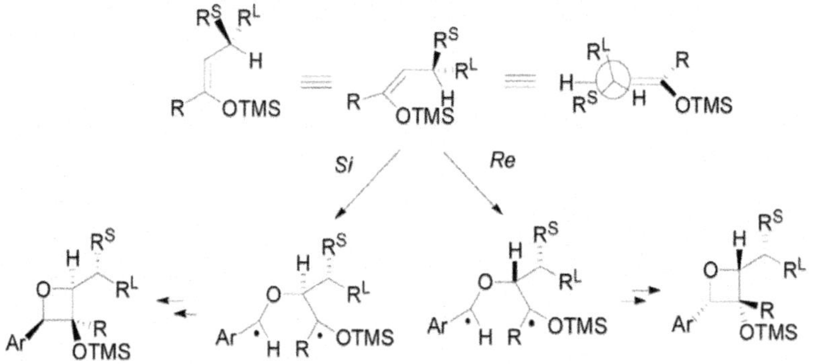

Scheme 14: Diastereoselectivity in chiral enol silyl ethers reaction with benzaldehyde.

In this case, a fluoride-promoted cleavage can be used to open the oxetane ring [85]. When the chiral center is in α position low facial diastereoselectivity was observed [86]. The best results were obtained by using both the silyl enol ether**19**, which gave the adducts with a 67/33 d.r., and the compound **20**, giving the adducts in 15/85 d.r. (Scheme 15) [87].

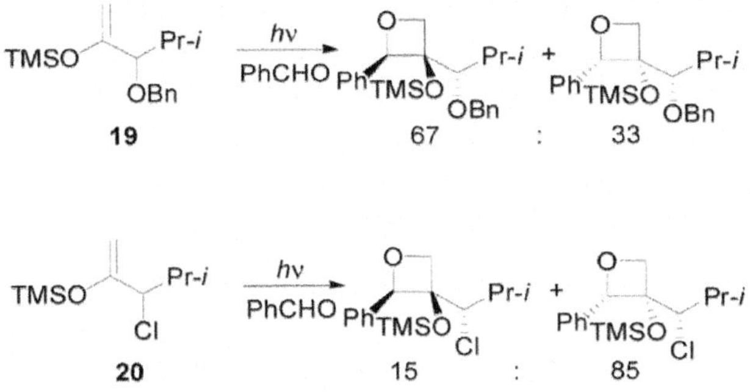

Scheme 15: Diastereoslectivity on the reaction of some enol ethers.

In the reaction of **19** with benzaldehyde two conformers of the biradical intermediate have been observed. These conformers showed almost the same energy (the conformer A is more stable than B for 3.11 kcal/mol): the conformers of the biradical are shown in Figure 3 [70].

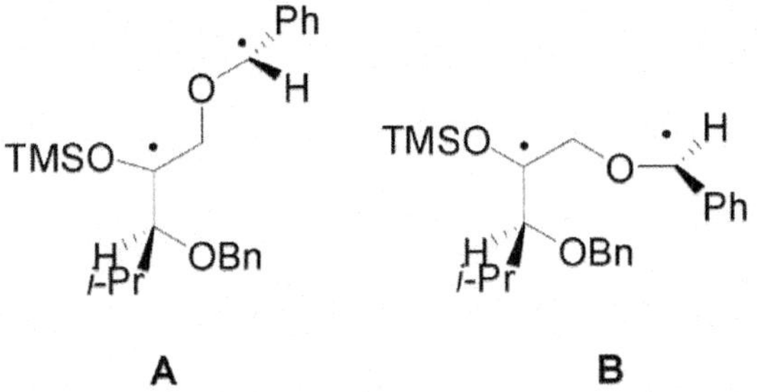

Figure 3: The conformers (**A** and **B**) of the biradical intermediate in the photoreaction between **19** and benzaldehyde.

Considering the reactive sites and the atomic coefficients at these sites, the coupling of the radical carbon atoms can occur only as depicted in the Scheme 16.

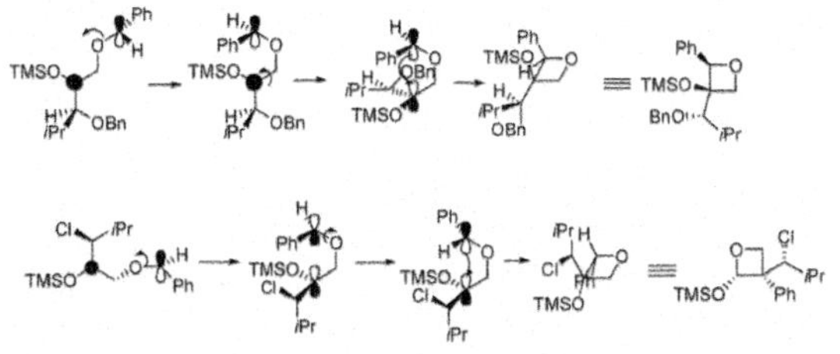

Scheme 16: Possible explanation of the observed diastereoselectivity in the reaction of chiral enol ethers.

The conformer **A** can give the major stereoisomer observed in the reaction. The results are in agreement with the experimental results showing that the course of this reaction is strictly frontier orbitals controlled. The other biradical conformer gave the other diastereoisomer. The observed diastereoisomer ratio (67:33) can be explained by the low difference between the energies of the conformers of the biradical intermediate (3.11 kcal/mol).

To confirm this result the behavior of **20** has been examined. In this case, the authors observed an inverse diastereoselectivity [86]. Also in this case two conformers of the biradical intermediate were possible. The energy difference between these two conformers was 4.6 kcal/mol. The coupling between the carbon atom considering the atomic coefficents allowed the formation of the observed diastereoisomers: the most stable conformer gave the main product. The larger diastereoisomeric ratio (85:15) here observed in comparison with that found in the other case is in agreement with the larger energy difference between the conformers of the biradical intermediate.

The synthetic utility of the oxetanes has been studied considering the presence of a removable protecting group in the side chain of the same oxetane. The possible intramolecular nucleophilic attack on the oxetane ring gave some interesting products (Scheme 17) [87].

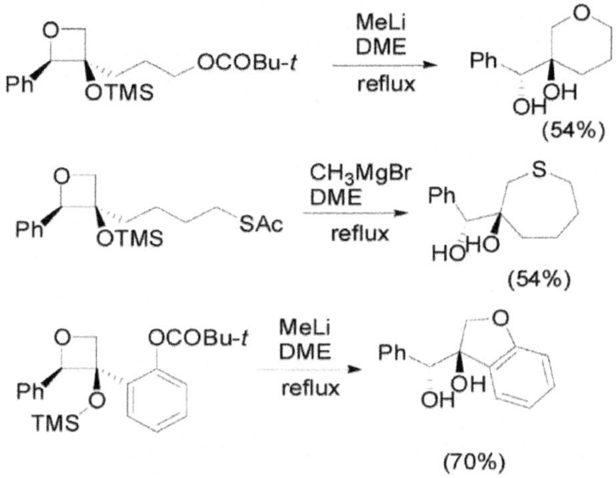

Scheme 17: Elaboration on the oxetane ring.

1,2-Diketones reacted with trimethylsilyl ketene acetals giving the corresponding oxetanes in low yields [88]. In the reaction of ketene silyl acetals with aromatic carbonyl compounds byproducts have been identified as oxetanes: in some cases they represent the main product [89]. Cyclic ketene silyl acetals reacted with 2-naphthaldehyde to give the corresponding adduct.

The treatment of the reaction mixture with water gave the aldol-type product with high stereoselectivity. The oxetane was obtained with *anti* stereochemistry [90,91]. The effect of both the solvent and the nature of the silyl group has been examined [92]. When silyl O,S-ketene acetals were used as alkene some interesting results were obtained [93].

The stereochemical behavior was explained by considering the capability of the sulfur atom to coordinate the oxygen atom of the carbonyl compound [93]. The presence of such an atom induced the attack of the carbonyl compound on the side where the sulfur atom is present (Scheme 18). The same regio-and stereoselectivity was observed when silyl O,Se-ketene acetals were used [94].

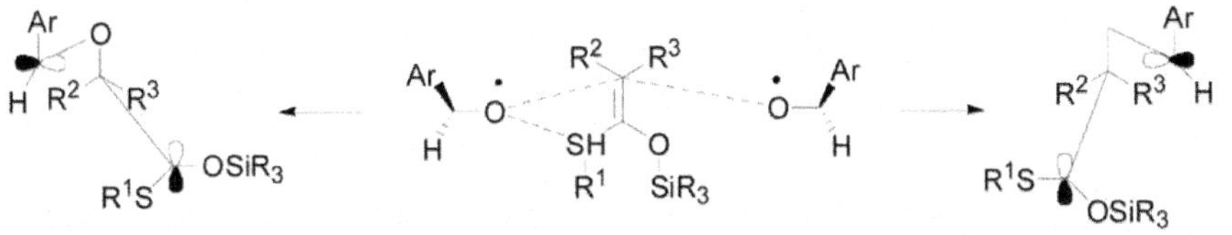

Scheme 18: The reaction O,S-ketene acetals with carbonyl compound. A possible explanation of the stereochemical behavior.

Oxetanes were obtained in the reaction of *N*-acyl enamines **21** and **22**, these compounds gave the corresponding adducts with high regio- and stereoselectivity (Scheme 19) [95,96,97,98]. The main product was the thermodynamically more stable isomer [99].

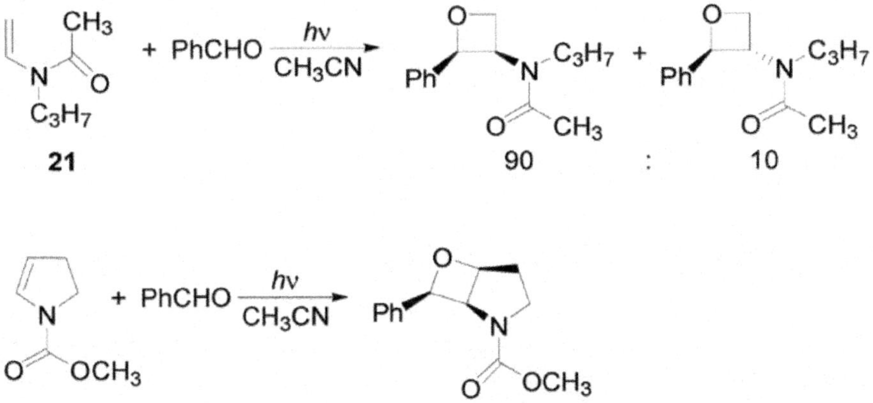

Scheme 19: Reacions of *N*-acyl enamines.

The optimized structure of the biradical intermediate is shown in Figure 4 with the HSOMO at −0.215 H and the LSOMO at −0.241 H [70].

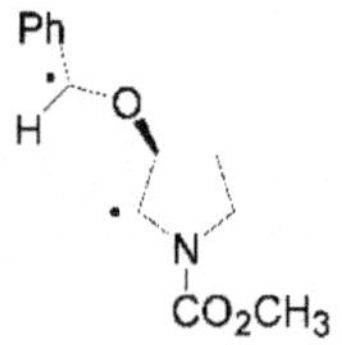

Figure 4: Optimized structure of the biradical intermediate in the reaction of compound **22**.

The coupling of the radical carbon atoms considering the atomic coefficients on the SOMOs allowed the formation of the *exo* isomer, in agreement with the experimental results (Scheme 20).

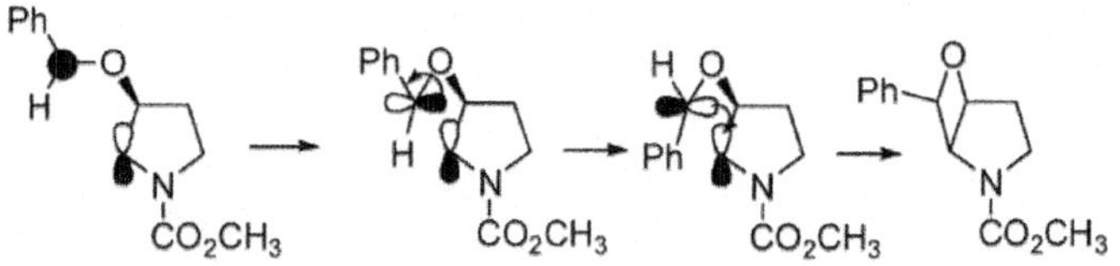

Scheme 20: Possible explanation of the stereoselctivity observed with *N*-acyl enamines.

Chiral enamines did not give the corresponding adduct with high diastereoselectivity. The only exception was found using the enamine **23**. It gave the corresponding adduct with 62% *de* (Scheme 21) [100,101].

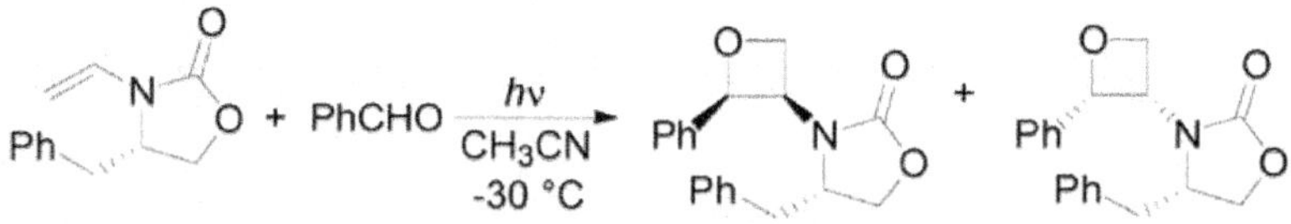

Scheme 21: Reaction with a chiral enamine.

N-Formyl protected oxetanes could be converted into the corresponding *syn*-1,2-amino alcohols through treatment with $LiAlH_4$ [97]. Alternatively, the treatment with TFA gave the corresponding oxazolidinone. Furthermore, the product can be converted into the corresponding *anti*-1,2-amino alcohol [97,102]. This approach has been used in the synthesis of both (±)-oxetin [103] and (+)-preussin [104,105].

The Paternò-Büchi reaction could be obtained also using alkenes bearing both electron donating groups (nitrogen atoms) and electron withdrawing groups. 2-Morpholinopropenenitrile gave in very low yields a product deriving from a Paternò-Büchi reaction when it reacted with naphthalene-1,4-dicarboxylic acid [106]. The same result was obtained when benzil was used as substrate. On the contrary, when methyl phenylglyoxylate was used as carbonyl compound, the corresponding adduct was isolated in good yields [107].

When benzil was irradiated in the presence of a 2-aminopropenenitrile derivative gave the corresponding adduct in variable yields [108]. The reaction showed a good stereoselectivity. When unsymmetrical benzil derivatives are used, both the isomers were obtained.

REACTIONS WITH HETEROCYCLIC COMPOUNDS

The reactivity of pentatomic aromatic heterocycles different from furan towards carbonyl compounds to give the corresponding oxetanes has been the object of other review articles [109]. These compounds showed lower reactivity than furan (see below). The reason of this behavior is not clear. It could be related to the different aromaticity of these compounds in comparison with that of furan, or could be due, as reported below, to the quenching properties of the heterocycles.

Thiophene did not react with benzophenone. The reaction occurred only if the irradiation is performed in the presence of BF_3 [110]. Probably, in this reaction BF_3 was able to catalyze the ring opening of the oxetane. The benzophenone BF_3complex, excited by light, leads to an exciplex whose excitation energy was lower that the lowest triplet energy level of thiophene, which under the circumstances cannot act as a quencher. On the contrary, 2,5-dimethylthiophene reacted with benzophenone at -10 °C yielding the corresponding cycloadduct in 62% yield (Scheme 22) [111,112].

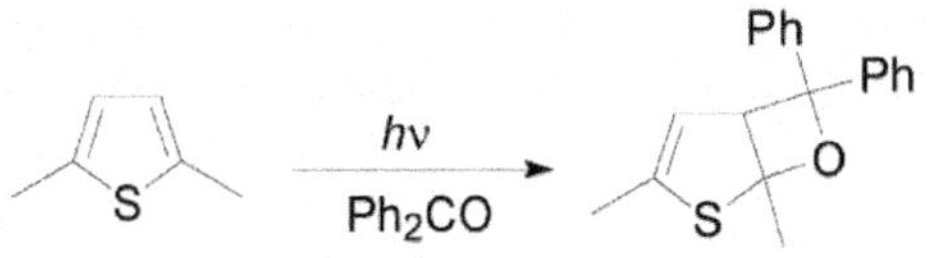

Scheme 22: The reaction of 2,5-dimethylthiophene with benzophenone.

The reaction product could be obtained also using 1-naphthaldehyde (50%), 2-, 3-, and 4-benzoyl- pyridine (62, 58, and 60%, respectively), and 2-benzoylthiophene (50%), while 2-naphthaldehyde, benzaldehyde, and acetophenone did not react [113]. 2,3-Dimethylthiophene also gave the corresponding oxetane when irradiated in the presence of benzophenone (60%) [114]. On the contrary, 2,3-dimethyl- and 2,3,5-trimethylthiophene did not react.

Pyrrole, such as thiophene, did not react with benzophenone and did not furnish the corresponding oxetane. However, pyrrole reacted with aliphatic aldehydes and ketones giving the corresponding 3-pyrryl carbinols. The alcohols derived from the cleavage of the corresponding oxetanes (Scheme 23) [115]. The yields increased when N-methylpyrrole was used as substrate, while the reactivity was depressed in the presence of subsituents on the pyrrole ring.

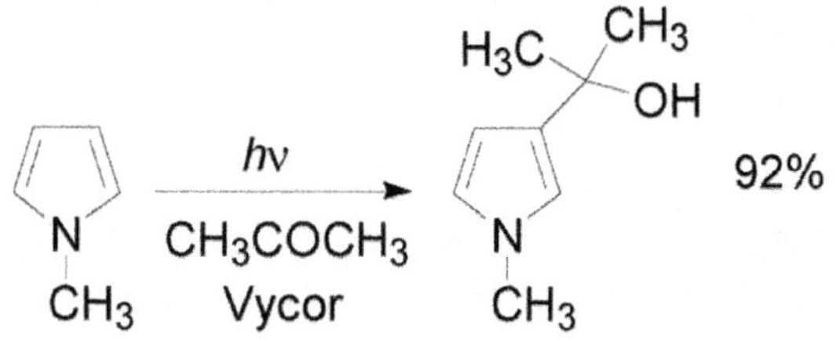

Scheme 23: The reaction of N-methylpyrrole with acetone.

Only when pyrrole, with an electron withdrawing group such as benzoyl on to the nitrogen atom, was irradiated in the presence of benzophenone, it was able to give the corresponding oxetane [116,117]. When N-phenylpyrrole was used as substrate, the corresponding 2-pyrryl carbinol was isolated [116].

Selenophene did not react with benzophenone [111]. On the contrary, the irradiation of 2-methylselenophene gave the corresponding adduct. The reaction occurred on the most hindered side of the molecule [118].

Imidazole, N-methylimidazole, and 1,2-dimethylimidazole reacted with aliphatic aldehydes and ketones. In all the cases, the product was the corresponding 4-imidazolyl carbinols (Scheme 24) [114,119]. On the contrary, the reaction of imidazole with benzophenone gave only 7.7% yield of the corresponding alcohol [119], while, the same reaction, performed in the presence of acetophenone as carbonyl compound, gave only 1.2% yield of the product [119].

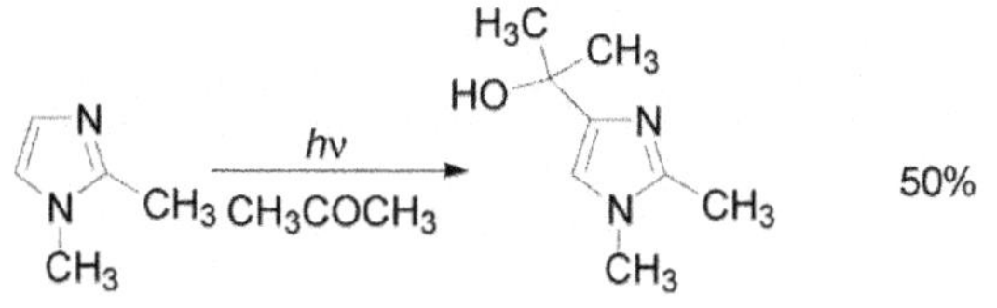

Scheme 24: Reaction of N-methyl-2-methylimidazole with acetone.

On the photochemical behavior of N-methylimidazole, there was not agreement between the work of Jones [115] and that of Matsuura [119]. While Jones reported that N-methylimidazole gave the corresponding imidazolyl carbinol in excellent yields, Matsuura reported that the same substrate gave the corresponding carbinol in 9% yield. They also reported that 2-methylimidazole gave a mixture of two products with an overall yield of 22.5% [119]. 1,2-Dimethylimidazole reacted with benzophenone. However, this compound gave a reaction on the methyl in the 2 position in low yield [120,121]. *N*-Benzylimidazole, when irradiated in the presence of benzophenone in *t*-BuOH, gave a reaction on the methylene group [120,121]. Only the irradiation of *N*-acetylimidazole allowed to obtain the corresponding oxetane when the reaction was performed in the presence of benzophenone (Scheme 25). The same result was obtained by using *N*-benzoylimidazole and *N,N'*-carbonyldiimidazole [120,121].

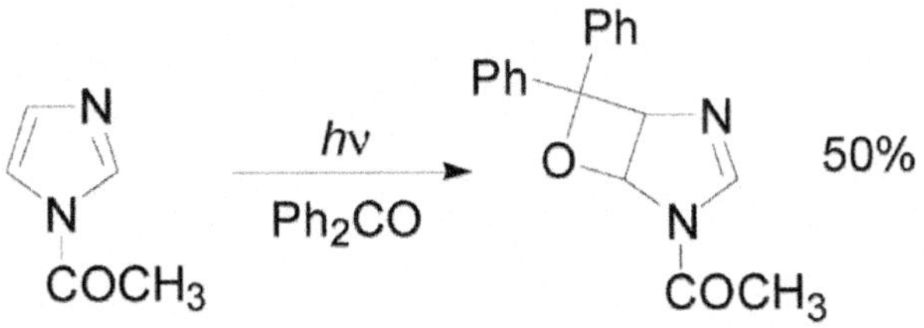

Scheme 25: The reaction of *N*-acetylimidazole with benzophenone.

The oxetane can be obtained also in the reaction of *N*-methyl-2,4,5-triphenylimidazole with aromatic ketones. It is noteworthy that, in this case, the reaction did not work in the presence of acetophenone and benzaldehyde [122].

The irradiation in the presence of benzophenone of 2,4-dimethylthiazole gave the corresponding oxetane, while the same reaction failed when acetophenone was used as carbonyl compound [123]. The irradiation in the presence of benzophenone of 3,5- and 4,5-dimethylisoxazole gave the corresponding oxetane in good yields, while 4-methylisothiazole gave a reaction on the methyl substituent [123].

Recently, the reactivity of isoxazole derivatives has been revisited, showing that some derivatives, such as 3,4,5-trimethylisoxazole, gave the corresponding oxetanes in very good yield (in the case reported in the Scheme 26, almost quantitative yields were obtained) and excellent diastereoselectivity (>99:1) [124].

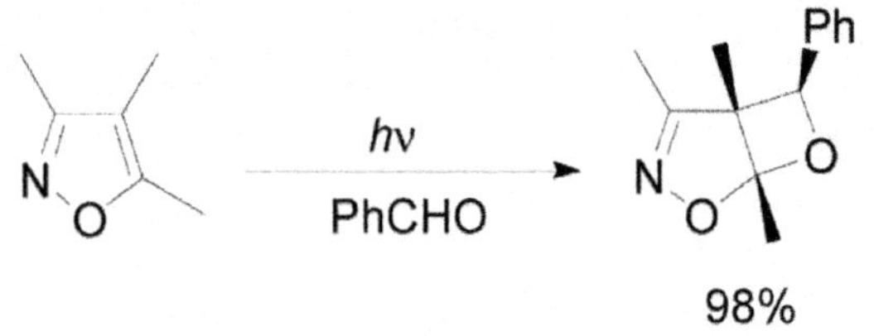

98%

Scheme 26: Reaction of an isoxazole derivative.

On the contrary, 3-phenyl substituted derivatives did not give the Paternò-Büchi reaction, allowing one to obtain only the ring contraction products [124]. In this case, the isoxazole was able to absorb the light giving the corresponding excited state able to give the isomerization reaction [125,126,127,128].

Aliphatic and aromatic carbonyl compounds reacted with oxazole derivatives with good to high exo diastereoselectivity, but low facial stereoselectivity. This reaction can be used in the synthesis of erythro-α-amino β-hydroxy carboxylic acid derivatives [129,130,131,132,133].

The irradiation of indole with benzophenone did not give the corresponding adduct. On the contrary, a benzoyl derivative of indole reacted with benzophenone, giving the corresponding oxetane [134]. It did not react with acetophenone, benzaldehyde, acetone, and propionaldehyde. When methyl pyruvate was used as carbonyl compound, the corresponding 3-indolyl carbinol was obtained [134].

The same reaction has been described on *N*-acetyl derivative of 7-azaindole. Although the reaction represents a method able to obtain a new class of compound, the low yields of the product (4%) prevent a synthetic use of this reaction [134].

Schenck reported that the irradiation of benzophenone in furan gave the corresponding adduct in 94% yield (Scheme 27) [135]. The structure was confirmed by Gagnaire *et al* [136].

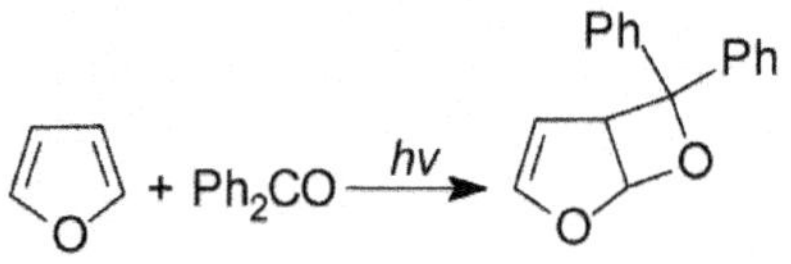

Scheme 27: Reaction of furan.

Furan and 2-methylfuran were found to react with propanal and benzaldehyde [111,137,138,139,140,141,142]. In this case, the *exo* stereochemistry at C-6 on the dioxabicyclo[3.2.0]heptene skeleton has been assigned [143]. Good regioselectivity was observed using silyl and stannyl furan derivatives: in this case, the reaction occurred on the less hindered side of the molecule [144]. When 2-silyloxyfuran was irradiated in the presence of aliphatic carbonyl compounds or with benzaldehyde, it gave a 1:1 mixture of regioisomeric products. On the contrary, when benzophenone was used, only the product deriving from the attack on the most hindered side of the molecule was recovered. The same result was obtained using acetone in a reaction when a low concentration of furan was used. In all the cases, an *exo* selectivity was observed [145]. On the contrary, 2-furylmethanol and the corresponding silyl ether gave low regioselectivity [145,146].

The high *exo* stereoselectivity of the reaction has been extensively studied: the formation of the product occurred on a triplet 1,4-biradical. The triplet biradical must be converted into the singlet one to give the product. In order to explain the pronounced *exo* stereoselectivity, a secondary orbital effect can be postulated: an interaction between the rather flexible α-oxy radical center and the allyloxy ring localized radical likely plays a major role (Figure 5) [147,148].

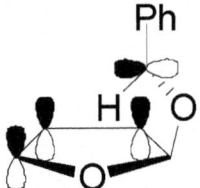

Figure 5: Structure of the biradical intermediate in the reaction of furan with benzaldehyde.

The regioisomeric biradical intermediates A and B resulting from the head-to-head and head-to-tail addition, respectively, have been examined (Figure 6) [69]. The biradical A is more stable than B by 16.5 kcal mol^{-1}. The biradical A exists as two conformers and only that conformer able to give the ring closure was considered.

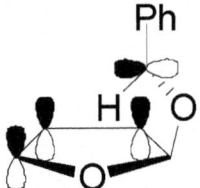

Figure 6: Possible biradical intermediates in the reaction of furan with benzaldehyde.

The HSOMO is mainly localized on the benzaldehyde fragment of the biradical while the LSOMO is mainly localized on the furanoid part of the molecule. The coupling between the radical carbons in these two orbitals, considering that the atomic orbitals involved can superimpose themselves, can give only the *exo* isomer, in agreement with the experimental results (Scheme 28).

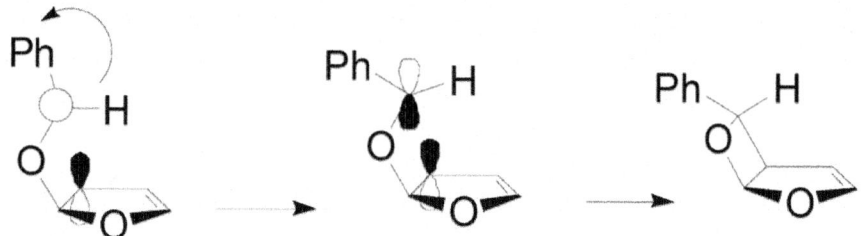

Scheme 28: Ring closure reaction in the formation of the *exo* isomer of the adduct between furan and benzaldehyde.

When the reaction was carried out in benzene, dimers can be obtained [55,149,150,151]. Furan reacted with chloral to give unexpectedly the corresponding 2-furyl carbinols [152]. Furthermore, 2-cyanofuran did not react, while 2-furfural diacetate and furfural ethylene acetal showed low reactivity [152].

The cycloaddition reaction can be performed on esters. In this case, the adducts can be obtained in a few cases. In most of the examples, they underwent a cycloreversion reaction to give the ring opening products (Scheme 29) [153]. More recently, this result has been questioned and a 95:5 mixture of stereoisomeric adducts was identified as the product [148]. Cycloreversion products were obtained also in the reaction of furan in the presence of furan-2-carboxyaldehyde [154].

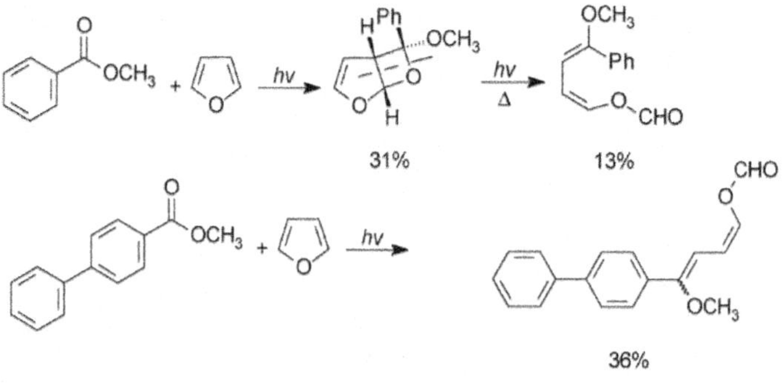

Scheme 29: Metathesis reactions on oxetanes obtained from furan derivatives.

Coupling products can be obtained carrying out the reaction between an amide and furan: also in this case the cycloadduct cannot be isolated, but the subsequent decomposition products can be isolated [155]. Furan quenches the fluorescence of the substrate, while a small new emission at 500 nm appears: this evidence is in agreement with a mechanism involving a reaction in the excited singlet state *via* the formation of an exciplex.

The reaction of some aromatic carbonyl compounds (benzophenone, benzaldehyde) with benzofuran has also been reported; when compounds showing high triplet energy were used, dimers of benzofuran were obtained. On the contrary, when carbonyl carbonyl with a low triplet energy were used, oxetanes were the products of the reaction [156]. As in the case of furan, Schenck obtained only one regioisomer. A reinvestigation of the reaction of benzofuran when it was irradiated in the presence of acetophenone or propiophenone, showed that, also in this case, the corresponding oxetanes were the products of the reaction [157].

The irradiation of 4-amino-2,7-dimethylbenzofuran in the presence of benzaldehyde in benzene gave the corresponding adduct in good yields [158]. It is interesting to note that the reaction allowed the formation of the *endo* isomer. The reaction between benzo[*b*]furan and benzaldehyde gave an adduct whose stereochemistry has not been described [156].

3-Furylcarbinol derivatives could be obtained through the treatment of the oxetanes with TsOH [142]. The formation of a protonated oxetane is in agreement with the high negative entropy of activation [159]. If the irradiation was performed in the presence of an acid, 3-furylcarbinol can be obtained in a one-step procedure in better yields [160]. This procedure has been used in the synthesis of perillaketone, a naturally occurring 3-substituted furan [160]. On the contrary, Lewis acids catalyzed a different behavior: the treatment with $BF_3 \cdot Et_2O$ gave only 3-substituted furan in THF and 89% of 2-substituted furan in acetonitrile [159].

The oxetane derivatives were treated also with $KMnO_4$ and the resulting *cis* diol reacted with acetone in the presence of an acid. This procedure allowed the synthesis of a carbohydrate derivative [161].

An epimerisation reaction occurred on the *trans* diol obtained when the oxetane was treated with MCPBA [161]. Schreiber studied the possible chemical modifications on the cycloadduct deriving from the reaction between an aldehyde and furan [162]. Hydrolytic ring opening, reductions, hetero Diels-Alder, the reaction with MCPBA, and hydroboration-oxidation were the reactions he studied. The reaction with MCPBA was used by Schreiber and coworkers in the synthesis of asteltoxin [163,164], and in that of avenaciolide, an antifungal metabolite [165]. The reaction of tributylstannylfuran with butyl glyoxylate was used in the synthesis of a ginkgolide B-kadsurenone hybrid of two inhibitors of a platelet activating factor [144].

In the synthesis of asteltoxin, the synthetic sequence implies the photochemical coupling of 3,4-dimethylfuran with a functionalised aldehyde to give the corresponding adduct in 63% yield (Scheme 30).

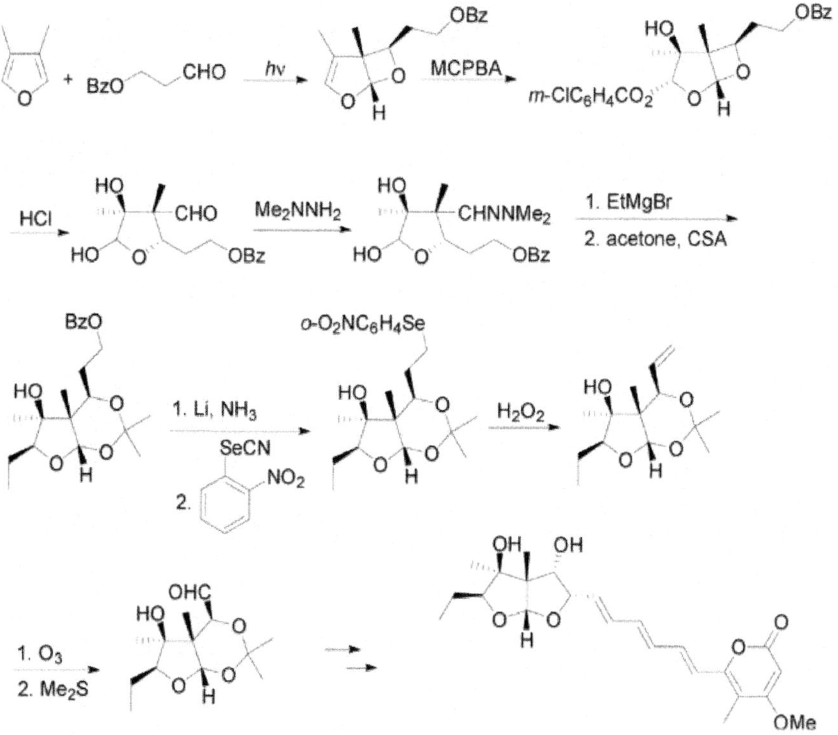

Scheme 30: Synthesis of asteltoxin.

The same research group reported a formal synthesis of avenaciolide, an antifungal metabolite. In this case the oxetane (obtained in multigram quantities in high yields and with complete stereochemical control) was treated with hydrogen to give the saturated compound. The key step in this synthetic procedure is a reaction with ozone followed by a base catalysed epimerisation with potassium carbonate and cyclization in acidic medium.

Furthermore, the cycloadduct obtained from the reaction between furan and an aldehyde can be treated with an excess of Schlosser's base (BuLi, t-BuOK). The reaction gave the corresponding anion which can react with carbonyl compounds or alkyl halides [166].

The most important target in this field was oxetanocin (**3**), a nucleoside isolated from *Bacillus megaterium* NK 84-0218 showing anti-HIV activity. An approach to the synthesis of this compound has been reported [167]. The treatment of oxetane obtained in the reaction between furan and benzaldehyde with *N*-iodosuccinimmide in the presence of methallyl alcohol gave the corresponding iodoacetals. The subsequent treatment with iodonium di-*sym*-collidine perchlorate (IDCP) gave a product with a structure related to that of **3** in low yields as a relatively unstable mixture of diastereoisomers. Oxetanocin was also obtained carrying out the reaction between 2-methylfuran and benzoyloxyacetaldehyde [168]. The corresponding adduct was treated with ozone and the product was reduced with $NaBH_4$. The obtained alcohols were protected. The product was treated with N-benzoyl-disilyladenine and $SnCl_4$ to give **3**.

The reaction of glyoxylates with furan can be performed also using chiral glyoxylates. In particular, the use of *R*-(−)-menthol, 2-octanol, and 2,2-dimethyl-3-butanol as chiral auxiliaries gave the corresponding oxetanes in high yields. These compounds can be converted into the corresponding 3-substituted furans. The furans showed low enantiomeric excess (Scheme 31) [169].

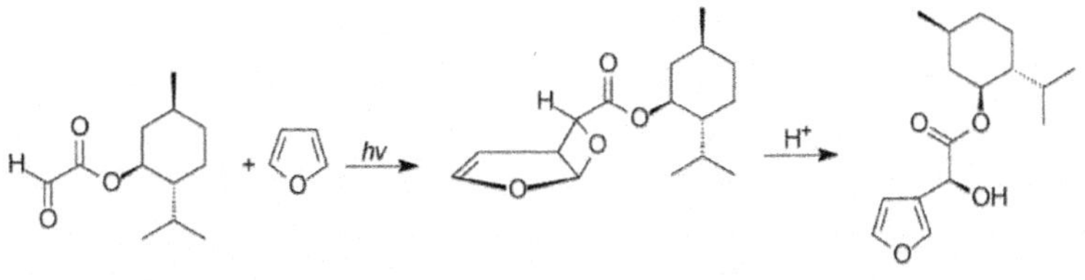

Scheme 31: Reaction of furan with a chiral glyoxylate.

The use of chiral phenylglyoxylate gave better results. The use of chiral alcohols gave diastereoisomeric excess in the range of 4%–80% [170,171,172,173].

An important variability of the diastereoisomeric excess in function of the temperature has been observed with the presence of an inversion temperature [174]. When the reaction is carried out on 2-methylfuran, a 2:1 regioisomeric mixture was obtained with a very high diastereoisomeric excess [175].

In order to justify the observed stereoselectivity the triplet biradical intermediates in the reaction of some phenylglyoxylate derivatives with furan could be considered [173]. Calculations on these biradical intermediates showed that **25** (the precursor of the observed product) was more stable than **26** by 0.73 kcal mol^{-1} (Figure 7).

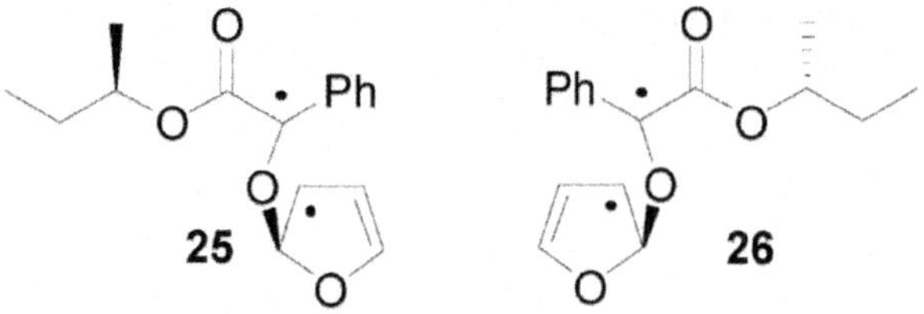

Figure 7: Radical intermediates in the reaction of chiral phenylglyoxylates with furan.

Neckers studied the reaction of ethyl phenylglyoxylates with several alkenes showing that it gave the cycloaddition reaction only when electron-rich alkenes were used. Furthermore, the main reaction observed when monosubstituted alkenes were used was the Norrish type II reaction and the same behavior was observed reducing the electron richness of the alkenes [176]. Methyl 2-thienylglyoxylate showed a transient absorption at 390 nm and a broad band around 600 nm. This transient band has been assigned to the triplet state. This compound reacted with 2,3-dimethyl-2-butene to give the corresponding adduct [177].

To improve diastereoselectivity, the reaction was carried out in the presence of different zeolites [178]. NaY was the best solid zeolite support for this reaction. The origin of this behavior can be explained considering the possible effect of the confinement on the reaction. In the Paternò-Büchi reaction the biradical intermediate can assume two possible conformations [30,68]. One has the radical carbons in almost *anti* conformation, while the other one shows a *syn* conformation (Scheme 32).

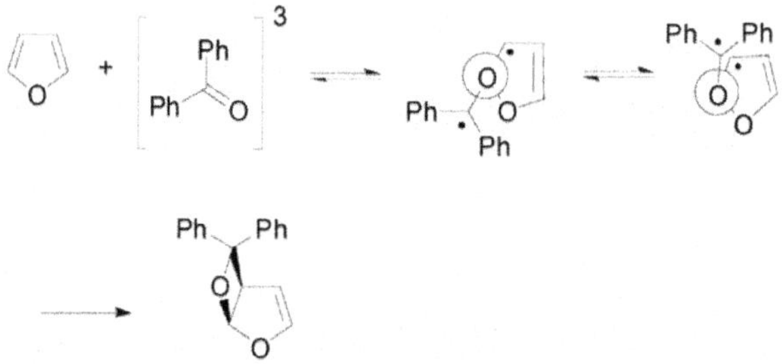

Scheme 32: The reaction of furan with benzophenone. Conformers of the biradical intermediate.

The *syn* conformation can undergo the following cyclization, after the intersystem crossing from the triplet to the singlet state. For the *anti* conformation the most favored process is the retrocleavage to the starting material. The *syn* conformations take up a smaller volume than the *anti* one. The preferential formation of a stereoisomer in the reaction within a zeolite could be explained: the *SS-syn* conformer was obtained preferentially, while the *RS* conformer was obtained preferentially in*anti* conformation. The latter conformer has to rotate along the C-O bond to give the *syn* conformation able to cyclize. Zeolites could act as agents able to reduce the conformational mobility of the intermediates. They inhibited the rotation of this conformer and, then, favored the formation of the product deriving from the *SS* biradical intermediate.

Recently, Abe reported an attempt to obtain an intramolecular version of the Paternò-Büchi reaction between furan and a carbonyl compound. Nevertheless, the irradiation of a phenylglyoxylate derivative allowed only the formation of a product deriving from an intermolecular reaction. Low yields were obtained [179].

The irradiation of furan in the presence of acyl cyanides yielded the corresponding oxetanes but both diastereoisomeric*endo*- and *exo*-oxetanes are formed. Low asymmetric induction was observed when chiral acyl cyanides were used [180]. Furan reacted also with chiral ketones. In this case, before the 2+2 cycloaddition, an α-cleavage reaction modified the expected products. A chiral product was obtained as 2:1 diastereoisomeric mixture where the most abundant product has 1*R*, 3*R* configuration when (−)-menthone was used as a substrate [181]. If the irradiation was performed on the ketonic functional group in a protected carbohydrate, a complex reaction mixture was obtained [182,183].

The reaction between 3,4-dimethylfuran and *R*-isopropylidene glyceraldehyde was performed in order to obtain a stereoselective Paternò-Büchi reaction. The coupling products were obtained with an overall yield of 35% as a 1.2:1 mixture of diastereoisomers [184].

This behavior suggested a mechanism that was insensitive to the substitution pattern of chiral aldehydes. Reaction between an excited aldehyde (singlet or triplet state) and furan proceeds with initial carbon-oxygen bond formation to produce either of the two biradical species. The stereocenter adjacent to the carbonyl is now in a 1,4-relationship to the newly formed stereocenter at the acetal carbon and is expected to exert little influence as a stereocontrol device [184]. This hypothesis is not in agreement with the stereochemical behavior of using phenyl glyoxylate esters where the chiral carbon atom was farther than in *R*-isopropylidene glyceraldehyde. The extensive racemization observed probably reflects the photolability of the aldehydes towards racemization under the conditions of the reaction [184]. Nevertheless, the product of this reaction was used in a chiral synthesis of the bicyclic part of asteltoxin confirming the assigned absolute configuration [184]. On the contrary, benzoin reacted with furan to give the corresponding adduct in acceptable yield (56%) and *de* > 98% [185].

The observed stereoselectivity was explained considering the relative stability of the biradical intermediates. Nevertheless, chiral ketones gave the Norrish Type II reaction as the only observed reaction. In order to avoid Norrish Type II reaction a substrate without γ-hydrogen was used [186].

In 1990 Griesbeck found that the reaction of benzaldehyde with homoallylic alcohols did not show diastereoselectivity [61]. Nevertheless, Adam showed that allylic alcohols reacted with benzophenone to give the corresponding adducts with high regio- and diastereo-selectivity (Scheme 33) [187,188,189,190].

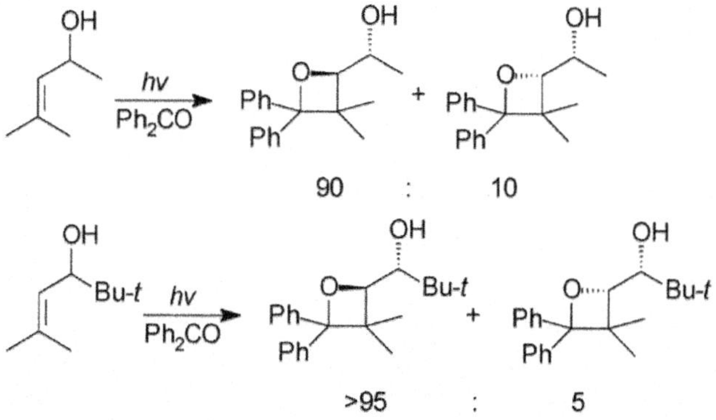

Scheme 33: Reaction of allylic alcohol derivatives with benzophenone.

In presence of protic methanol, the diastereoselectivity dropped drastically, while it totally disappeared using the corresponding silyl ethers. These data are in agreement with the presence of a hydroxyl directing effect in the Paternò-Büchi reaction. The formation of a hydrogen bond between triplet excited benzophenone and the substrate in the exciplex favored the formation of the *threo* isomer. On the contrary, the formation of the *erythro* stereoisomer would be less favored due to allylic strain.

The formation of a hydrogen bond to direct the Paternò-Büchi reaction has been considered by other researchers. Diastereoselective cycloaddition has been obtained using chiral enamide [191,192], or in the reaction of allylic alcohols with naphthalene rings [193]. When unsymmetrical carbonyl partners such as acetophenone or benzaldehyde were used, the corresponding *cis* isomer was observed with high diastereoselectivity. The regioselectivity was high with acetophenone but lower with benzaldehyde [188]. *Cis* diastereoselectivity can be explained by using the Griesbeck rule on the possible triplet biradicals formed in the reaction. Steric interactions are minimized when the biradical assumes the optimal conformation and this conformation accounts for the formation of the observed stereoisomer [190].

When chiral allylic alcohols were used as substrates in the reaction *cis* diastereoisomers were formed. Furthermore, also in this case, a pronounced *threo* diastereoselectivity was observed, in agreement with a less pronounced hydroxyl directing effect when acetophenone and benzaldehyde were used [188,190]. Chiral allyl ether gave the corresponding adduct with high diastereoselectivity [98]. We have to note that this result is not in agreement with previous reported data on the reactivity of silyl ethers of allylic alcohols [187].

The reaction of 2,3-dihydrofuran-3-ol derivatives (a particular type of allylic alcohol) with benzophenone gave the corresponding adducts. In methanol, a *trans* relationship between the oxetane ring and the hydroxyl group was observed, while, in benzene, the *cis* isomer prevailed. The Eyring plot showed that the *trans* isomer increased with a not linear behavior upon decreasing the temperature [194].

The reaction of allylic alcohols with carbonyl compounds was tested also on 2-furylmethanol derivatives. The presence of large substituents on the carbon bearing the alcoholic function allows a high regioselectivity (Scheme 34) [195].

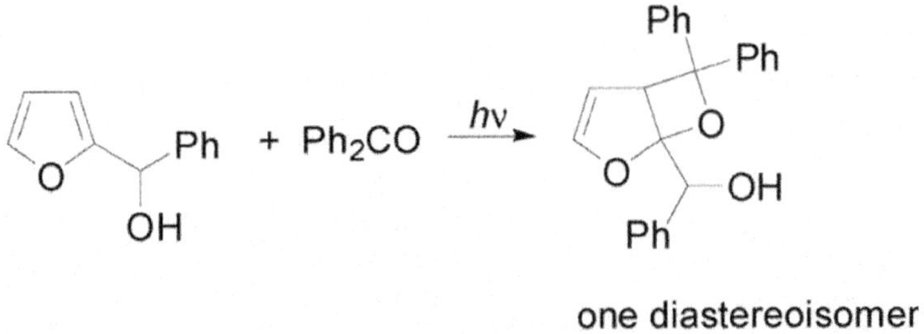

one diastereoisomer

Scheme 34: Reaction of 2-furylmethanol derivatives.

5-Methyl-2-furyl derivatives were used as substrates showing a different regioselectivity. This type of substrates gave a 1:1 mixture of regioisomers, when irradiated in the presence of benzophenone, and a single regioisomer in the presence of benzaldehyde [196]. In agreement with the results obtained with 2-furyl derivatives, the products deriving from the attack on the side bearing the alcoholic function were obtained as a single diastereoisomer, while those deriving from the attack on the side bearing the methyl group were obtained as a mixture of diastereoisomers.

The above described reactions showed that on furan two possible regioisomers could be obtained. The experimental results showed that, in some cases, the reactions occurred mainly on the side when the hydroxyl group was present, while other reactions showed a completely different regioselectivity. The reason of this behavior can be found in kinetic factors depending on the different stability of the biradical intermediates. These considerations are in agreement with the results obtained irradiating geraniol in the presence of benzophenone [197].

In this case, two different double bonds were present: a double bond able to give a tertiary radical intermediate, and an allylic alcoholic moiety. The authors found that at −75 °C the product deriving from the attack on the first double bond was obtained in 44% yields while the product deriving from the attack on the allylic double bond was obtained only in 13% yields. Furthermore, at 20 °C the yields were 29% and 15%, respectively. In this case, then, the attack on the terminal double bond was favored.

The reaction of 2-furylmethanol derivatives with aliphatic aldehydes and ketones gave the corresponding adducts with high regioselectivity: the reaction occurred on the most hindered side of the substrate. However, no diastereoselectivity was observed [198]. The relative stability of the biradical intermediates was able to explain the regioselectivity of the reaction. A computational study (DFT) showed that the biradical obtained on the most hindered side of the molecule was more stable than the other one [196].

The diastereoselectivity of the reaction between 2-furylmethanol derivatives and aromatic carbonyl compounds clearly showed that it increased in relation to the nature of the substituents on the carbon bearing the alcoholic function as described by Adam. However, while Adam considered the allylic strain with a methyl group in β-position as the driving force for the diastereoselectivity, in this case, a methyl group on the C-3 of the furan ring was not present.

In order to have more data to explain the observed stereoselectivity the photochemical behavior of tertiary 2-furylcarbinols was studied [199]. The irradiation of 1-methyl-1-phenyl-1-(2-furyl)methanol with benzaldehyde gave a mixture of two regioisomeric products. The regioisomer on the most hindered side of the molecule was obtained in low yield but it showed a complete diastereoisomeric control. On the contrary the main product was a mixture of four diastereoisomeric products. The reaction of the same compound with benzophenone gave only the product deriving from the attack on the most hindered side of molecule. This compound was obtained with 48% diastereoisomeric excess. The regioselectivity was explained, as described above, considering the relative stability of the biradical intermediates. In this case, in the reaction of the 1-methyl-1-phenyl-1-(2-furyl)methanol with benzaldehyde the biradical obtained on the less hindered side of the substrate was more stable than the other one by 18.03 kJ mol⁻¹. Recently, some authors reported that no stereoselectivity was observed using cyclic 2-furyl methanol derivatives [200]. However, in another work, a stereoselective behavior of the same substrates was assumed [201].

On the basis of these results an explanation of the stereochemical behavior was attempted [199]. 1-Methyl-1-phenyl-1-(2-furyl)methanol showed three conformations. All three conformers were in the range of 1.97 kJ mol⁻¹ and they did not show a preference. The directing effect exerted by the hydroxyl group is due to the formation of a hydrogen bond between the hydroxyl group and the oxygen of the excited carbonyl compound, or it is due to the formation of a complex. This type of interaction could favor the formation of a preferential conformation in the biradical intermediate where the hydroxyl group and the oxygen of the carbonyl compound are near. These conformations could have different energies for different diastereoisomeric biradicals, giving an explanation of the observed behavior. In the case of 1-methyl-1-phenyl-1-(2-furyl)methanol, if the hydroxyl group drove the attack of the oxygen of the carbonyl group, the conformations of the biradical intermediate represented in Scheme 35 were obtained. **B** and **D** were the preferential conformations: calculations on these conformations showed that there was a difference of 13.26 kJ mol⁻¹ between the energies of these two conformations. This difference can account for the observed complete diastereoselectivity of the reaction. In the reaction of the same substrate with benzophenone, the corresponding conformers **B** and **D** showed a difference energy of 7.79 kJ mol⁻¹: this difference is in agreement with the observed diastereoselectivity.

The same approach can be used to justify the stereochemical behavior of the reaction of allylic alcohols with benzophenone [202]. The irradiation of 3-furylmethanol derivatives in the presence of benzophenone gave the corresponding adduct with a very high stereoselectivity [203].

The above reported results represent all the available data on the Paternò-Büchi reaction on pentaatomic heterocycles. We can see that, with the exception of furan, there are very few data: in particular, 1. most of the unsubstituted tested compounds different from furan did not react, 2. only few substituted derivatives showed a significative reactivity towards excited carbonyl compounds.

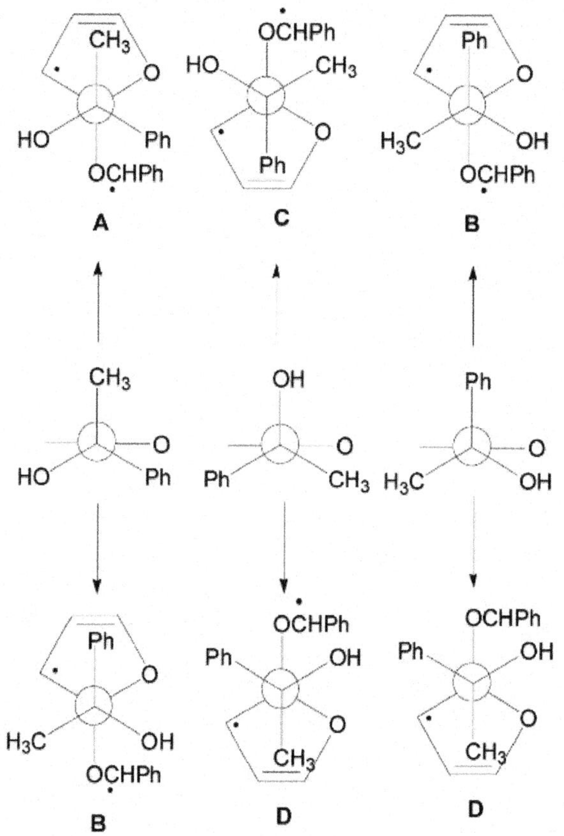

Scheme 35: Possible conformations of the biradical intermediate from the reaction of 1-methyl-1-phenyl-1-(2-furyl)methanol with benzaldehyde.

This behavior may be due to different reasons. First, the different aromaticity of the compounds could play an important role in order to define the reactivity of the compounds. Furan is the lowest aromatic pentaatomic heterocyclic compound known, while the other compounds show higher aromaticity. However, this type of explanation cannot justify why thiophene does not react while simple dimethylthienyl derivatives react and why some dimethylthienyl derivatives react while some others do not show any reactivity.

A different explanation was identified in the quenching properties of these heterocycles. Thiophene and monomethyl derivatives are efficient quenchers of triplet benzophenone. The Stern-Volmer plot showed a linear relationship [204,205]. On the contrary, 2,5-dimethylthiophene (a compound able to give the cycloaddition reaction) is not a good quencher of the triplet benzophenone [206]. In this case, the Stern-Volmer plot is not linear. This situation is commonly encountered when the quencher employed quenches two excited states. It seems reasonable that pyrrole acts as quencher of both triplet benzophenone and the exciplex between triplet benzophenone and pyrrole.

On the basis of these investigations, the common five-membered heterocycles may be classified in two categories in regard to their quenching properties: those with electron-donating groups, which give Stern-Volmer plots in the shape of straight lines, and those substituted with halogens and electron attracting groups, such as *N*-benzoylpyrrole, which give parabola-shaped curves. Those in the first category give oxetanes when they are bad quenchers. Since these compounds do not presumably form exciplexes, they should go from starting materials to products through a biradical intermediate. On the other hand, those in the second category most likely give oxetanes when they are good quenchers, as may be shown for 2,5-dibromothiophene, and furthermore they go from starting materials to products through an exciplex [207].

The (6-4) photoproduct is an adduct of two pyrimidines that occupy adjacent sites on the same DNA strand. It is the second major lesion induced in DNA by UV radiation. Although statistically four times less frequent than CPD lesion, (6-4) photoproducts are believed to be severely mutagenic. Its formation is believed to occur via an initial Paternò-Büchi type cycloaddition to form an oxetane intermediate. Subsequent C4-O bond cleavage gave the observed (6-4) photoproducts (Scheme 36).

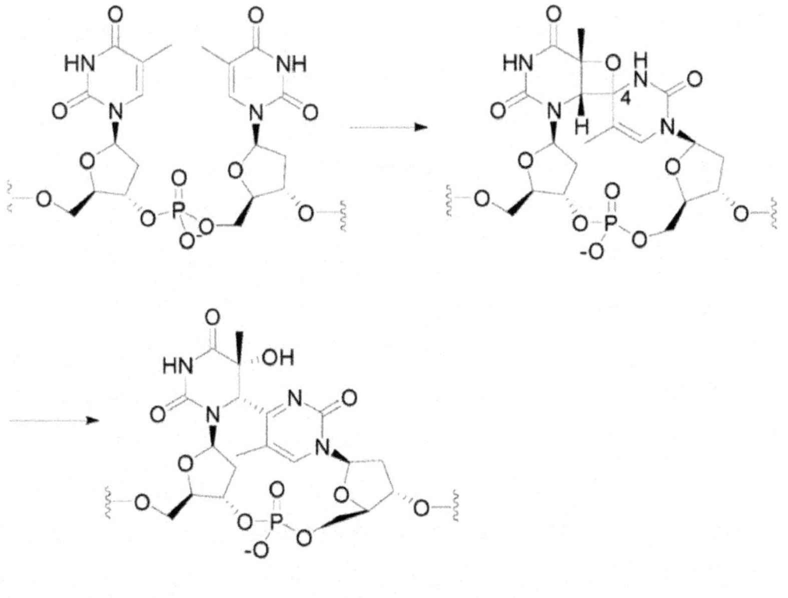

Scheme 36: Photochemical reaction between DNA pyrimidine derivatives.

The (6-4) photoproduct is one of the major mutagenic classes of DNA photoproducts and is involved in the etiology of skin cancer. The oxetane **27** was prepared in a triplet reaction and both electron donors and acceptors substituents were found to be able to photosensitize the splitting reaction (Scheme 37) [208,209,210,211]. Only one regioisomer (**27**) was observed. The other regioisomer **28** was rarely observed [212].

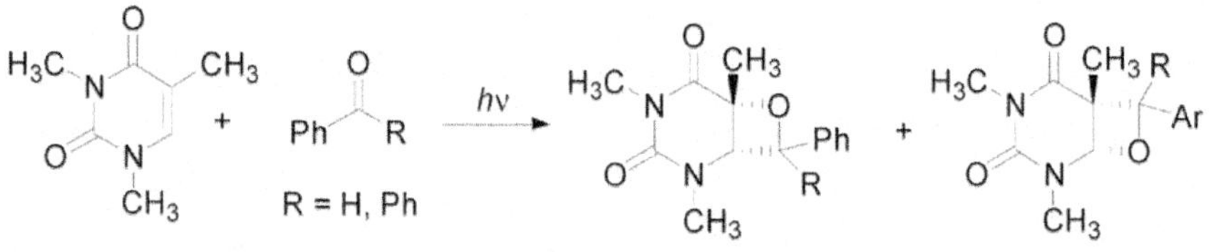

Scheme 37: Reaction of uracil derivative with benzophenone and benzaldehyde.

The temperature can determine the regiochemistry of the reaction. At −38 °C the regioisomer **27** dominates (61:39). On the other hand, at 70 °C the regioisomeric behavior of the reaction is inverted [212,213]. A change of the selectivity-determining step has been determined considering a non-linear Eyring plot. Two different cases can be assumed: the situation where the conformational changes of the triplet intermediates were slower than ISC at low temperature and the case where the conformational changes exceed the ISC process [213]. When the conformational interchange is faster than the ISC process, the population of high potential energy conformations decreases, while the population of a lower potential energy conformer increases.

A weak solvent effect was observed on the regioselectivity of the reaction [214]. Substituents on benzophenone modified the regiochemistry of the reaction [215,216,217]. The quantum yields correlated with the energy gap between SOMO of benzophenone derivatives and the HOMO of the uracil. Recently, a computational approach to this reaction appeared [218]. The authors sudied the biradical intermediates. They found that one of them is more stable than the other, and that the formation of the first is faster than the other one. Then, 28 can be considered as a thermodynamic product, while 27 a kinetic one.

Recently, it has been observed that, due to the energy barriers between the two stable conformers, the equilibrium was more favorable for the formation of the oxetane 28, rather than oxetane 27 at a higher temperature. Triplet benzophenones with a short lifetime would give rise to a less efficient Paternò-Büchi reaction [217]. The oxetane was obtained also in the reaction between benzophenone and benzophenone-derived drugs and thymidine [211,219]. An enantioselectivity factor for triplet deactivation was found using enantiopure ketoprofen. The use of enantiopure ketoprofen showed that thimidine was able to give an enantioselective quenching of the chiral ketoprofen triplet state. This quenching was related to the formation of C-O bond, the first step of oxetane formation [67]. An intramolecular reaction beween thymidine esterified by ketoprofen has been reported [220]. The irradiation of 29 gave 30 deriving from the (6–4) photoadduct 31 (Scheme 38) [221].

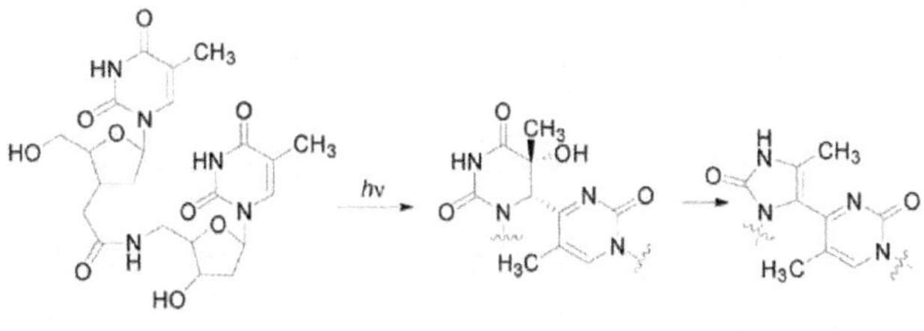

Scheme 38: Irradiation of 29.

A reaction between 5-fluoro-1,3-dimethyluracil with 1,5-dimethoxynaphthalene has been reported [222]. The reaction gave product where an aromatic ring of naphthalene is broken. The presence of the intermediate adduct has been proposed.

OTHER INTERMOLECULAR REACTIONS

5-Substituted adamantan-2-ones gave the corresponding cycloadducts via the n, π^* excited singlet state. This substrate gave a regiospecific reaction with a low stereoselectivity (the best result was 59:41 ratio in favor of the *anti* isomer) [223,224,225,226]. Methyleneadamantane reacted with acetone. However, the Paternò-Büchi adduct was obtained in only 5% yield [223]. The reaction of biacetyl with benzvalene gave the corresponding adduct, while benzophenone gave, as the only product, benzene [44]. When homobenzvalene was used as the alkene in the reaction with ethyl phenylglyoxylate, the irradiation gave the corresponding adduct in 70% yield with high stereoselectivity (Scheme 39) [227].

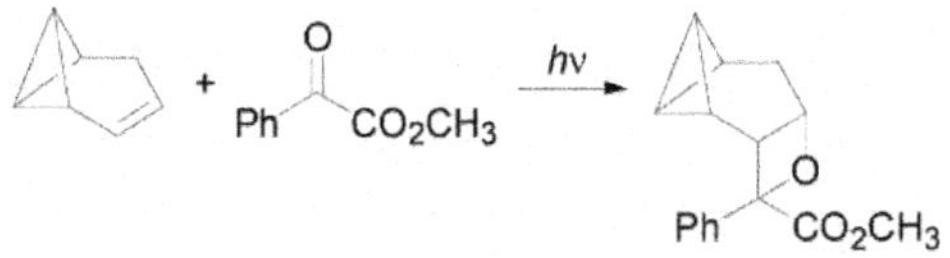

Scheme 39: Reaction of homobenzvalene.

Esters could be used as carbonyl compounds in the reaction with chiral alkyl cyanobenzoate derivatives (Scheme 40) [228,229].

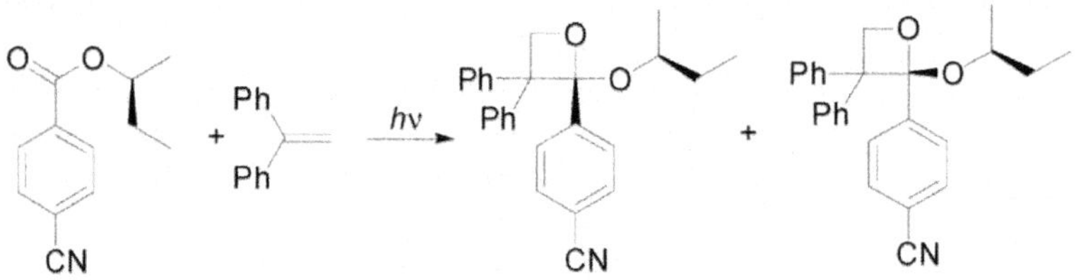

Scheme 40: Esters as carbonyl compounds in the Paternò-Büchi reaction.

The reaction occurred through the first excited singlet state. The authors showed the presence of a charge transfer band at 300–330 nm. Irradiating at 254–290 nm, a good diastereoselectivity was observed (*de* 77%). The irradiation at 330 nm afforded the formation of an epimeric mixture. The photocyclization was likely to proceed *via* a short lived 1,4-biradical mechanism. Then, the diastereo-face-differentiating complexation of chiral alkyl cyanobenzoate with the olefin donor is the responsible for the observed *de* values. Calculations showed that the conformational population in the charge-transfer complex is almost the same in agreement with the observed lack in diastereoselectivity.

A stereoselective behavior has been described in the reaction between benzophenone and cyclooctene. At −95 °C the *cis*adduct was formed with a very high diastereoselectiviy from *cis*-cyclooctene. On the contrary, at −20 °C both diastereoisomers were generated. At higher temperatures, the *trans* adduct dominated. Over a broad temperature range (−80 to +60 °C), the more strained *trans*-cyclooctene gave nearly exclusively the *trans* adduct [93,94]. Triplet ketone could attack the double bond to give two biradical intermediates and the authors suggested that temperature-dependent conformational changes of the biradical competed with the cyclization to the oxetane and the retro cleavage to the *cis*-cyclooctene. The activation barrier between two conformers of the biradical intermediate could explain the increasing of *trans*-oxetane with the temperature [94].

1-Acetylisatin reacted with styrene and furan derivatives to give the *endo* adduct with high stereoselectivity [230,231].

Quinones react with stilbene derivatives has been described [73,232]. In this case, a charge-transfer absorption band was observed at 480 nm. Time-resolved (ps) absorption spectrum showed an absorption at 480 nm (chloranil anion radical) and a broad band around 760 nm. After 50 ps a transient absorption typical for triplet chloranil at 510 nm was observed. The reaction occurred through selective excitation at 480 nm. Quinones reacted with diarylacetylene giving a product deriving from the ring opening of a transient oxetene. Time-resolved (ps) absorption spectrum was in agreement with an electron transfer mechanism [233]. When 2-chloro-5-methoxybenzoquinone reacted with arylacetylenes, CIDNP experiments suggested a mechanism where the attack of the triplet quinone on the acetylene gives a biradical intermediate [92]. The same behavior has been observed in the reaction between quinone and quadricyclane and norbornadiene [234,235]. The adduct was found also in the reaction between quinones and acenaphthylene [77]. Benzophenone can react also with a paracyclophane derivative **32** (Scheme 41) [54].

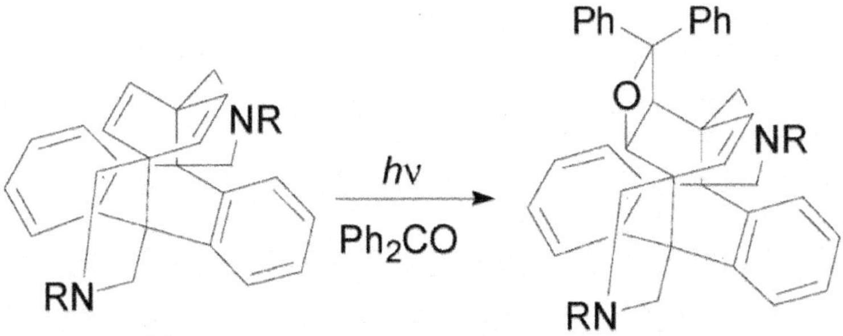

Scheme 41: Reaction of a paracyclophane derivative.

The Paternò-Büchi reaction can also be applied to other unsaturated systems other than alkenes, such as dienes [10], allenes [236], acetylenes [237], and ketenimines [236]. When tetramethylallene reacted with aromatic ketones the main products resulted from bis-addition of the carbonyl compounds to give dioxaspiroheptanes [238,239]. On the contrary, 1,1-dimethylallene gave only the monoaddition 2:1 mixture of the adducts in a reaction with aliphatic aldehydes [240]. A higher regioselectivity was obtained using an allene derivative. The irradiation of isoquinoline-1,3,4-trione with alkynes gave a product deriving from an initial Paternò-Büchi reaction on the carbonyl at C-4 to give the corresponding oxetene [241]. DFT calculations on the relative stability of the biradical intermediate showed that the more stable one was the precursor of the main product of the reaction.

INTRAMOLECULAR REACTIONS

An intramolecular Paternò-Büchi reaction on a paracyclophane derivative has been reported giving the corresponding adduct in quantitative yield [90]. An intramolecular Paternò-Büchi reaction of **33** to give **34** has been used in the synthesis of 2,7,9-trimethylenetricyclo[4.3.0.0^{3,8}]nonane **35** (Scheme 42) [242].

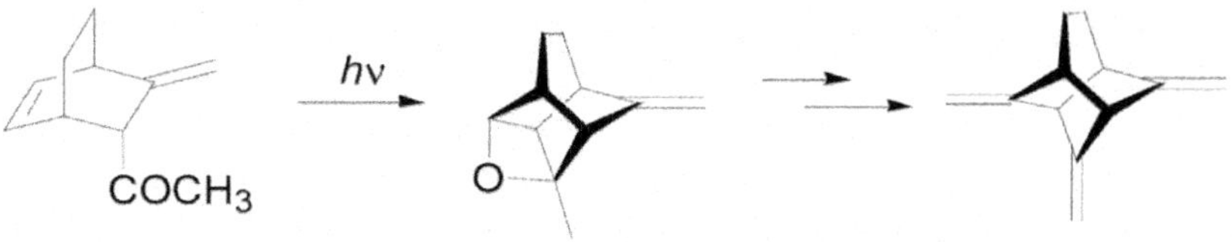

Scheme 42: Intramolecular oxetane formation in the synthesis of 2,7,9-trimethylenet ricyclo[4.3.0.0^{3,8}]nonane.

The same synthetic scheme was used in the synthesis of 2,7,9-trimethylenetricyclo[4.3.0.0^{3,8}]non-4-ene [243]. The preparation of some stelladiones such as tricyclo[3.3.0.0^{3,7}]octane-2,4-dione or tricyclo[3.3.0.0^{3,7}]-octane-2,6-dione [244]. This type of intramolecular reaction was the key step used in the synthesis of diquinanes and triquinanes [245,246,247].

The synthesis of 1,13-herbertenediol was performed using an intramolecular Paternò-Büchi reaction between an aldehydic group and an α-alkyl substituted styrene moiety [248]. An intramolecular stereoselective Paternò-Büchi reaction was the key step also in the synthesis of some derivatives of R-(+)-sclareolide [249]. A synthesis of the scaffold of merrilactone A involved also an intramolecular 2+2 cycloaddition to give the adduct [250].

On the other hand, a project devoted to obtain tromboxane analogs using an intramoelcular reaction of a carbonyl group with an enol ether, failed [251].

An intramolecular reaction on allyl cyclopentanone derivatives has been reported [227,252,253]. In this case both *straight* 37 and *crossed* 38 oxetanes can be obtained (Scheme 43). Using 2-allylcyclopentanone nearly equal amounts of these isomers were obtained. However, the use of the more rigid starting material 36 allowed the preferential formation the *straight* isomer.

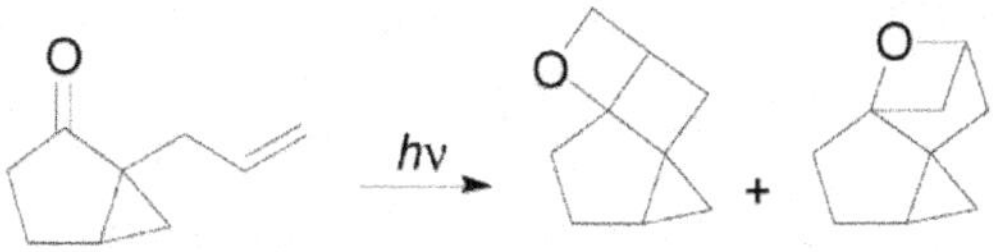

Scheme 43: Intramolecular oxetane formation on allyl cyclopentanone derivatives.

An intramolecular cycloaddition reaction was found in the reaction of 39 (Scheme 44) [254].

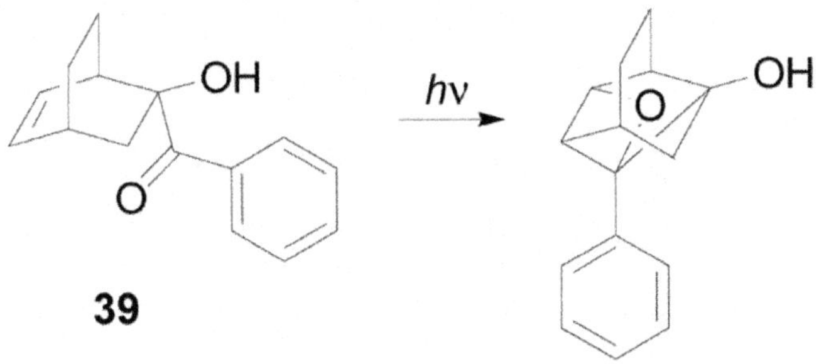

Scheme 44: Intramolecular oxetane formation in the reaction of **39**.

Compound **40** gave in quantitative yield the oxetane through an intramolecular reaction. The oxetane thus obtained could be converted into **41** via fluoride desilylation (Scheme 45) [255].

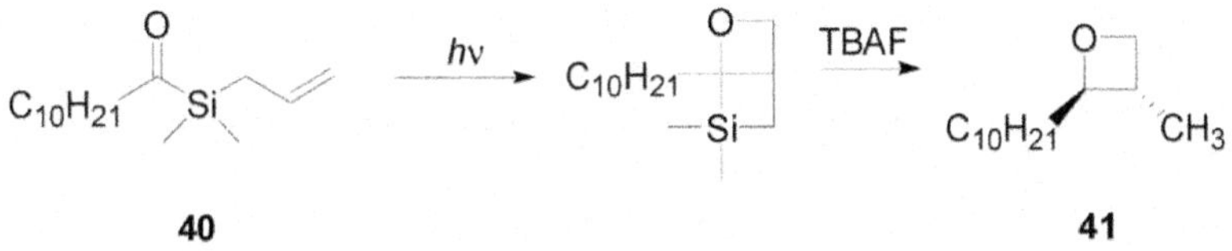

Scheme 45: The use of silyl derivatives in the intramolecular synthesis of oxetanes.

When alkenyl phenylglyoxylates were used as substrates, Norrish type II is the main reaction in most of the cases [256,257].

N-Isopropyl-*N*-tigloylbenzoylformamides gave an intramolecular cycloaddition reaction in high yields (50-99%) to give the *syn* adduct (*syn/anti* = 2.1) [258,259].

When the ketone **42** was irradiated in the solid state, the only possible reaction was a Norrish Type I reaction. The reaction gave an alkene and a carbonyl compounds, able to react in an intermolecular reaction giving the corresponding adduct (Scheme 46). The reaction occurred with high stereoselectivity, and, when performed on a chiral salt gave the product with *ee* in the range 42%–52% [234].

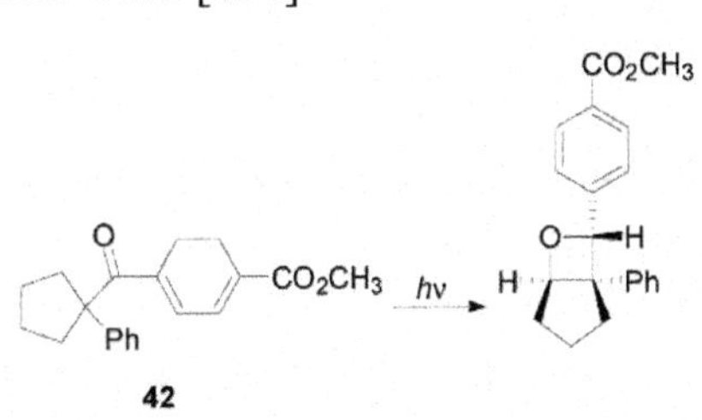

Scheme 46: Chiral intramolecular oxetane synthesis.

THE FUTURE OF THE PATERNÒ-BÜCHI REACTION

One of the most important field to be developed in the near future to obtain a more sustainable organic chemistry is connected to the flow chemistry. Several attempts have been made to apply flow techniques to photochemical reactions [260]. The photochemical reaction of benzophenone with prenyl alcohol to give the corresponding oxetane in flux conditions has been attempted with success [261].

Another field to be developed in the future is the use of photochemical reactions to functionalize polymeric materials. On this subject, two interesting results has been obtained. Oxetanes were obtained by using a photochemical coupling of aromatic aldehydes on a polymeric materials with alkenes [262]. On the other hand, also the coupling of polyvinyl compounds with aromatic aldehydes has been performed [263].

CONCLUSIONS

The aim of this review was to furnish an overview of the latest results obtained by using the Paternò-Büchi reaction. The reaction can be used as a suitable synthetic method to obtain an uncommon structural moiety such as the oxetane ring. The oxetane can be modified allowing the synthesis of several biological active compounds. Furthermore, the reaction can be accomplished in a stereoselective manner, allowing new sustainable chemical processes.

REFERENCES

1.   Kingston, D.G.I. The chemistry of taxol. *Pharmacol. Ther.* 1991, *52*, 1–34.

2.   Huang, J.; Yokoyama, R.; Yang, C.; Fukuyama, Y. Merrilactone A, a novel neurotrophic sesquiterpene dilactone from*Illicium merrillianum. Tetrahedron Lett.* 2000, *41*, 6111–6114.

3.   Wang, Y.; Fleet, G.W.J.; Storer, R.; Myers, P.L.; Wallis, C.J.; Doherty, O.; Watkin, D.J.; Vogt, K.; Witty, D.R.; Wilson, F.X.; Peach, J.M. Synthesis of the potent antiviral oxetane nucleoside epinoroxetanocin from D-lyxonolactone. *Tetrahedron Asymmetry* 1990, *1*, 527–530.

4.   Yang, C.O.; Kurz, W.; Eugui, E.M.; Mc Roberts, M.J.; Verheyden, J.P.H.; Kurz, L.J.; Walker, K.A.M. 4′-Substituted nucleosides as inhibitors of HIV: An unusual oxetane derivative. *Tetrahedron Lett.* 1992, *33*, 41–44.

5.   Kawahata, Y.; Takatsuto, S.; Ikekawa, N.; Murata, M.; Omura, S. Synthesis of a new amino acid-antibiotic, oxetin and its three stereoisomers. *Chem. Pharm. Bull.* 1986, *34*, 3102–3110.

6. Paternò, E. Sintesi in chimica organica per mezzo della luce. Nota II. Composti degli idrocarburi non saturi con aldeidi e chetoni. In *Synthesis in Organic Chemistry by Means of Light*; D'Auria, M., Ed.; Società Chimica Italiana: Rome, Italy, 2009; pp. 85–105.

7. Büchi, G.; Inman, C.G.; Lipinsky, E.S. Light-catalyzed organic reactions. I. The reaction of carbonyl compounds with 2-methyl-2-butene in the presence of ultraviolet light. *J. Am. Chem. Soc.* 1954, *76*, 4327–4331.

8. Arnold, D.R. The photocycloaddition of carbonyl compounds to unsaturated systems: The syntheses of oxetanes. *Adv. Photochem.* 1968, *6*, 301–423.

9. Jones, G. Synthetic applications of the Paternò–Büchi reaction. *Org. Photochem.* 1981, *5*, 1–122.

10. Carless, H.A.J. Photochermical synthesis of oxetanes. In *Synthetic Organic Photochemistry*; Horspool, W.M., Ed.; Plenum: New York, NY, USA, 1984; pp. 425–487.

11. Porco, J.A.; Schreiber, S.L. The Paternò–Büchi reaction. In *Comprehensive Organic Synthesis*; Trost, B.M., Fleming, I., Paquette, L.A., Eds.; Plenum: New York, NY, USA, 1991; Volume 5, pp. 151–192.

12. Griesbeck, A.G. Oxetane formation stereocontrol. In *Handbook of Photochemistry and Photobiology*; Horspool, W.A., Song, P.-S., Eds.; CRC Press: Boca Raton, FL, USA, 1994; pp. 522–535, 550–559.

13. Bach, T. Stereoselective intermolecular [2+2]-photocycloaddition reactions and their application in synthesis. *Synthesis* 1998, *1998*, 683–703.

14. D'Auria, M.; Emanuele, L.; Racioppi, R.; Romaniello, G. The Paternò–Büchi reaction on furan derivatives. *Curr. Org. Chem.* 2003, *7*, 1443–1459.

15. D'Auria, M. Paternò-Büchi reaction on furan: Regio- and stereochemistry. In *Targets in Heterocyclic Systems, Chemistry and Properties*; Attanasi, O.A., Spinelli, D., Eds.; Italian Society of Chemistry: Rome, Italy, 2003; Volume 7, pp. 157–173.

16. Griesbeck, A.G.; Bondock, S. Oxetane formation: Stereocontrol. In *Handbook of Photochemistry and Photobiology*; Horspool, W.A., Lenci, F., Eds.; CRC Press: Boca Raton, FL, USA, 2004; pp. 59:1–59:9.

17. Griesbeck, A.G.; Bondock, S. Oxetane formation: Intermolecular additions. In *Handbook of Photochemistry and Photobiology*; Horspool, W.A., Lenci, F., Eds.; CRC Press: Boca Raton, FL, USA, 2004; p. 60.

18. Abe, M. Photochemical oxetane formation: Addition to heterocycles.

In *Handbook of Photochemistry and Photobiology*; Horspool, W.A., Lenci, F., Eds.; CRC Press: Boca Raton, FL, USA, 2004; p. 62.

19. D'Auria, M.; Emanuele, L.; Racioppi, R. 1,2-Cycloaddition reaction of carbonyl compounds and pentaatomic heterocyclic compounds. In *Advances in Photochemistry*; Neckers, D.C., Jenks, W.S., Wolff, T., Eds.; John Wiley & Sons: Hoboken, New Jersey, NJ, USA, 2005; Volume 28, pp. 81–127.

20. D'Auria, M.; Emanuele, L.; Racioppi, R. 2+2 Cycloaddition on furan derivatives. In *Photochemistry Research Progress*; Sanchez, A., Gutierrez, S.J., Eds.; Nova Science Publishers Inc.: Hauppage, NY, USA, 2008; pp. 373–438.

21. D'Auria, M.; Racioppi, R. Concepts of stereoselective photochemistry and a case study: The Paternò-Büchi reaction. *Curr. Org. Chem.* 2009, *13*, 939–954.

22. Abe, M. Formation of a 4-membered ring III (oxetanes). In *Handbook of Synthetic Photochemistry*; Albini, A., Fagnoni, M., Eds.; Wiley: New York, NY, USA, 2010; pp. 217–239.

23. Griesbeck, A.G. Photocycloadditions of alkenes to excited carbonyls. In *Synthetic Organic Photochemistry*; Griesbeck, A.G., Mattay, J., Eds.; Marcel Dekker: New York, NY, USA, 2005; pp. 89–139.

24. D'Auria, M. Paternò–Büchi Reaction. In *CRC Handbook of Organic Photochemistry and Photobiology*, 3rd Edition; Griesbeck, A., Oelgemöller, M., Ghetti, F., Eds.; CRC Press: Boca Raton, FL, USA, 2012; pp. 653–681.

25. Kopecký, J. *Organic Photochemistry*; VCH: New York, NY, USA, 1992; p. 126.

26. Turro, N.J.; Dalton, J.C.; Dawes, K.; Farrington, G.; Hautala, R.; Morton, D.; Niemczyk, M.; Schore, N. Molecular photochemistry. L. Molecular photochemistry of alkanones in solution. α-Cleavage, hydrogen abstraction, cycloaddition, and sensitization reactions. *Acc. Chem. Res.* 1972, *5*, 92–101.

27. Freilich, S.C.; Peters, K.S. Observation of the 1,4 biradical in the Paterno-Buchi reaction. *J. Am. Chem. Soc.* 1981, *103*, 6255–6257.

28. Caldwell, R.A.; Majima, T.; Pac, C. Some structural effects on triplet biradical lifetimes. Norrish II and Paterno-Buchi biradicals. *J. Am. Chem. Soc.* 1982, *104*, 629–630.

29. Freilich, S.C.; Peters, K.S. Picosecond dynamics of the Paterno-Buechi reaction. *J. Am. Chem. Soc.* 1985, *107*, 3819–3822.

30. Abe, M.; Kawakami, T.; Ohata, S.; Nozaki, K.; Nojima, M. Mechanism of stereo- and regioselectivity in the Paternò–Büchi reaction of furan derivatives with aromatic carbonyl compounds: Importance of the conformational distribution in the intermediary triplet 1,4-diradicals. *J. Am. Chem. Soc.* 2004, *126*, 2838–2846.

31. Wilson, R.M.; Wunderly, S.W.; Walsh, T.F.; Musser, A.K.; Outcalt, R.; Geiser, F.; Gee, S.K.; Brabender, W.; Yerino, L., Jr.; Conrad, T.T.; *et al.* Laser photochemistry: Trapping of quinone-olefin preoxetane intermediates with molecular oxygen and chemistry of the resulting 1,2,4-trioxanes. *J. Am. Chem. Soc.* 1982, *104*, 4429–4446.

32. Adam, W.; Kliem, U.; Lucchini, V. Preparative UV-VIS laser photochemistry; molecular oxygen trapping of the Paterno-Büchi triplet diradicals derived from 1,4-dioxene. *Tetrahedron Lett.* 1986, *27*, 2953–2956.

33. Adam, W.; Kliem, U.; Mosandl, T.; Peters, E.-M.; Peters, K.; von Schnering, H.G. Preparative visible-laser photochemistry: Qinghaosu-type 1,2,4-trioxanes by molecular oxygen trapping of Paterno-Buechi triplet 1,4-diradicals derived from 3,4-dihydro-4,4-dimethyl-2*H*-pyran-2-one and quinones. *J. Org. Chem.* 1988, *53*, 4986–4992.

34. Adam, W.; Kliem, U.; Lucchini, V. Preparative UV-VIS laser photochemistry: Photocycloadditions of methylenelactones with benzophenone and p-benzoquinone. Oxygen trapping of paterno-Büchi triplet 1,4-diradicals as model reactions for quinghaosu-type 1,2,4-trioxanes. *Liebigs Ann.* 1988, 869–875.

35. Pinter, B.; de Proft, F.; Veszprémi, T.; Geerlings, P. Regioselectivity in the [2 + 2] cyclo-addition reaction of triplet carbonyl compounds to substituted alkenes (Paterno-Büchi reaction): A spin-polarized conceptual DFT approach. *J. Chem. Sci.* 2005, *117*, 561–571.

36. Noorizadeh, S. Minimum electrophilicity principle in photocycloaddition formation of oxetanes. *J. Phys. Org. Chem.* 2007, *20*, 514–524.

37. Palmer, I.J.; Ragazos, I.N.; Bernardi, F.; Olivucci, M.; Robb, M.A. An MC-SCF Study of the (Photochemical) Paterno-Buchi Reaction. *J. Am. Chem. Soc.* 1994, *116*, 2121–2132.

38. Kutateladze, A.G. Conformational analysis of singlet-triplet state mixing in Paternò-Büchi diradicals. *J. Am. Chem. Soc.* 2001, *123*, 9279–9282.

39. Kutateladze, A.G.; McHale, W.A., Jr. Toward parameterization of spin-orbit coupling in triplet organic diradicals separated by a partially conjugated spacer. *Arkivoc* 2005, 88–101.

40. Minaev, B.F. Spin–orbit coupling in oxygen containing diradicals. *J.*

Mol. Struct. 1998, *434*, 193–206.

41. Minaev, B.F.; Ågren, H. The role of one-center spin-orbit coupling in organic chemical reactions. *EPA Newslett.* 1999,*65*, 7–38.

42. Mattay, J. Charge transfer and radical ions in photochemistry. *Angew. Chem. Int. Ed. Engl.* 1987, *26*, 825–845.

43. Paternò, E.; Chieffi, G. Sintesi in chimica organica per mezzo della luce. Nota V. Comportamento degli acidi e degli eteri col benzofenone. *Gazz. Chim. Ital.* 1910, *40*, 321–331.

44. Gáplovský, A.; Donovalová, J.; Toma, S.; Kubinec, R. Ultrasound effects on photochemical reactions, Part 1: Photochemical reactions of ketones with alkenes. *Ultrasonics Sonochem.* 1997, *4*, 109–115.

45. Schroeter, S.H.; Orlando, C.M. Photocycloaddition of various ketones and aldehydes to vinyl ethers and ketene diethyl acetal. *J. Org. Chem.* 1969, *34*, 1181–1187.

46. Khan, N.; Morris, T.H.; Smith, E.H.; Walsh, R. Alkenyl sulphides and ketene S,S-dithioacetals as olefin components in the Paterno-Büchi reaction: A regioselective synthesis of oxetanes. *J. Chem. Soc. Perkin Trans. 1* 1991, *1991*, 865–870.

47. Mattay, J.; Buchkremer, K. Thermal and photochemical oxetane formation. A contribution to the synthesis of Branched-Chain aldonolactone. *Helv. Chim. Acta* 1988, *71*, 981–987.

48. Mattay, J.; Buchkremer, K. Thermal and photochemical oxetane formation with α-Ketoesters. *Heterocycles* 1988, *27*, 2153–2166.

49. Buhr, S.; Griesbeck, A.G.; Lex, J.; Mattay, J.; Schröer, J. Stereoselectivity in the Paternò-Büchi reaction of 2,2-diisopropyl-1,3-dioxol with methyl trimethylpyruvate. *Tetrahedron Lett.* 1996, *37*, 1195–1196.

50. Abe, M.; Taniguchi, K.; Hayashi, T. Exo-selective formation of bicyclic oxetanes in the photocycloaddition reaction of carbonyl compounds with vinylene carbonate: The important role of intermediary triplet diradicals in the stereoselectivity. *Arkivoc* 2007, *2007*, 58–65.

51. Gan, C.Y.; Lambert, J.N. The tandem intermolecular Paternò–Büchi reaction: Formation of tetrahydrooxepins. *J. Chem. Soc. Perkin Trans. 1* 1998, *1998*, 2363–2372.

52. Park, S.-K.; Lee, S.-J.; Baek, K.; Yu, C.-M. Diastereoselective routes in the Paterno-Büchi reactions of cyclic enol ortho ester with aldehydes. *Bull. Korean Chem. Soc.* 1998, *19*, 35–36.

53. Netto-Ferreira, J.C.; Silva, M.T.; Puget, F.P. Photochemistry of cyclic vicinal tricarbonyl compounds: [2+2] Photocycloaddition of

1,2,3-indanetrione to electron rich olefins. *J. Photochem. Photobiol. A* 1998, *119*, 165–170.

54. Silva, M.T.; Braz-Filho, R.; Netto-Ferreira, J.C. Photochemistry of cyclic vicinal tricarbonyl compounds. Photochemical reaction of 1,2,3-indanetrione with 2,3-Dimethyl-2-butene: Hydrogen abstraction and photocycloaddition. *J. Braz. Chem. Soc.* 2000, *11*, 479–485.

55. Ogata, M.; Watanabe, H.; Kanō, H. Photochemical cycloaddition of benzophenone to furans. *Tetrahedron Lett.* 1967, *8*, 533–537.

56. Carless, H.A.; Haywood, D.J. Photochemical syntheses of 2,6-dioxabicyclo[3.2.0]heptanes. *J. Chem. Soc. Chem. Commun.* 1980, 1067–1068.

57. Bondock, S.; Griesbeck, A.G. Spin-dependent diastereoselectivity in the photocycloaddition of aldehydes to 2,2-dimethyl-2,3-dihydrofuran. *Int. J. Photoenergy* 2005, *7*, 23–25.

58. Griesbeck, A.G.; Fiege, M.; Bondock, S.; Gudipati, M.S. Spin-directed stereoselectivity of carbonyl-alkene photocycloadditions. *Org. Lett.* 2000, *2*, 3623–3625.

59. Griesbeck, A.G. Stereoselective and spin-selective photochemical reactions. *J. Photoscience* 2003, *10*, 49–60.

60. Griesbeck, A.G. Spin-selectivity in photochemistry: A tool for organic synthesis. *Synlett* 2003, *2003*, 451–472.

61. Griesbeck, A.G.; Abe, M.; Bondock, S. Selectivity control in electron spin inversion processes: Regio- and stereochemistry of Paternò–Büchi photocyclo- additions as a powerful tool for mapping intersystem crossing processes. *Acc. Chem. Res.* 2004, *37*, 919–928.

62. Griesbeck, A.G.; Bondock, S.; Gudipati, M.S. Temperature and viscosity dependence of the spin-directed stereoselectivity of the carbonyl-alkene photocycloaddition. *Angew. Chem. Int. Ed.* 2001, *40*, 4684–4687.

63. D'Auria, Maurizio. University of Basilicata, Potenza, Italy, Unpublished work. 2013.

64. Griesbeck, A.G.; Stadtmüller, S. Photocycloaddition of benzaldehyde to cyclic olefins: Electronic control of endo stereoselectivity. *J. Am. Chem. Soc.* 1990, *112*, 1281–1283.

65. Griesbeck, A.G.; Stadtmüller, S. Electronic control of stereoselectivity in photocycloaddition reactions. 4. Effects of methyl substituents at the donor olefin. *J. Am. Chem. Soc.* 1991, *113*, 6923–6928.

66. Griesbeck, A.G.; Stadtmüller, S. Regio- und stereoselektive photocycloadditionen aromatischer aldehyde an furan und

2,3-dihydrofuran. *Chem. Ber.* 1990, *123*, 357–362.

67. Lhiaubet-Vallet, V.; Encinas, S.; Miranda, M.A. Excited state enantiodifferentiating interactions between a chiral benzophenone derivative and nucleosides. *J. Am. Chem. Soc.* 2005, *127*, 12774–12775.

68. Abe, M. Recent progress regarding regio-, site-, and stereoselective formation of oxetanes in Paternò-Büchi reactions.*J. Chin. Chem. Soc.* 2008, *55*, 479–486.

69. D'Auria, M.; Emanuele, L.; Racioppi, R. Regio- and stereoselectivity in the Paterno-Buchi reaction between 2,3-dihydrofuran and furan with benzaldehyde. *Lett. Org. Chem.* 2006, *3*, 244–246.

70. Griesbeck, A.G.; Mauder, H.; Peters, K.; Peters, E.-M.; von Schnering, H.G. Photocycloadditionen mit α- und β-naphthaldehyd: Vollständige umkehr der diastereoselektivität als konsequenz unterschiedlich konfigurierter elektronischer Zustände. *Chem. Ber.* 1991, *124*, 407–410.

71. D'Auria, M.; Racioppi, R. A DFT study of 1,4-biradical intermediates involved in stereoselective Paternò–Büchi reactions. *Eur. J. Org. Chem.* 2010, 3831–3836.

72. Smith, A.B., III; Sulikowski, G.A.; Sulikowski, M.M.; Fujimoto, K. Applications of an asymmetric [2 + 2]-photocycloaddition. Total synthesis of (−)-echinosporin. Construction of an advanced 11-deoxyprostaglandin intermediate. *J. Am. Chem. Soc.* 1992, *114*, 2567–2576.

73. Thopate, S.R.; Kulkarni, M.G.; Puranik, V.G. The Paternò–Büchi reaction of l-ascorbic acid. *Angew. Chem. Int. Ed. Engl.*1998, *37*, 1110–1112.

74. Perez-Ruiz, R.; Miranda, M.A.; Alle, R.; Meerholz, K.; Griesbeck, A.G. An efficient carbonyl-alkene metathesis of bicyclic oxetanes: Photoinduced electron transfer reduction of the Paternò–Büchi adducts from 2,3-dihydrofuran and aromatic aldehydes. *Photochem. Photobiol. Sci.* 2006, *5*, 51–55.

75. Rodina, L.L.; Galkina, O.S.; Supurgibekov, M.B.; Grigov'ev, Y.M.; Utsal, V.A. Photolysis of regioisomeric α,α-diphenyl-substituted diazotetrahydrofuranones: Primary and secondary photochemical processes. *Russian J. Org. Chem.* 2010,*46*, 1542–1545.

76. Vasudeva, S.; Brock, C.P.; Watt, D.S.; Morita, H. Diastereoselectivity in the Paterno-Buechi reaction of enol acetates and benzaldehydes. *J. Org. Chem.* 1994, *59*, 4677–4679.

77. Kollenz, G.; Terpetschnig, E.; Sterk, H.; Peters, K.; Peters, E.-M. Regio- and stereoselective photocycloadditions of heterocyclic 2,3-diones— Evidence for an unexpected 1,2-aroyl migration. *Tetrahedron* 1999, *55*,

2973–2984.

78. Fleming, S.A.; Gao, J.J. Stereocontrol of Paterno-Büchi photocycloadditions. *Tetrahedron Lett.* 1997, *38*, 5407–5410.

79. Bach, T.; Kather, K. The β-silicon effect as a control element for the regioselective ring opening of oxetanes. *Tetrahedron*1994, *50*, 12319–12328.

80. Bach, T. 3-Trimethylsilyloxy-oxetanes via a highly selective Paterno-Büchi reaction. *Tetrahedron Lett.* 1991, *32*, 7037–7038.

81. Vogt, F.; Jödicke, K.; Schröder, J.; Bach, T. Paternò-Büchi reactions of silyl enol ethers and enamides. *Synthesis* 2009, 4268–4273.

82. Bach, T. Silyl enol ethers in Paternò-Büchi reactions. Functional group tolerance and the effect of β-alkyl substitution.*Liebigs Ann.* 1995, 855–865.

83. Bach, T. Regioselective ring opening of oxetanes by hydrogenolysis—A convenient method for the carbohydroxylation of enol ethers. *Liebigs Ann.* 1995, 1045–1053.

84. Bach, T.; Jödicke, K.; Kather, K.; Fröhlich, R. 1,3-Allylic strain as a control element in the Paternò-Büchi reaction of chiral silyl enol ethers: Synthesis of diastereomerically pure oxetanes containing four contiguous stereogenic centers.*J. Am. Chem. Soc.* 1997, *119*, 2437–2445.

85. Bach, T.; Jödicke, K.; Kather, K.; Hecht, J. Facial diastereoselectivity in the Paternò–Büchi reaction of chiral silyl enol ethers. *Angew. Chem. Int. Ed. Engl.* 1995, *34*, 2271–2273.

86. Bach, T.; Jödicke, K.; Wibbeling, B. Reversal of the facial diastereoselectivity in the Paternò-Büchi reaction of silyl enol ethers carrying a chiral substituent in α-position. *Tetrahedron* 1996, *52*, 10861–10878.

87. Bach, T.; Kather, K.; Krämer, O. Synthesis of five-, six-, and seven-membered heterocycles by intramolecular ring opening reactions of 3-Oxetanol derivatives. *J. Org. Chem.* 1998, *63*, 1910–1918.

88. Cho, D.W.; Lee, H.-Y.; Oh, S.W.; Choi, J.H.; Park, H.J.; Mariano, P.S.; Yoon, U.C. Photoaddition reactions of 1,2-diketones with silyl ketene acetals. Formation of β-hydroxy-γ-ketoesters. *J. Org. Chem.* 2008, *73*, 4539–4547.

89. Yoon, U.C.; Kim, M.J.; Moon, J.J.; Oh, S.W.; Kim, H.J.; Mariano, P.S. Photoaddition reactions of silyl ketene acetals with aromatic carbonyl compounds: A new procedure for β-hydroxyester synthesis. *Bull. Korean Chem. Soc.* 2002, *23*, 1218–1228.

90. Abe, M.; Ikeda, M.; Shirodai, Y.; Nojima, M. Regio- and stereo-selective formation of 2-siloxy-2-alkoxyoxetanes in the photoreaction of cyclic ketene silyl acetals with 2-naphthaldehyde and their transformation to aldol-type adducts.*Tetrahedron Lett.* 1996, *37*, 5901–5904.

91. Abe, M.; Ikeda, M.; Nojima, M. A stereoselective, tandem [2+2] photocycloaddition-hydrolysis route to aldol-type adducts. *J. Chem. Soc. Perkin Trans. 1* 1998, *1998*, 3261–3266.

92. Abe, M.; Shirodai, Y.; Nojima, M. Regioselective formation of 2-alkoxyoxetanes in the photoreaction of aromatic carbonyl compounds with β,β-dimethyl ketene silyl acetals: Notable solvent and silyl group effects. *J. Chem. Soc., Perkin Trans. 1* 1998, *1998*, 3253–3260.

93. Abe, M.; Fujimoto, K.; Nojima, M. Notable sulfur atom effects on the regio- and stereoselective formation of oxetanes in Paternò–Büchi photocycloaddition of aromatic aldehydes with silyl O,S-ketene Acetals. *J. Am. Chem. Soc.* 2000, *122*, 4005–4010.

94. Abe, M.; Tachibana, K.; Fujimoto, K.; Nojima, M. Regioselective formation of 3-selanyl-3-siloxyoxetanes in the Paternò-Büchi reaction of silyl O,Se-Ketene acetals (O,Se-SKA). *Synthesis* 2001, *2001*, 1243–1247.

95. Bach, T. N-Acyl enamines in the Paternò-Büchi reaction: Stereoselective preparation of 1,2-amino alcohols by CC bond formation. *Angew. Chem. Int. Ed. Engl.* 1996, *35*, 884–886.

96. Bach, T.; Schröder, J. Photocycloaddition of N-Acyl enamines to aldehydes and its application to the synthesis of diastereomerically pure 1,2-amino alcohols. *J. Org. Chem.* 1999, *64*, 1265–1273.

97. Bach, T.; Bergmann, H.; Brummerhop, H.; Lewis, W.; Harms, K. The [2+2]-photocycloaddition of aromatic aldehydes and ketones to 3,4-dihydro-2-pyridones: Regioselectivity, diastereoselectivity, and reductive ring opening of the product oxetanes. *Chem. Eur. J.* 2001, *7*, 4512–4521.

98. Bach, T. The Paternò-Büchi reaction of 3-heteroatom-substituted alkenes as a stereoselective entry to polyfunctional cyclic and acyclic molecules. *Liebigs Ann. Recueil* 1997, 1627–1634.

99. Bach, T.; Schröder, J. The Paternò-Büchi reaction of α-alkyl-substituted enecarbamates and benzaldehyde. *Synthesis*2001, *2001*, 1117–1124.

100. Bach, T.; Schröder, J.; Brandl, T.; Hecht, J.; Harms, K. Facial diastereoselectivity in the photocycloaddition of chiral N-acyl enamines to benzaldehyde. *Tetrahedron* 1998, *54*, 4507–4520.

101. Bach, T. The Paternò-Büchi reaction of N-acyl enamines and aldehydes—The development of a new synthetic method and its application to total synthesis and molecular recognition studies. *Synlett* 2000, *2000*, 1699–1707.

102. Bach, T.; Schröder, J. Synthesis of syn- and anti-1,2-amino alcohols by regioselective ring opening reactions of cis-3-aminooxetanes. *Tetrahedron Lett.* 1997, *38*, 3707–3710.

103. Bach, T.; Schröder, J. A Short Synthesis of (±)-Oxetin. *Liebigs Ann. Recueil* 1997, 2265–2267.

104. Bach, T.; Brummerhop, H. Unprecedented facial diastereoselectivity in the Paternò–Büchi reaction of a chiral dihydropyrrole—A short total synthesis of (+)-preussin. *Angew. Chem. Int. Ed.* 1998, *37*, 3400–3402.

105. Bach, T.; Brummerhop, H.; Harms, K. The Synthesis of (+)-preussin and related pyrrolidinols by diastereoselective Paternò–Büchi reactions of chiral 2-substituted 2,3-dihydropyrroles. *Chem. Eur. J.* 2000, *6*, 3838–3848.

106. Kugelberg, A.; Döpp, D.; Görner, H. Photoadditions of 2-morpholinopropenenitrile to naphthalenecarboxylic acids. *J. Inf. Recording* 2000, *25*, 187–194.

107. Van Wolven, C.; Döpp, D.; Fischer, M.A. α-Cyanoenamines in the Paterno-Buchi reaction. *J. Inf. Recording* 2000, *25*, 209–214.

108. Döpp, D.; Fischer, M.-A. [2+2] Photoaddition of 2-aminopropenenitriles to diarylethanediones. A product study. *Recl. Trav. Chim. Pays-Bas* 1995, *114*, 498–503.

109. Rivas, C.; Vargas, F. Oxetane formation: Addition to heterocycles. In *CRC Handbook of Organic Photochemistry and Photobiology*; Horspool, W.M., Song, P.-S, Eds.; CRC Press: Boca Raton, FL, USA, 1995; p. 536.

110. Vargas, F.; Rivas, C.; Navarro, M.; Alvarado, Y. Synthesis of an elusive oxetane by photoaddition of benzophenone to thiophene in the presence of a Lewis acid. *J. Photochem. Photobiol. A Chem.* 1996, *93*, 169–171.

111. Rivas, C.; Bolivar, R.A.; Cucarella, M. Photoaddition of carbonyl compounds to five-membered ring heterocycles. *J. Heterocyclic Chem.* 1982, *19*, 529–535.

112. Rivas, C.; Velez, M.; Crescente, O. Synthesis of an oxetan by photoaddition of benzophenone to a thiophen derivative. *J. Chem. Soc. Chem. Commun.* 1970, 1474.

113. Rivas, C.; Bolivar, R.A. Synthesis of oxetanes by photoaddition of carbonyl compounds to 2,5-dimethylthiophene. *J. Heterocyclic*

Chem. 1973, *10*, 967–971.

114. Rivas, C.; Pacheco, D.; Vargas, F.; Ascanio, J. Synthesis of oxetanes by photoaddition of carbonyl compounds to thiophene derivatives. *J. Heterocyclic Chem.* 1981, *18*, 1065–1067.

115. Jones, G.; Gilow, H.M.; Low, J. Regioselective photoaddition of pyrroles and aliphatic carbonyl compounds. A new synthesis of 3(4)-substituted pyrroles. *J. Org. Chem.* 1979, *44*, 2949–2951.

116. Rivas, C.; Velez, M.; Cucarella, M.; Bolivar, R.A.; Flores, S.E. Synthesis of an oxetane by photoaddition of benzophenone to a pyrrole derivative. *Acta Cient. Venezolana* 1971, *22*, 145–146.

117. Rivas, C.; Bolivar, R.A. Synthesis of oxetanes by photoaddition of carbonyl compounds to pyrrole derivatives. *J. Heterocyclic Chem.* 1976, *13*, 1037–1040.

118. Rivas, C.; Pacheco, D.; Vargas, F. Synthesis of an oxetane by photoaddition of benzophenone to a selenophene derivative. *Acta Sud Am. Quim.* 1982, *2*, 1–3.

119. Matsuura, T.; Banba, A.; Ogura, K. Photoinduced reactions—XLV: Photoaddition of ketones to methylimidazoles.*Tetrahedron* 1971, *27*, 1211–1219.

120. Nakano, T.; Rivas, C.; Perez, C.; Larrauri, J.M. Photoaddition of ketones to imidazoles: Synthesis of oxetanes. *J. Heterocyclic Chem.* 1976, *13*, 173–174.

121. Nakano, T.; Rodriguez, W.; de Roche, S.Z.; Larrauri, J.M.; Rivas, C.; Perez, C. Photoaddition of ketones to imidazoles, thiazoles, isothiazoles and isoxazoles. Synthesis of their oxetanes. *J. Heterocyclic Chem.* 1980, *17*, 1777–1780.

122. Ito, Y.; Ji-Ben, M.; Suzuki, S.; Kusunaga, Y.; Matsuura, T.; Fukuyama, K. Efficient photochemical oxetane formation from 1-methyl-2,4,5-triphenylimidazole and benzophenones. *Tetrahedron Lett.* 1985, *26*, 2093–2096.

123. Julian, D.R.; Tringham, G.D. Photoaddition of ketones to indoles: Synthesis of oxeto[2,3-*b*]indoles. *J. Chem. Soc. Chem. Commun.* 1973, 13.

124. Griesbeck, A.G.; Franke, M.; Neudörfl, J.; Kotaka, H. Photocycloaddition of aromatic and aliphatic aldehydes to isoxazoles: Cycloaddition reactivity and stability studies. *Beilstein J. Org. Chem.* 2011, *7*, 127–134.

125. D'Auria, M. Unified Approach to the photochemical isomerization of five-membered heteroaromatic compounds. In*Targets in Heterocyclic*

Systems, Chemistry and Properties; Attanasi, O.A., Spinelli, D., Eds.; Italian Society of Chemistry: Rome, Italy, 1999; Volume 2, pp. 233–279.

126. D'Auria, M. Towards a unitary description of the photochemical isomerization reactions in pentaatomic aromatic heterocycles: The case of furan, thiophene, pyrrole, isoxazole, imidazole, and pyrazole derivatives. *Heterocycles* 1999,*50*, 1115–1136.

127. D'Auria, M. Photochemical isomerization of pentaatomic heterocycles. *Adv. Heterocycl. Chem.* 2001, *79*, 41–48.

128. D'Auria, M. Photochemical and photophysical behavior of thiophene. *Adv. Heterocycl. Chem.* 2011, *104*, 127–390.

129. Griesbeck, A.G.; Fiege, M.; Lex, J. Oxazole–carbonyl photocycloadditions: Selectivity pattern and synthetic route to *erythro* α-amino, β-hydroxy ketones. *Chem. Commun.* 2000, 589–590.

130. Griesbeck, A.G.; Bondock, S.; Lex, J. Synthesis of erythro-α-amino β-hydroxy carboxylic acid esters by diastereoselective photocycloaddition of 5-methoxyoxazoles with aldehydes. *J. Org. Chem.* 2003, *68*, 9899–9906.

131. Griesbeck, A.G.; Bondock, S.; Lex, J. Stereoselective generation of vicinal stereogenic quaternary centers by photocycloaddition of 5-methoxy oxazoles to α-keto esters: Synthesis of *erythro* β-hydroxy dimethyl aspartates. *Org. Biomol. Chem.* 2004, *2*, 1113–1115.

132. Bondock, S.; Griesbeck, A.G. Diastereoselective photochemical synthesis of α-amino-β-hydroxyketones by photocycloaddition of carbonyl compounds to 2,5-dimethyl-4-isobutyloxazole. *Monatsh. Chem.* 2006, *137*, 765–777.

133. Griesbeck, A.G.; Bondock, S. Photocycloaddition of 5-methoxyoxazoles to aldehydes and α-keto esters: A comprehensive view on stereoselectivity, triplet biradical conformations, and synthetic applications of Paternò–Büchi adducts. *Aust. J. Chem.* 2008, *61*, 573–580.

134. Nakano, T.; Santana, M. Photoaddition of benzophenone to azaindole, synthesis of the oxetane of 7-azaindole. *J. Heterocyclic Chem.* 1976, *13*, 585–587.

135. Schenck, G.O.; Hartman, W.; Steinmetz, R. Vierringsynthesen durch photosensibilisierte Cycloaddition von Dimethylmaleinsäureanhydrid an Olefine. *Chem. Ber.* 1963, *96*, 498–508.

136. Gagnaire, D.; Payo-Subiza, E. Nuclear magnetic resonance coupling of methyl protons over five bonds: Heterocyclic vinyl ethers. *Bull. Soc. Chim. Fr.* 1963, 2623–2631.

137. Toki, S.; Shima, K.; Sakurai, H. Organic photochemical reactions. I. The synthesis of substituted oxetanes by the photoaddition of aldehydes to furans. *Bull Chem. Soc. Jpn* 1965, *38*, 760–762.

138. Nakano, T.; Rivas, C.; Perez, C. Configuration and stereochemistry of photoproducts by application of the nuclear Overhauser effect. Adducts of benzophenone with methyl-substituted furans and 2,5-dimethylthiophen, and of methyl-substituted maleic anhydrides with thiophen and its methyl derivatives, and benzo-[*b*]thiophen. *J. Chem. Soc. Perkin Trans. 1* 1973, 2322–2327.

139. Rivas, C.; Payo, E. Synthesis of oxetanes by photoaddition of benzophenone to furans. *J. Org. Chem.* 1967, *32*, 2918–2920.

140. Shima, K.; Sakurai, H. Organic photochemical reactions. IV. Photoaddition reactions of various carbonyl compounds to furan. *Bull. Chem. Soc. Jpn* 1966, *39*, 1806–1808.

141. Toki, S.; Sakurai, H. Kinetic studies of the photoreaction of benzophenone with furan. *Bull. Soc. Chim. Jpn* 1967, *40*, 2885–2889.

142. Zamojski, A.; Koźluk, T. Synthesis of 3-substituted furans. *J. Org. Chem.* 1977, *42*, 1089–1090.

143. Whipple, E.B.; Evanega, G.R. The assignment of configuration to the photoaddition products of unsymmetrical carbonyls to furan using pseudocontact shifts. *Tetrahedron* 1968, *24*, 1299–1310.

144. Schreiber, S.L.; Desmaele, D.; Porco, J.A. On the use of unsymmetrically substituted furans in the furan-carbonyl photocycloaddition reaction: Synthesis of a kadsurenone-ginkgolide hybrid. *Tetrahedron Lett.* 1988, *29*, 6689–6693.

145. Schreiber, S.L.; Desmaele, D.; Porco, J.A. On the use of unsymmetrically substituted furans in the furan-carbonyl photocycloaddition reaction: Synthesis of a kadsurenone-ginkgolide hybrid. *Tetrahedron Lett.* 1988, *29*, 6689–6693.

146. Carless, H.A. J.; Halfhide, A.F. Highly regioselective [2 + 2] photocycloaddition of aromatic aldehydes to acetylfurans.*J. Chem. Soc., Perkin Trans. 1* 1992, 1081–1082.

147. Griesbeck, A.G.; Mauder, H.; Stadtmüller, S. Intersystem crossing in triplet 1,4-biradicals: Conformational memory effects on the stereoselectivity of photocycloaddition reactions. *Acc. Chem. Res.* 1994, *27*, 70–76.

148. Griesbeck, A.G.; Buhr, S.; Fiege, M.; Schmickler, H.; Lex, J. Stereoselectivity of triplet photocycloadditions: Diene−carbonyl reactions and solvent effects. *J. Org. Chem.* 1998, *63*, 3847–3854.

149. Leitich, J. Comments on the paper "Photochemical cycloaddition of benzophenone to furans". *Tetrahedron Lett.* 1967, 1937–1939.

150. Evanega, G.R.; Whipple, E.B. The photochemical addition of benzophenone to furan. *Tetrahedron Lett.* 1967, 2163–2168.

151. Toki, S.; Sakurai, H. On the structure of the 2:1-adduct of benzophenone and furan. *Tetrahedron Lett.* 1967, 4119–4122.

152. Sekretar, S.; Rudā, J.; Štibranyi, L. Photochemical reactions of 2-substituted furans with some carbonyl compounds.*Coll. Czech. Chem. Commun.* 1984, *49*, 71–77.

153. Cantrell, T.S.; Allen, A.; Ziffer, H. Photochemical 2 + 2 cycloaddition of arenecarboxylic acid esters to furans and 1,3-dienes. 2 + 2 Cycloreversion of oxetanes to dienol esters and ketones. *J. Org. Chem.* 1989, *54*, 140–145.

154. D'Auria, M.; Racioppi, R.; Viggiani, L. Paternò–Büchi reaction between furan and heterocyclic aldehydes: Oxetane formation vs. metathesis. *Photochem. Photobiol. Sci.* 2010, *9*, 1134–1138.

155. Kubo, Y.; Suto, M.; Tojo, S.; Araki, T. Photochemical reaction of N-methylnaphthalene-1,8-dicarboximide with alkenes, dienes, and furans. Cyclobutane and oxetane formation. *J. Chem. Soc. Perkin Trans. 1* 1986, 771–779.

156. Krauch, C.H.; Metzner, W.; Schenck, G.O. Photochemische C4- und C3O-Cycloadditionen an Cumaron. *Chem. Ber.*1966, *99*, 1723–1731.

157. Farid, S.; Hartman, S.E.; DeBoer, C.D. Reversible energy transfer and oxetane formation in the photoreactions of carbonyl compounds with benzofurans. *J. Am. Chem. Soc.* 1975, *97*, 808–812.

158. Capozzo, M.; D'Auria, M.; Emanuele, L.; Racioppi, R. Synthesis and photochemical reactivity towards the Paternò–Büchi reaction of benzo[*b*] furan derivatives: Their use in the preparation of 3-benzofurylmethanol derivatives. *J. Photochem. Photobiol. A: Chem.* 2007, *185*, 38–43.

159. Jarosz, S.; Zamojski, A. Rearrangement of 6-substituted 2,7-dioxabicyclo[3.2.0]hept-3-enes to furans. *J. Org. Chem.*1979, *44*, 3720–3723.

160. Kitamura, T.; Kawakami, Y.; Imagawa, T.; Kawanisi, M. One-pot synthesis of 3-substituted furan: A synthesis of perillaketone. *Synth. Commun.* 1977, *7*, 521–528.

161. Kozluk, T.; Zamojski, A. The synthesis of 3-deoxy-dl-streptose. *Tetrahedron* 1983, *39*, 805–810.

162. Schreiber, S.L.; Hoveyda, A.H.; Wu, H.-J. A photochemical route to the

formation of threo aldols. *J. Am. Chem. Soc.*1983, *105*, 660–661.

163. Schreiber, S.L.; Satake, K. Application of the furan carbonyl photocycloaddition reaction to the synthesis of the bis(tetrahydrofuran) moiety of asteltoxin. *J. Am. Chem. Soc.* 1983, *105*, 6723–6724.

164. Schreiber, S.L.; Satake, K. Total synthesis of (±)-asteltoxin. *J. Am. Chem. Soc.* 1984, *106*, 4186–4188.

165. Schreiber, S.L.; Hoveyda, A.H. Synthetic studies of the furan-carbonyl photocycloaddition reaction. A total synthesis of (±)-avenaciolide. *J. Am. Chem. Soc.* 1984, *106*, 7200–7202.

166. Schreiber, S.L.; Porco, J.A. Studies of the furan-carbonyl photocycloaddition reaction: Vinylic substitution reactions. *J. Org. Chem.* 1989, *54*, 4721–4723.

167. Hambalek, R.; Just, G. Trisubstituted oxetanes from 2,7-dioxa-bicyclo-[3,2,0]-hept-3-enes. *Tetrahedron Lett.* 1990, *31*, 4693–4696.

168. Hambalek, R.; Just, G. A short synthesis of (±)-oxetanocin. *Tetrahedron Lett.* 1990, *31*, 5445–5448.

169. Jarosz, S.; Zamojski, A. Asymmetric photocycloaddition between furan and chiral alkyl glyoxylates. *Tetrahedron* 1982,*38*, 1447–1451.

170. Pelzer, R.; Jütten, P.; Scharf, H.-D. Chirale Induktion bei photochemischen Reaktionen, IX. Isoselektivität bei der asymmetrischen Paterno-Büchi-Reaktion unter Verwendung von Kohlenhydraten als chirale Auxiliare. *Chem. Ber.*1989, *122*, 487–491.

171. Pelzer, R.; Scharf, H. –D.; Buschmann, H.; Runsink, J. Chirale Induktion bei Photochemischen Reaktionen, XI. Derivate der (+)-Menthyl-glyoxylate in der Paternò-Büchi-Reaktion. Einfluß von Substituenten im Glyoxylsäurerest auf die Diastereoselektivität. *Chem. Ber.* 1989, *122*, 1187–1192.

172. Hu, S.; Neckers, D.C. Photocycloaddition and ortho-hydrogen abstraction reactions of methyl arylglyoxylates: Structure dependent reactivities. *J. Chem. Soc. Perkin Trans. 2* 1999, 1771–1778.

173. D'Auria, M.; Emanuele, L.; Racioppi, R. On the Paternò–Büchi reaction of chiral phenylglyoxylate esters with furan derivatives. *Photochem. Photobiol. Sci.* 2003, *2*, 904–913.

174. Buschmann, H.; Scharf, H.-D.; Hoffmann, N.; Plath, M.W.; Runsink, J. Chiral induction in photochemical reactions. 10. The principle of isoinversion: A model of stereoselection developed from the diastereoselectivity of the Paterno-Buechi reaction. *J. Am. Chem. Soc.* 1989, *111*, 5367–5373.

175. Buschmann, H.; Scharf, H.-D.; Hoffmann, N.; Esser, P. The isoinversion principle—A general model of chemical selectivity. *Angew. Chem. Int. Ed. Engl.* 1991, *30*, 477–515.

176. Hu, S.; Neckers, D.C. Rapid regio- and diastereoselective Paternò–Büchi reaction of alkyl phenylglyoxylates. *J. Org. Chem.* 1997, *62*, 564–567.

177. Kaneko, Y.; Hu, S.; Neckers, D.C. Photochemical reactivities of alkyl thiopheneglyoxylates and alkyl furanylglyoxylates. *J. Photochem. Photobiol. A* 1998, *114*, 173–179.

178. D'Auria, M.; Emanuele, L.; Racioppi, R. Stereoselectivity in the reaction of chiral phenylglyoxylate esters with furan within zeolites and cyclodextrin. *Lett. Org. Chem.* 2008, *5*, 249–256.

179. Arimura, J.; Mizuta, T.; Hiraga, Y.; Abe, M. Formation of macrocyclic lactones in the Paternò-Büchi dimerization reaction. *Beilstein J. Org. Chem.* 2011, *7*, 265–269.

180. Žagar, C.; Scharf, H.-D. Chiral induction in photochemical reactions, XIII. The Paternó-Büchi reaction of achiral and chiral acyl cyanides with furan. *Chem. Ber.* 1991, *124*, 967–969.

181. Jarosz, S.; Zamojski, A. Asymmetric photocycloaddition between furan and optically active ketones. *Tetrahedron* 1982, *38*, 1453–1456.

182. Tronchet, J.M.J.; Baehler, B. Derives C-glycosyliques. XXII. Glycosyl-6-dioxa-2,7-bicyclo[3,2,0]heptenes-3. *J. Carbohydrates Nucleosides Nucleotides* 1974, *1*, 449–459.

183. Schreiber, S.L. [2 + 2] photocycloadditions in the synthesis of chiral molecules. *Science* 1985, *227*, 857–863.

184. Schreiber, S.L.; Satake, K. Studies of the furan-carbonyl photocycloaddition reaction: The determination of the absolute stereostructure of asteltoxin. *Tetrahedron Lett.* 1986, *27*, 2575–2578.

185. D'Auria, M.; Emanuele, L.; Racioppi, R. Diastereoselectivity in the Paternò–Büchi reaction on furan derivatives. *Tetrahedron Lett.* 2004, *45*, 3877–3880.

186. D'Auria, M.; Emanuele, l.; Pace, V.; Racioppi, R. Diastereoselective Paternò-Büchi reaction on furan derivatives - reaction with asymmetric ketones. *Lett. Org. Chem.* 2006, *3*, 350–355.

187. Adam, W.; Peters, K.; Peters, E.M.; Stegmann, V.R. Hydroxy-directed regio- and diastereoselective [2+2] photocycloaddition (Paternò–Büchi reaction) of benzophenone to chiral allylic alcohols. *J. Am. Chem. Soc.* 2000, *122*, 2958–2959.

188. Adam, W.; Stegmann, V.R. Hydroxy-group directivity in the regioselective

and diastereoselective [2+2] photocycloaddition (Paternò–Büchi reaction) of aromatic carbonyl compounds to chiral and achiral allylic substrates: The preparation of oxetanes with up to three stereogenic centers as synthetic building blocks. *Synthesis* 2001, *2001*, 1203–1214.

189. Adam, W.; Prein, M. The Schenck ene reaction: Diastereoselective oxyfunctionalization with singlet oxygen in synthetic applications. *Angew. Chem. Int. Ed. Engl.* 1996, *35*, 477–494.

190. Griesbeck, A.G.; Bondock, S. Paternò–Büchi reactions of allylic alcohols and acetates with aldehydes: Hydrogen-bond interaction in the excited singlet and triplet states? *J. Am. Chem. Soc.* 2001, *123*, 6191–6192.

191. Bach, T.; Schröder, J.; Harms, K. Diastereoselective photocycloaddition of an axial chiral enamide. *Tetrahedron Lett.* 1999, *40*, 9003–9004.

192. Bach, T.; Bergmann, H.; Harms, K. High facial diastereoselectivity in the photocycloaddition of a chiral aromatic aldehyde and an enamide induced by intermolecular hydrogen bonding. *J. Am Chem. Soc.* 1999, *121*, 10650–10651.

193. Yokohama, A.; Mizuno, K. Stereoselective photocycloaddition of alkenes to naphthalene rings assisted by hydrogen bonding. *Org. Lett.* 2000, *2*, 3457–3459.

194. Abe, M.; Terazawa, M.; Nozaki, K.; Masuyama, A.; Hayashi, T. Notable temperature effect on the stereoselectivity in the photochemical [2+2] cycloaddition reaction (Paternò–Büchi reaction) of 2,3-dihydrofuran-3-ol derivatives with benzophenone. *Tetrahedron Lett.* 2006, *47*, 2527–2530.

195. D'Auria, M.; Racioppi, R.; Romaniello, G. The Paternò-Büchi reaction of 2-furylmethanols. *Eur. J. Org. Chem.* 2000, 3265–3272.

196. D'Auria, M.; Racioppi, R. Paterno-Büchi reaction on 5-methyl-2-furylmethanol derivatives. *Arkivoc* 2000, *1*, 133–140.

197. Hisamoto, K.; Hiraga, Y.; Abe, M. Hydroxy-group effect on the regioselectivity in a photochemical oxetane formation reaction (the Paternò-Büchi Reaction) of geraniol derivatives. *Photochem. Photobiol. Sci.* 2011, *10*, 1469–1473.

198. D'Auria, M.; Emanuele, L.; Poggi, G.; Racioppi, R.; Romaniello, G. On the stereoselectivity of the Paternò–Büchi reaction between carbonyl compounds and 2-furylmethanol derivatives. The case of aliphatic aldehydes and ketones. *Tetrahedron* 2002, *58*, 5045–5051.

199. D'Auria, M.; Emanuele, L.; Racioppi, R. The stereoselectivity of the Paternò–Büchi reaction between tertiary 2-furylmethanol derivatives and aromatic carbonyl compounds: On the nature of the hydroxy directing

effect.*Photochem. Photobiol. Sci.* 2004, *3*, 927–932.

200. Yabuno, Y.; Hiraga, Y.; Abe, M. Site- and stereoselectivity in the photochemical oxetane formation reaction (Paternò–Büchi reaction) of tetrahydrobenzofuranols with benzophenone: Hydroxy-directed diastereoselectivity? *Chem. Lett.*2008, *37*, 822–823.

201. Yabuno, Y.; Hiraga, Y.; Takagi, R.; Abe, M. Concentration and temperature dependency of regio- and stereoselectivity in a photochemical [2 + 2] cycloaddition reaction (the Paternò–Büchi Reaction): Origin of the hydroxy-group directivity. *J. Am. Chem. Soc.* 2011, *133*, 2592–2604.

202. D'Auria, M.; Emanuele, L.; Racioppi, R. Stereoselectivity in the Paterno-Buchi reaction on chiral allylic alcohols, for a discussion of the hydroxy directing effect. *Lett. Org. Chem.* 2005, *2*, 132–135.

203. D'Auria, M.; Emanuele, L.; Racioppi, R.; Valente, A. Paternò–Büchi reaction between aromatic carbonyl compounds and 1-(3-furyl) alkanols. *Photochem. Photobiol. Sci.* 2008, *7*, 98–103.

204. Bolivar, R.A.; Rivas, C. Quencher effect of thiophene and its derivatives on the photoreduction of carbonyl compounds. *J. Photochem.* 1981, *17*, 91.

205. Bolivar, R.A.; Rivas, C. Quencher effect of thiophene and its monomethyl derivatives on photo-reduction and photocycloaddition reactions of ketones. *J. Photochem.* 1982, *19*, 95–99.

206. Bolivar, R.A.; Machado, R.; Montero, L.; Vargas, F.; Rivas, C. Quenching effect of five-membered ring heterocycles on the photoreduction and photocycloaddition reactions of benzophenone: II. *J. Photochem.* 1983, *22*, 91–95.

207. Vargas, F.; Rivas, C. Photochemical formation of oxetanes derived from aromatic ketones and substituted thiophenes and selenophenes. *Int. J. Photoenergy* 2000, *2*, 97–101.

208. Von Wilucki, I.; Matthaus, H.; Krauch, C.H. Photosensibilisierte Cyclodimerisation von Thymin in Lösung.*Photochem. Photobiol.* 1967, *6*, 497–500.

209. Prakash, G.; Falvey, D.E. Model studies of the (6–4) photoproduct DNA photolyase: Synthesis and photosensitized splitting of a thymine-5,6-oxetane. *J. Am. Chem. Soc.* 1995, *117*, 11375–11376.

210. Joseph, A.; Prakash, G.; Falvey, D.E. Model studies of the (6–4) photoproduct photolyase enzyme: Laser flash photolysis experiments confirm radical ion intermediates in the sensitized repair of thymine oxetane adducts. *J. Am. Chem. Soc.* 2000, *122*, 11219–11225.

211. Lhiaubet-Vallet, V.; Belmadoui, N.; Climent, M.J.; Miranda, M.A. The long-lived triplet excited state of an elongated ketoprofen derivative and its interactions with amino acids and nucleosides. *J. Phys. Chem. B* 2007, *111*, 8277–8282.

212. Nakatani, K.; Yoshida, T.; Saito, I. Photochemistry of benzophenone immobilized in a major groove of DNA: Formation of thermally reversible interstrand cross-link. *J. Am. Chem. Soc.* 2002, *124*, 2118–2119.

213. Hei, X.-M.; Song, Q.-H.; Li, X.-B.; Tang, W.-J.; Wang, H.-B.; Guo, Q.-X. Origin of a large temperature dependence of regioselectivity observed for [2+2] photocycloaddition (Paternò-Büchi reaction) of 1,3-dimethylthymine with benzophenone and its derivatives: Conformational property of the intermediary triplet 1,4-diradicals. *J. Org. Chem.* 2005, *70*, 2522–2527.

214. Zhai, B.-C.; Luo, S.-W.; Kong, F.-F.; Song, Q.-H. Notable solvent effects on the regioselectivity in the Paternò–Büchi reaction of 1,3-dimethylthymine with 4-methoxybenzophenone. *J. Photochem. Photobiol. A* 2007, *187*, 406–409.

215. Song, Q.-H; Wang, H.-B.; Li, X.-B.; Hei, X.-M.; Guo, Q.-X.; Yu, S.-Q. Notable substituent and temperature effects on the regioselectivity and the efficiency in Paternò–Büchi reaction of 4,4′-disubstituted benzophenones with 1,3-dimethyluracil and 1,3-dimethylthymine. *J. Photochem. Photobiol. A* 2006, *183*, 198–204.

216. Song, Q.-H.; Zhai, B.-C.; Hei, X.-M.; Guo, Q.-X. The Paternò–Büchi reaction of 1,3-dimethyluracil and 1,3-dimethylthymine with 4,4′-disubstituted benzophenones. *Eur. J. Org. Chem.* 2006, 1790–1800.

217. Kong, F.-F.; Wang, J.-B.; Song, Q.-H. Heavy atom effects in the Paternò-Büchi reaction of pyrimidine derivatives with 4,4'-disubstituted benzophenones. *Beilstein J. Org. Chem.* 2011, *7*, 113–118.

218. Kong, F.-F.; Zhai, B.-C.; Song, Q.-H. Substituent effects on the regioselectivity of the Paternò-Büchi reaction of 5- or/and 6-methyl substituted uracils with 4,4′-disubstituted benzophenones. *Photochem. Photobiol. Sci.* 2008, *7*, 1332–1336.

219. Encinas, S.; Belmadoui, N.; Climent, M.J.; Gil, S.; Miranda, M.A. Photosensitization of thymine nucleobase by benzophenone derivatives as models for photoinduced DNA damage: Paterno-Büchi vs energy and electron transfer processes. *Chem. Res. Toxicol.* 2004, *17*, 857–858.

220. Belmadoui, N.; Encinas, S.; Climent, M.J.; Gil, S.; Miranda, M.A. Intramolecular interactions in the triplet excited states of benzophenone-thymine dyads. *Chem. Eur. J.* 2006, *12*, 553–561.

221. Thomas, M.; Guillaume, D.; Fourrey, J.-L.; Clivio, P. Further insight in the photochemistry of DNA: Structure of a 2-imidazolone (5–4) pyrimidone adduct derived from the mutagenic pyrimidine (6–4) pyrimidone photolesion by UV irradiation. *J. Am. Chem. Soc.* 2002, *124*, 2400–2401.

222. Seki, K.-I.; Aizawa, K.; Sugaoi, T.; Kimura, M.; Ohkura, K. Synthesis of highly conjugated arylpropenylidene-1,3-diazin-2-ones via Paterno-Büchi reaction by photoreaction of 5-fluoro-1,3-dimethyluracil with 1-methoxynaphthalenes. *Chem. Lett.* 2008, *37*, 872–873.

223. Chung, W.-S.; Turro, N.J.; Srivastava, S.; Li, H.; le Noble, W.J. Hyperconjugation as a factor in face selectivity during cycloaddition. *J. Am. Chem. Soc.* 1988, *110*, 7882–7883.

224. Chung, W.-S.; Turro, N.J.; Srivastava, S.; le Noble, W.J. Stereochemistry of photocycloaddition of (E)-1,2-dicyano- and (Z)-1,2-diethoxyethylene to 5-substituted adamantanones. *J. Org. Chem.* 1991, *56*, 5020–5025.

225. Chung, W.-S.; Wang, N.-J.; Liu, Y.-D.; Leu, Y.-J.; Chiang, M.Y. Photocycloaddition of fumaronitrile to adamantan-2-ones and modification of face selectivity by inclusion in β-cyclodextrin and its derivatives. *J. Chem. Soc. Perkin Trans.* 2 1995, 307–313.

226. Chung, W.-S.; Liu, Y.-D.; Wang, N.-. J Face selectivity in the Paterno-Büchi reactions of methacrylonitrile to 5-substituted adamantan-2-ones. *J. Chem. Soc. Perkin Trans.* 2 1995, 581–586.

227. Kossanyi, J.; Jost, P.; Furth, B.; Daccord, G.; Chaquin, P. Intramolecular photoaddition of the excited carbonyl group of cycloakanones to a non-conjugated ethylene double bond. *J. Chem. Res. Synop.* 1980, 368–369.

228. Matsumura, K.; Mori, T.; Inoue, Y. Wavelength control of diastereodifferentiating Paternò–Büchi reaction of chiral cyanobenzoates with diphenylethene through direct versus charge-transfer excitation. *J. Am. Chem. Soc.* 2009, *131*, 17076–17077.

229. Matsumura, K.; Mori, T.; Inoue, Y. Solvent and temperature effects on diastereodifferentiating Paternó-Büchi reaction of chiral alkyl cyanobenzoates with diphenylethene upon direct versus charge-transfer excitation. *J. Org. Chem.* 2010, *75*, 5461–5469.

230. Xue, J. Photoinduced [2+2] cycloadditions (the Paterno-Büchi reaction) of 1H-1-acetylindole-2,3-dione with alkenes. *J. Chem. Soc. Perkin Trans.* 1 2001, *2001*, 183–191.

231. Zhang, Y.; Xue, J.; Gao, Y.; Fun, H.-K.; Xu, J.-H. Photoinduced [2+2] cycloadditions (the Paternò–Büchi reaction) of 1-acetylisatin with enol ethers—regioselectivity, diastereoselectivity and acid catalysed transformations of the spirooxetane products. *J. Chem. Soc. Perkin*

Trans. 1 2002, *2002*, 345–353.

232. Bosch, E.; Hubig, S.M.; Kochi, J.K. Paterno-Büchi coupling of (diaryl) acetylenes and quinone via photoinduced electron transfer. *J. Am. Chem. Soc.* 1998, *120*, 386–395.

233. Christl, M.; Braun, M. [2+2] Photocycloadditions of homobenzvalene. *Liebigs Ann. Recueil* 1997, 1135–1141.

234. Kang, T.; Scheffer, J.R. An Unexpected Paternò-Büchi reaction in the crystalline state. *Org. Lett.* 2001, *3*, 3361–3364.

235. Adam, W.; Stegmann, V.R. Unusual temperature dependence in the cis/trans-oxetane formation discloses competitive syn versus anti attack for the Paternò–Büchi reaction of triplet-excited ketones with cis- and trans-cylooctenes. Conformational control of diastereoselectivity in the cyclization and cleavage of preoxetane diradicals. *J. Am. Chem. Soc.* 2002, *124*, 3600–3607.

236. Horspool, W.; Armesto, D. *Organic Photochemistry: A Comprehensive Treatment*; Prentice Hall: London, UK, 1992.

237. Büchi, G.; Kofron, J.T.; Koller, E.; Rosenthal, D. Light catalyzed organic reactions. V. 1 The addition of aromatic carbonyl compounds to a disubstituted acetylene. *J. Am. Chem. Soc.* 1956, *78*, 876–877.

238. Arnold, D.; Glick, A. The photocycloaddition of carbonyl compounds to allenes. *Chem. Commun.* 1966, 813–814.

239. Gotthardt, J.K.; Steinmetz, R.; Hammond, G.S. Mechanisms of photochemical reactions in solution. LIII. Cycloaddition of carbonyl compounds to allenes. *J. Org. Chem.* 1968, *33*, 2774–2780.

240. Howell, A.R.; Fan, R.; Truong, A. Preparation of 2-alkylidene oxetanes: An investigation of the Paterno-Büchi reaction between aliphatic aldehydes and allenes. *Tetrahedron Lett.* 1996, *37*, 8651–8654.

241. Yu, H.; Li, J.; Kou, Z.; Du, X.; Wei, Y.; Fun, H.-K.; Xu, J.; Zhang, Y. Photoinduced tandem reactions of isoquinoline-1,3,4-trione with alkynes to build aza-polycycles. *J. Org. Chem.* 2010, *75*, 2989–3001.

242. Gleiter, R.; Herb, T.; Borzyk, O.; Hyla-Kryspin, I. 2,7,9-Trimethylenetricyclo[4.3.0.0^{3,8}]nonane and 2,9-dimethylenetricyclo[4.3.0.0^{3,8}]non-4-ene. Synthesis and properties. *Liebigs Ann.* 1995, 357–364.

243. Herb, T.; Gleiter, R. 2,7,9-Trimethylenetricyclo[4.3.0.0^{3,8}]-non-4-ene. *Angew. Chem. Int. Ed. Engl.* 1996, *35*, 2368–2369.

244. Gleiter, R.; Gaa, B.; Sigwart, C.; Lange, H.; Borzyk, O.; Rominger, F.; Irngartinger, H.; Oeser, T. Preparation and Properties of Stelladiones. *Eur.*

J. Org. Chem. 1998, 171–176.

245. Rawal, V.H.; Dufour, C. A general strategy for increasing molecular complexity: Photocycloaddition-fragmentation route to functionalized di- and triquinanes. *J. Am. Chem. Soc.* 1994, *116*, 2613–2614.

246. Dvorak, C.A.; Dufour, C.; Iwasa, S.; Rawal, V.H. Rapid synthesis of di- and triquinanes by direct reductive fragmentation of Paterno–Büchi-derived oxetanes. *J. Org. Chem.* 1998, *63*, 5302–5303.

247. Reddy, T.J.; Rawal, V.H. Expeditious syntheses of (±)-5-oxosilphiperfol-6-ene and (±)-silphiperfol-6-ene. *Org. Lett.* 2000,*2*, 2711–2712.

248. Boxall, R.J.; Ferris, L.; Grainger, R.S. Synthesis of C-13 oxidised cuparene and herbertane sesquiterpenes via a Paternò–Büchi photocyclisation–oxetane fragmentation strategy: Total synthesis of 1,13-herbertenediol. *Synlett* 2004, 2379–2381.

249. De la Torre, M.; Garca, I.; Sierra, M.A. Photochemical access to tetra- and pentacyclic terpene-like products from R-(+)-sclareolide. *J. Org. Chem.* 2003, *68*, 6611–6618.

250. Iriondo-Alberdi, J.; Perea-Buceta, J.; Greaney, M.F. A Paternò-Büchi approach to the synthesis of merrilactone A. *Org. Lett.* 2005, *7*, 3969–3971.

251. Carless, H.A. J.; Fekarurhobo, G.K. A photochemical route to a thromboxane A$_2$ ring analogue. *J. Chem. Soc. Chem. Commun.* 1984, 667–668.

252. Furth, B.; Daccord, G.; Kossanyi, J. Photochimie en solution. IX. Reactivite des allyl-2 cyclanones. *Tetrahedron Lett.*1975, 4259–4262.

253. Kirschberg, T.; Mattay, J. Photoinduced electron transfer reactions of α-cyclopropyl- and α-epoxy ketones. Tandem fragmentation–cyclization to bi-, tri-, and spirocyclic ketones. *J. Org. Chem.* 1996, *61*, 8885–8896.

254. Gescheidt, G.; Neshchadin, D.; Rist, G.; Borer, A.; Dietliker, K.; Misteli, K. Stereocontrolled photo-reaction pathways of endo/exo-2-benzoyl-substituted bicyclo[2.2.2]oct-5-en-2-ol: Paternò-Büchi reaction versus α-cleavage. *Phys. Chem. Chem. Phys.* 2003, *5*, 1071–1077.

255. Hammaecher, C.; Portella, C. New 6-oxa-2-silabicyclo[2.2.0]hexanes by photochemical conversion of acyl(allyl)(dimethyl)silanes. *Chem. Commun.* 2008, 5833–5835.

256. Hu, S.; Neckers, D.C. Photochemical reactions of alkenyl phenylglyoxylates. *J. Org. Chem.* 1997, *62*, 6820–6826.

257. Rochat, S.; Minardi, C.; de Saint Laumer, J.-Y.; Herrmann, A. Controlled release of perfumery aldehydes and ketones by Norrish type-II

photofragmentation of α-keto esters in undegassed solution. *Helv. Chim. Acta* 2000, *83*, 1645–1671.

258. Sakamoto, M.; Aoyama, H.; Omota, Y. Photochemical reactions of N-benzoylformyl α,β-unsaturated amides. *J. Chem. Soc. Perkin 1* 1986, 1759–1762.

259. Sakamoto, M.; Takahashi, M.; Fujita, T.; Watanabe, S.; Nishio, T.; Iida, I.; Aoyama, H. Solid state photochemical reaction of N-(α,β-unsaturated carbonyl)benzoylformamides. *J. Org. Chem.* 1997, *62*, 6298–6308.

260. Shvydkiv, O.; Yavorskyy; Nolan, A.K.; Oelgemöller, M. Microflow photochemistry—An advantageous combination of synthetic photochemistry and microreactor technology. *EPA Newsletter* 2012, *83*, 65–69.

261. Fukuyama, T.; Kajihara, Y.; Hino, Y.; Ryu, I. Continuous microflow [2+2] photocycloaddition reactions using energy-saving compact light sources. *J. Flow Chem.* 2011, *1*, 40–45.

262. Conradi, M.; Junkers, T. Photoinduced conjugation of aldehyde functional polymers with olefins via [2+2]-cycloaddition. *Macromolecules* 2011, *44*, 7969–7976.

263. Ethirajan, A.; Baeten, L.; Conradi, M.; Ranieri, K.; Conings, B.; Boyen, H.-G.; Junkers, T. UV-induced functionalization of poly(divinylbenzene) nanoparticles *via* efficient [2+2]-photocycloadditions. *Polym. Chem.* 2013, *4*, 4010–4016.

a

Chapter 4

MANNICH-TYPE REACTIONS OF ALDIMINES AND HETERO DIELS-ALDER REACTIONS OF ALDEHYDES CATALYZED BY ANION-TYPE LEWIS BASES DERIVED FROM A SINGLE MOLECULE

Kaori Ishimaru, Daiki Maeda, Kaori Ono, Yuya Tanimura

Department of Chemistry, National Defense Academy, Yokosuka, Japan

ABSTRACT

Mannich-type reactions of aldimines with silyl enolates and hetero Diels-Alder reactions of aldehydes with Danishefsky's diene in the presence of anion catalysts derived from proline were performed to afford the corresponding products in high yields.

INTRODUCTION

Organocatalytic reactions using natural amino acids such as proline or its derivatives have recently received much attention in organic synthesis [1-9].Despite the many reports on proline-derived catalysts, the generation of iminium ions or enamine intermediates was necessary in most reactions [10-30]. Unlike these catalysts, we envisioned that anion-type Lewis base catalysts (Scheme 1) prepared from a single molecule, i.e., proline, would have a wide range of Lewis basicities and activate the various silyloxy compounds.

Yamaguchi et al. have first reported the enantioselective Michael addition of a simple malonate to enones and enals in the presence of rubidium salt of proline [31-34]. Recent reports have shown that the simple anion catalysts were useful for various reactions [35-50] including Mannich-type reactions [51-55] and hetero Diels-Alder reactions [56]. The combination of proline and amine also has been developed [57-58], however, the anioncatalyzed reactions using proline-derivedcompounds are still challenging. Here we report two different reactions (the Mannich-type reactions and hetero Diels-Alder reactions) using the anion-type Lewis base catalysts derived from a single molecule.

RESULTS AND DISCUSSION

Mannich-Type Reactions of Aldimines

We first prepared various proline-derived compounds1a- 1c according to the literature [59-61]. In our initial studies, the Mannich-type reaction of aldimine 2a and silyl enolate 3a was performed in DMF solution using lithium salt of 1a as a Lewis base catalyst which was prepared from 1a and MeLi in THF [50] just before use (Table 1). The use of 25 mol% of 1a and 25 mol% of MeLi resulted in low yields of the corresponding β-amino ester 4a (entry 1 in Table 1). Since a small excess of MeLi would decompose the product, 0.5 equiv of MeLi relative to the proline-derived compound was used for the reaction (entries 2 - 6). We found that 2.5 mol% of Lewis base derived from 1b was effective to afford the corresponding β-amino ester in 74% yield (entry 5). However, attempt to use aldimine 2e having p-chlorophenyl group gave <20% yield of the product under the same conditions,

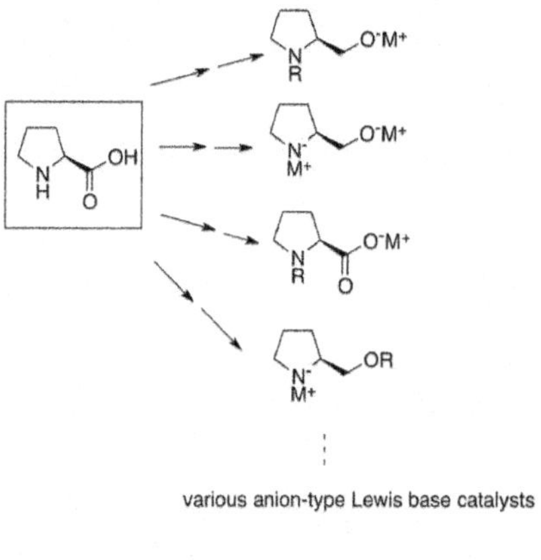

various anion-type Lewis base catalysts

Scheme 1: Anion-type Lewis base catalysts derived from a single molecule.

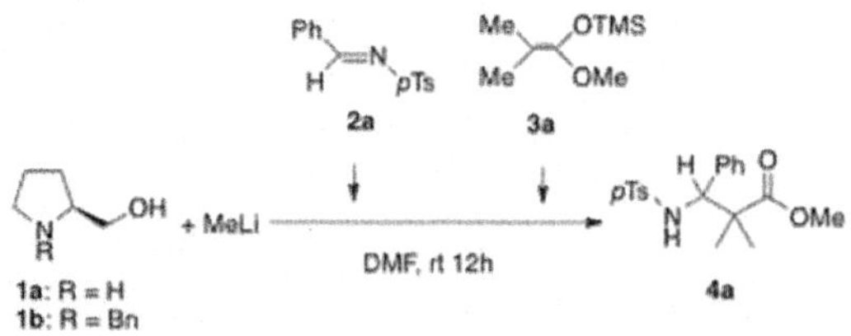

Table 1: Mannich-type reaction of aldimine 2a and silyl enolate 3a in the presence of Lewis base catalysts[a]

Entry	1 (equiv to aldimine)	MeLi (equiv to aldimine)	Yield(%)[b]
1	**1a** (0.25)	0.25	38
2	**1b** (1.5)	0.75	0
3	**1b** (0.25)	0.125	60
4	**1b** (0.1)	0.05	71
5	**1b** (0.05)	0.025	74
6	**1b** (0.025)	0.0125	61

a: Reaction conditions: aldimine (1 mmol), silyl enolate (1.5 mmol) in DMF (2 ml). Reactions were carried out at rt for 12 h; b: Isolated yields after column chromatography.

indicating that the reaction depended on the substituent of the aldimines. After some experiments, 30 mol% of the anion catalyst 1c in the reaction of 2a with 3a also gave a high yield of the product 4a (entry 1 in Table 2). Under these optimized conditions, the Mannich-type reactions of various aldimines (2b - 2e) were carried out (entries 2 - 5). To our delight, the reaction of 2e also proceeded in high yield (entry 5). Electron-poor (entries 4 and 5) and electron-rich (entry 2) aromatic aldehydes reacted with equal facility. We also carried out the reaction with 3b having a phenyl group to afford corresponding β-amino esters in high yields (entries 6 - 8 in Table 2).

Hetero Diels-Alder Reactions of Aromatic Aldehydes

With these results in hand, we next examined the hetero Diels-Alder reaction of aromatic aldehyde and Danishefsky's diene (Table 3) using the anion-type catalysts derived from the same molecule (proline), in which the substrates were quite different from those in Mannich-type reactions. After some experiments, the anion catalyst prepared from 1b and BuLi was suitable for the reaction. We found that 20 mol% of the catalyst promoted the reaction of benzaldehyde in 84 % yield (entry 1), however, the low yield was observed with o-tolualdehyde. Additional experiments were performed and the use of 30 mol% of the catalyst has been successfully applied for the substituted benzaldehydes (entries 2-5). The reaction of p-anisaldehyde having an electron-donating group gave moderate yield of the product

Table 2: Mannich-type reactions of aldimines with silyl enolates in the presence of Lewis base catalyst derived from 1c[a]

Entry	2	R^1	3	R^2	Product	Yield (%)[b]
1	2a	Ph	3a	Me	4a[c]	97
2	2b	p-MeOPh	3a	Me	4b[c]	95
3	2c	o-tolyl	3a	Me	4c	92
4	2d	p-BrPh	3a	Me	4d	91
5	2e	p-ClPh	3a	Me	4e[c]	88
6	2a	Ph	3b	Ph	4f	95
7	2b	p-MeOPh	3b	Ph	4g	92
8	2c	o-tolyl	3b	Ph	4h	90

a: Reaction conditions: anion catalyst (0.3 mmol), aldimine (1 mmol), silyl enolate (1.5 mmol) in DMF (2 ml). Reactions were carried out at rt for 12h; b: Isolated yields after column chromatography; c: The spectral data were consistent with those in ref 55.

Table 3: Hetero diels-alder reactions of aromatic aldehydes with danishefsky's diene[a]

Entry	Ar	Catalyst (mol%)	Product[c]	Yield (%)[b]
1	Ph	20	4a	84
2	o-MePh	30	4b	93
3	m-MePh	30	4c	83
4	p-MePh	30	4d	90
5	p-BrPh	30	4e	88
6	p-MeOPh	30	4f	65

a: Reaction conditions: aldehyde (0.5 mmol), Danishefsky's diene (1.0 mmol) in DMF (total 3 ml). Reactions were carried out at rt for 12 h; b: Isolated yields after column chromatography; c: The spectral data of the products were consistent with that in ref. [62] for 4c and 4e, ref. [63] for 4a, 4d, and 4f, and ref. [64] for 4b.

(entry 6): In all cases, addition of water resulted in low yields of products with starting materials.

CONCLUSION

In conclusion, Mannich-type reactions of aldimines with silyl enolates and hetero Diels-Alder reactions of aldehydes in the presence of anion catalysts derived from proline were performed to afford the corresponding products in high yields. Further applications using the various anions of the proline are now in progress.

EXPERIMENTAL

General

All reactions were carried out under an inert atmosphere and in dried glassware. Anhydrous THF and DMF were used for the all reactions. Flash column chromatography was performed on silica gel (particle size 0.063 - 0.200 mm, Merck silica gel 60). The ^{1}H NMR spectra were recorded with a JEOL JNM-AL300 BK1 spectrometer at 300 MHz with chemical shift values (d) reported in ppm relative to an internal standard (TMS). High resolution mass spectra were measured with a JEOL SX-102A spectrometer.

Typical Experimental Procedure for Mannich-Type Reactions

General Procedure for Preparation of N-Benzylproline Lithium Salt.

To a stirred solution of 1c (0.64 g, 3.1 mmol) in THF (25 ml) was slowly added 1.55 mmol of MeLi (1.09 M diethylether solution) at 0°C under Ar. The reaction mixture was stirred at 30°C for 20 min and cooled to r.t. The catalyst solution was used without further purification.

General Procedure for Mannich-Type Reactions

The catalyst solution prepared as mentioned above (5.1 ml, 0.3 mmol) was transferred to a two-necked flask, and THF was evaporated in vacuo. Dry DMF (1 ml) was added to the flask, and 1 mmol of aldimine in dry DMF (1 ml) and silyl enolate (1.5 mmol) were added successively at r.t. The reaction mixture was stirred at r.t. for 12 h, and quenched with saturated aqueous NH_4Cl. The mixture was extracted with AcOEt, and the organic layers were dried over sodium sulfate. After filtration, the solvent was evaporated to give the crude product. The crude product was purified by flash column chromatography (hexane:EtOAc:CH$_2$Cl$_2$ = 3:1:1).

Identification of the Products

Methyl 2,2-Dimethyl-3-(2-tolyl)-3-(tosylamino) propanoate (4c). Colorless oil; ^{1}H NMR (300 MHz, CDCl$_3$) δ = 1.06 (s, 3H), 1.35 (s, 3H), 2.27 (s, 6H), 3.65 (s, 3H) 4.72 (d, 1H, J = 9.2 Hz), 6.24 (d, 1H, J = 9.2 Hz), 6.82 - 6.99 (m, 6H), 7.27 - 7.35 (m, 2H); HRMS-FAB(M + H)$^+$m/z calcd for C$_{20}$H$_{26}$O$_4$NS 376.1599, found 376.1663.

Methyl 3-(4-Bromophenyl)-2,2-dimethyl-3-(tosylamino) propanoate (4d).

Colorless oil; ^{1}H NMR (300 MHz, CDCl$_3$) δ = 1.06 (s, 3H), 1.31 (s, 3H), 2.33 (s, 3H), 3.61(s, 3H), 4.31 (d, 1H, J = 9.5 Hz), 6.27 (d, 1H, J = 9.5 Hz), 6.77 (d, 2H, J = 8.4 Hz), 7.00 (d, 2H, J = 8.1 Hz), 7.13 - 7.16 (m, 2H), 7.38 (d, 2H, J = 8.4 Hz); HRMSFAB(M+H)$^+$m/z calcd for C$_{19}$H$_{23}$O$_4$NSBr 440.0549, found 440.0560.

Phenyl 2,2-Dimethyl-3-phenyl-3-(tosylamino) propanoate (4f).

Colorless oil; ^{1}H NMR (300 MHz, CDCl$_3$) δ = 1.17 (s, 3H), 1.36 (s, 3H), 2.16 (s, 3H), 4.54 (d, 1H, J = 10.3 Hz), 6.18 (d, 1H, J = 10.3 Hz), 6.80 - 7.32 (m, 14H); HRMS-FAB(M+H)$^+$m/z calcd for C$_{24}$H$_{26}$O$_4$NS 424.1599, found 424.1614.

Phenyl 3-(4-Methoxyphenyl)-2,2-dimethyl-3- (tosylamino)pro-panoate (4g).

Colorless oil; ^{1}H NMR (300 MHz, CDCl$_3$) δ = 1.16 (s, 3H), 1.27 (s, 3H), 2.15 (s, 3H), 3.62 (s, 3H), 4.57 (d, 1H, J = 10.4 Hz), 6.36 (d, 1H, J = 10.4 Hz), 6.44 - 6.45 (m, 2H), 6.47-6.48 (m, 4H), 6.94-6.95 (m, 2H), 7.13 - 7.17 (m, 1H), 7.23 - 7.34 (m, 4H); HRMS-FAB(M+H)$^+$m/z calcd for C$_{25}$H$_{28}$O$_5$NS 454.1705, found 454.1723.

Phenyl 2,2-Dimethyl-3-(2-tolyl)-3-(tosylamino)propanoate (4h).

Colorless oil; ^{1}H NMR (300 MHz, CDCl$_3$) δ = 1.23 (s, 3H), 1.46 (s, 3H), 2.23 (s, 3H), 2.31 (s, 3H), 4.99 (d, 1H, J = 9.8 Hz), 6.29 (d, 1H, J = 9.8 Hz), 6.85 - 7.03 (m, 8H), 7.21 - 7.40 (m, 5H); HRMS-FAB(M+H)$^+$ m/z calcd for C$_{25}$H$_{28}$O$_4$NS 438.1756, found 438.1764.

Typical Experimental Procedure for Hetero Diels-Alder Reactions

General Procedure for Preparation of Anion Catalysts for Hetero Diels-Alder Reaction

To a stirred solution of 1b (0.17 g, 0.9 mmol) in THF (2 ml) was slowly added 0.45 mmol of n-BuLi (1.6 M nhexane solution) at 0°C under Ar. The reaction mixture was stirred at r.t. for 15 min and the solvent was evaporated in vacuo. Dry DMF (3 ml) was added to the flask under Ar, and the catalyst solution was used without further purification.

General Procedure for Hetero Diels-Alder Reactions

The catalyst solution prepared as mentioned above (1 ml, 0.15 mmol) was transferred to a two-necked flask under Ar. To the stirred solution, DMF (2 ml), 0.5 mmol of aldehyde and Danishefsky's diene (1 mmol) were added successively at r.t. The reaction mixture was stirred at r.t. for 12 h, and quenched with saturated aqueous NH_4Cl. The mixture was extracted with AcOEt, and the organic layers were washed with water and brine, and dried over sodium sulfate. After filtration, the solvent was evaporated. To the crude mixture, diethylether (5 ml) and trifluoroacetic acid (0.45 ml) were added. The reaction mixture was stirred at r.t. for 15 min and quenched with saturated aqueous $NaHCO_3$. The mixture was extracted with AcOEt and the organic layers were dried over sodium sulfate. After filtration, the solvent was evaporated to give the crude product. The crude product was purified by flash column chromatography (hexane:EtOAc = 10:1). The spectral data of the products were consistent with that in ref. [62] for 4c and 4e, ref. [63] for 4a, 4d, and 4f, and ref. [64] for 4b.

REFERENCES

1.   G. Guillena, C. Nájera and D. J. Ramón, "Enantioselective Direct Aldol Reaction: The Blossoming of Modern Organocatalysis," Tetrahedron: Asymmetry, Vol. 18, No. 19, 2007, pp. 2249-2293. doi:10.1016/j. tetasy.2007.09.025

2.   A.Ting and S. E. Schaus, "Organocatalytic Asymmetric Mannich Reactions: New Methodology, Catalyst Design, and Synthetic Applications," European Journal of Organic Chemistry, Vol. 2007, No. 35, 2007, pp. 5797-5815. doi:10.1002/ejoc.200700409

3.   H. Kotsuki, H. Ikishima and A. Okuyama, "Organocatalytic Asymmetric Synthesis Using Proline and Related Molecules. Part 1," Heterocycles, Vol. 75, No. 3, 2008, pp. 493-529.doi:10.3987/REV-07-620

4. H. Kotsuki, H. Ikishima and A. Okuyama, "Organocatalytic Asymmetric Synthesis Using Proline and Related Molecules. Part 2," Heterocycles, Vol. 75, No. 4, 2008, pp. 757-797.doi:10.3987/REV-07-621

5. P. Melchiorre, M. Marigo, A. Carlone and G. Bartoli, "Asymmetric Aminocatalysis—Gold Rush in Organic Chemistry," Angewandte Chemie International Edition, Vol. 47, No. 33, 2008, pp. 6138-6171. doi:10.1002/anie.200705523

6. M. Gruttadauria, F. Giacalone and R. Noto, "Supported proline and proline-derivatives as recyclable organocatalysts," Chemical Society Reviews, Vol. 37, No. 8, 2008, pp. 1666-1688. doi:10.1039/b800704g

7. A.-N. Alba, X. Companyó, M. Viciao and R. Rios, "Organocatalytic Domino Reactions," Current Organic Chemistry, Vol. 13, No. 14, 2009, pp. 1432-1474.doi:10.2174/138527209789055054

8. Š. Toma, M. Mečiarová, R. Šebesta, "Are Ionic Liquids Suitable Media for Organocatalytic Reactions?" European Journal of Organic Chemistry, Vol. 2009, No. 3, 2009, pp. 321-327.doi:10.1002/ejoc.200800809

9. H. Pellissier, "Asymmetric Organocatalytic Cycloadditions," Tetrahedron, Vol. 68, No. 10, 2012, pp. 2197- 2232. doi:10.1016/j.tet.2011.10.103

10. B. Simmons, A. M. Walji and D. W. C. MacMillian, "Cycle-Specific Organocascade Catalysis: Application to Olefin Hydroamination, Hydro-oxidation, and Aminooxidation, and to Natural Product Synthesis," Angewandte Chemie International Edition, Vol. 48, No. 24, 2009, pp. 4349-4353. doi:10.1002/anie.200900220

11. C. Chandler, P. Galzerano, A. Michrowska and B. List, "The Proline-Catalyzed Double Mannich Reaction of Acetaldehyde with N-Boc Imines," Angewandte Chemie International Edition, Vol. 48, No. 11, 2009, pp. 1978- 1980. doi:10.1002/anie.2008060s49

12. S. Chercheja, T. Rothenbücher and P. Eilbracht, "Tandem Metal and Organocatalysis in Sequential Hydroformylation and Enantioselective Mannich Reactions," A Advanced Synthesis & Catalysis, Vol. 351, No. 3, 2009, pp. 339-344. doi:10.1002/adsc.200800720

13. A. Odedra and P. H. Seeberger, "5-(Pyrrolidin-2-yl)- tetrazole-Catalyzed Aldol and Mannich Reactions: Acceleration and Lower Catalyst Loading in a ContinuousFlow Reactor," Angewandte Chemie International Edition, Vol. 48, No. 15, 2009, pp. 2699-2702.doi:10.1002/anie.200804407

14. X. Ding, H.-L. Jiang, C.-J. Zhu and Y.-X. Cheng, "Direct Asymmetric α-Amination of Aldehydes with azodicarboxylates in Ionic Liquids Catalyzed by Imidazolium IonTagged Proline Organocatalyst," Tetrahedron Letters, Vol. 51, No. 47, 2010, pp. 6105-6107.doi:10.1016/j.

tetlet.2010.09.036

15. V. Rawat, P. V. Chouthaiwale, V. B. Chavan, G. Suryavanshi and A. Sudalai, "A Facile Enantioselective Synthesis of (S)-N-(5-Chlorothiophene-2-sulfonyl)-β,β-diethyla-laninol via Proline-Catalyzed Asymmetric α-Aminooxylation and α-Amination of Aldehyde," Tetrahedron Letters, Vol. 51, No. 50, 2010, pp. 6565-6567.doi:10.1016/j.tetlet.2010.10.029

16. F. Kelleher, S. Kelly, J. Watts and V. McKee, "Structure—Reactivity Relationships of L-Proline Derived Spirolactams and α-Methyl Prolinamide Organocatalysts in the Asymmetric Michael Addition Reaction of Aldehydes to Nitroolefins," Tetrahedron, Vol. 66, No. 19, 2010, pp. 3525-3536. doi:10.1016/j.tet.2010.03.002

17. W.-H. Wang, T. Abe, X.-B. Wang, K. Kodama, T. Hirose and G.-Y. Zhang, "Self-Assembled Proline-Amino Thioureas as Efficient Organocatalysts for the Asymmetric Michael Addition of Aldehydes to Nitroolefins," Tetrahedron: Asymmetry, Vol. 21, No. 24, 2010, pp. 2925-2933. doi:10.1016/j.tetasy.2010.11.025

18. S. P. Panchgalle, H. B. Bidwai, S. P. Chavan and U. R. Kalkote, "Organocatalytic Asymmetric Synthesis of (−)- δ-Coniceine Based on Sequential Proline-Catalyzed Asymmetric α-Amination—HWE Olefination," Tetrahedron: Asymmetry, Vol. 21, No. 19, 2010, pp. 2399-2401. doi:10.1016/j.tetasy.2010.08.009

19. A. K. Sharma and R. B. Sunoj, "Enamine versus Oxazolidinone: What Controls Stereoselectivity in ProlineCatalyzed Asymmetric Aldol Reactions?" Angewandte Chemie International Edition, Vol. 49, No. 36, 2010, pp. 6373-6377. doi:10.1002/anie.201001588

20. N. Győrffy, A. Tungler and M. Fodor, "Stereodifferentiation in Heterogeneous Catalytic Hydrogenation. Kinetic Resolution and Asymmetric Hydrogenation in the Presence of (S)-Proline: Catalyst-Dependent Processes," Journal of Catalysis, Vol. 270, No. 1, 2010, pp. 2-8. doi:10.1016/j.jcat.2009.10.018

21. M. Cui, H. Song, A. Feng, Z. Wang and Q. Wang, "Asymmetric Synthesis of (R)-Antofine and (R)-Cryptopleurine via Proline-Catalyzed Sequential α-Aminoxylation and Horner-Wadsworth-Emmons Olefination of Aldehyde," The Journal of Organic Chemistry, Vol. 75, No. 20, 2010, pp. 7018-7021. doi:10.1021/jo101510x

22. H. Suga, T. Arikawa, K. Itoh, Y. Okumura, A. Kakehi and M. Shiro, "Asymmetric 1,3-Dipolar Cycloaddition Reactions of Azomethine Imines with Acrolein Catalyzed by L-Proline and Its Derivatives," Heterocycles, Vol. 81, No. 7, 2010, pp. 1669-1688.doi:10.3987/COM-10-11967

23. E. Ververková, J. Štrasserová, R. Šebesta and Š. Toma, "Asymmetric Mannich Reaction Catalyzed by N-Arylsulfonyl-L-proline Amides," Tetrahedron: Asymmetry, Vol. 21, No. 1, 2010, pp. 58-61. doi:10.1016/j. tetasy.2009.12.013

24. M. Lu, Y. Lu, P. K. A. Tan, Q. Y. Lau and G. Zhong, "Highly Enantioselective Synthesis of Fluorinated b-Amino Ketones via Asymmetric Organocatalytic Mannich Reactions: A Case Study of Unusual Reversal of Regioselectivity," Synlett, 2011, pp. 477-480. doi:10.1055/s-0030-1259513

25. M .Penhoat, D. Barbry and C. Rolando, "Direct Asymmetric Aldol Reaction Co-Catalyzed by L-Proline and Group 12 Elements Lewis Acids in the Presence of Water," Tetrahedron Letters, Vol. 52, No. 1, 2011, pp. 159- 162. doi:10.1016/j.tetlet.2010.11.014

26. A. Rai, A. K. Singh, S. Singh and L. D. S. Yadav, "Chiral Amine-Triggered Triple Cascade Reactions: A New Approach to Functionalized Decahydroquinolines," Synlett, No. 3, 2011, pp. 335-340. doi:10.1055/s-0030-1259320

27. J. G. Hernández and E. Juaristi, "Asymmetric Aldol Reaction Organocatalyzed by (S)-Proline-Containing Dipeptides: Improved Stereoinduction under Solvent-Free Conditions," The Journal of Organic Chemistry, Vol. 76, No. 5, 2011, pp. 1464-1467.doi:10.1021/jo1022469

28. D. E. Siyutkin, A. S. Kucherenko, L. L. Frolova, A. V. Kuchin and S. G. Zlotin, "2-Hydroxy-3-[(S)-prolinamido]pinanes as Novel Bifunctional Organocatalysts for Asymmetric Aldol Reactions in Aqueous Media," Tetrahedron: Asymmetry, Vol. 22, No. 12, 2011, pp. 1320-1324. doi:10.1016/j.tetasy.2011.07.013

29. S. Fotaras, C. G. Kokotos, E. Tsandi and G. Kokotos, "Prolinamides Bearing Thiourea Groups as Catalysts for Asymmetric Aldol Reactions," European Journal of Organic Chemistry, Vol. 2011, No. 7, 2011, pp. 1310-1317. doi:10.1002/ejoc.201001417

30. J. G. Hernández, V. García-López and E. Juaristi, "Solvent-Free Asymmetric Aldol Reaction Organocatalyzed by (S)-Proline-Containing Thiodipeptides under BallMilling Conditions," Tetrahedron, Vol. 68, No. 1, 2012, pp. 92-97. doi:10.1016/j.tet.2011.10.093

31. M. Yamaguchi, T. Shiraishi and M. Hirama, "A Catalytic Enantioselective Michael Addition of a Simple Malonate to Prochiral α,β-Unsaturated Ketoses and Aldehyde," Angewandte Chemie International Edition, Vol. 32, No. 8, 1993, pp. 1176-1178.doi:10.1002/anie.199311761

32. M. Yasmaguchi, Y. Igarashi, R. S. Reddy, T. Shiraishi and M. Hirama,

"Catalytic Asymmetric Michael Addition of Nitroalkane to Enone and Enal," Tetrahedron Letters, Vol. 35, No. 44, 1994, pp. 8233-8236. doi:10.1016/0040-4039(94)88290-8

33. M. Yamaguchi, T. Shiraishi and M. Hirama, "Asymmetric Michael Addition of Malonate Anions to Prochiral Acceptors Catalyzed by L-Proline Rubidium Salt," The Journal of Organic Chemistry, Vol. 61, No. 10, 1996, pp. 3520-3530. doi:10.1021/jo960216c

34. M. Yamaguchi, Y. Igarashi, R. S. Reddy, T. Shiraishi and M. Hirama, "Asymmetric Michael Addition of Nitroalkanes to Prochiral Acceptors Catalyzed by Proline Rubidium Salts," Tetrahedron, Vol. 53, No. 32, 1997, pp. 11223-11236. doi:10.1016/S0040-4020(97)00379-7

35. H. Fujisawa and T. Mukaiyama, "A Catalytic Aldol Reaction between Ketene Silyl Acetals and Aldehydes Promoted by Lithium Amide under Non-Acidic Conditions," Chemistry Letters, Vol. 31, No. 2, 2002, pp. 182- 183. doi:10.1246/cl.2002.182

36. H. Fujisawa and T. Mukaiyama, "Lithium Pyrrolidone Catalyzed Aldol Reaction between Aldehyde and Trimethylsilyl Enolate," Chemistry Letters, Vol. 31, No. 8, 2002, pp. 858-859. doi:10.1246/cl.2002.858

37. T. Mukaiyama H. Fujisawa and T. Nakagawa, "Lewis Base Catalyzed Aldol Reaction of Trimethylsilyl Enolates with Aldehydes," Helvetica Chimica Acta, Vol. 85, No. 12, 2002, pp. 4518-4531.doi:10.1002/hlca.200290025

38. T. Mukaiyama, T. Nakagawa and H. Fujisawa, "Lewis Base Catalyzed Michael Reaction between Ketene Silyl Acetals and α,β-Unsaturated Carbonyl Compounds," Chemistry Letters, Vol. 32, No. 1, 2003, pp. 56-57. doi:10.1246/cl.2003.56

39. T. Nakagawa, H. Fujisawa and T. Mukaiyama, "Lithium Acetate-Catalyzed Aldol Reaction between Aldehyde and Trimethylsilyl Enolate," Chemistry Letters, Vol. 32, No. 5, 2003, pp. 462-463. doi:10.1246/cl.2003.462

40. T. Nakagawa, H. Fujisawa and T. Mukaiyama, T. "Lithium Acetate Catalyzed Aldol Reaction between Aldehyde and Trimethylsilyl Enolate in a Dimethylformamide—H_2O Solvent," Chemistry Letters, Vol. 32, No. 8, 2003, pp. 696-697. doi:10.1246/cl.2003.696

41. T. Nakagawa, H. Fujisawa and T. Mukaiyama, "SelfPromoted Aldol Reaction between Aldehyde Having Lewis Base Moiety and Trimethylsilyl Enolate," Chemistry Letters, Vol. 33, No. 2, 2004, pp. 92-93. doi:10.1246/cl.2004.92

42. T. Nakagawa, H. Fujisawa, Y. Nagata and T. Mukaiyama, "Lithium

Acetate-Catalyzed Michael Reaction between Trimethylsilyl Enolate and α,β-Unsaturated Carbonyl Compound," *Chemistry Letters*, Vol. 33, No. 8, 2004, pp. 1016-1017.doi:10.1246/cl.2004.1016

43.   T. Mukaiyama, T. Tozawa and H. Fujisawa, "Lithium Alkoxide-Promoted Michael Reaction between Silyl Enolates and α,β-Unsaturated Carbonyl Compounds," *Chemistry Letters*, Vol. 33, No. 11, 2004, pp. 1410-1411. doi:10.1246/cl.2004.1410

44.   E. Takahashi, H. Fujisawa, T. Yanai and T. Mukaiyama, "Lewis Base-Catalyzed Strecker-Type Reaction between Trimethylsilyl Cyanide and N-Tosylimines in Watercontaining DMF," *Chemistry Letters*, Vol. 34, No. 3, 2005, pp. 318-319. doi:10.1246/cl.2005.318

45.   Y. Sato, H. Fujisawa and T. Mukaiyama, "Lewis Base -Catalyzed [2,3]-Wittig Rearrangement of Silyl Enolates Generated from α-Allyloxy Ketones," *Chemistry Letters*, Vol. 34, No. 4, 2005, pp. 588-589. doi:10.1246/cl.2005.588

46.   E. Takahashi, H. Fujisawa, T. Yanai and T. Mukaiyama, "Lewis Base-Catalyzed Diastereoselective Strecker-Type Reaction between Trimethylsilyl Cyanide and Chiral Sulfinimines," *Chemistry Letters*, Vol. 34, No. 4, 2005, pp. 604-605.doi:10.1246/cl.2005.604

47.   Y. Kawano, H. Fujisawa and T. Mukaiyama, "Lithium Acetate-Catalyzed Crossed Aldol Reaction between Aldehydes and Trimethylsilyl Enolates Generated from Other Aldehydes," *Chemistry Letters*, Vol. 34, No. 4, 2005, pp. 614-615.doi:10.1246/cl.2005.614

48.   Y. Kawano, N. Kaneko and T. Mukaiyama, "Lewis BaseCatalyzed Cyanomethylation of Carbonyl Compounds with (Trimethylsilyl) Acetonitrile," *Chemistry Letters*, Vol. 34, No. 11, 2005, pp. 1508-1509. doi:10.1246/cl.2005.1508

49.   M. Hatano, T. Ikeno, T. Miyamoto and K. Ishihara, "Chiral Lithium Binaptholate Aqua Complex as a Highly Effective Asymmetric Catalyst for Cyanohydrin Synthesis," *Journal of the American Chemical Society*, Vol. 127, No. 31, 2005, pp. 10776-10777.doi:10.1021/ja051125c

50.   Y. Kawano, N. Kaneko and T. Mukaiyama, "Lewis BaseCatalyzed Perfluoroalkylation of Carbonyl Compounds and Imines with (Perfluoroalkyl)Trimethylsilane," *Bulletin of the Chemical Society of Japan*, Vol. 79, No. 7, 2006, pp. 1133-1145.doi:10.1246/bcsj.79.1133

51.   H. Fujisawa, E .Takahashi, T. Nagasawa and T. Mukaiyama, "Lewis Base-Catalyzed Mannich-Type Reaction between Aldimine and Trimethylsilyl Enolate," *Chemistry Letters*, Vol. 32, No. 11, 2003, pp. 1036-1037. doi:10.1246/cl.2003.1036

52. E. Takahashi, H. Fujisawa and T. Mukaiyama, "Lithium Acetate-Catalyzed Mannich-Type Reaction between Trimethylsilyl Enolates and Aldimines in a Water-Containing DMF," Chemistry Letters, Vol. 33, No. 7, 2004, pp. 936-937. doi:10.1246/cl.2004.936

53. E. Takahashi, H. Fujisawa and T. Mukaiyama, "Lewis Base-Catalyzed Anti-Selective Mannich-Type Reaction between Trimethylsilyl Enolates and Aldimines," Chemistry Letters, Vol. 34, No. 1, 2005, pp. 84-85. doi:10.1246/cl.2005.84

54. E. Takahashi, H. Fujisawa, T. Yanai and T. Mukaiyama, "One-Pot Synthesis of β-Lactams from Aldimines and Ketene Silyl Acetals by Tandem Lewis Base-Catalyzed Mannich-Type Addition and Cyclization," Chemistry Letters, Vol. 34, No. 2, 2005, pp. 216-217. doi:10.1246/cl.2005.216

55. H. Fujisawa, E. Takahashi and T. Mukaiyama, "Lewis Base Catalyzed Mannich-Type Reactions between Trimethylsilyl Enol Ethers and Aldimines," Chemistry—A European Journal, Vol. 12, No. 19, 2006, pp. 5082-5093. doi:10.1002/chem.200500821

56. T. Mukaiyama, T. Kitazawa and H. Fujisawa, "A Lewis Base-Catalyzed Hetero Diels—Alder Reaction between Aldehydes and the Danishefsky's Diene," Chemistry Letters, Vol. 35, No. 3, 2006, pp. 328-329. doi:10.1246/cl.2006.328

57. S. Hanessian and V. Pham, "Catalytic Asymmetric Conjugate Addition of Nitroalkanes to Cycloalkenones," Organic Letters, Vol. 2, No. 19, 2000, pp. 2975-2978. doi:10.1021/ol000170g

58. S. Hanessian, Z. Shao and J. S. Warrier, "Optimization of the Catalytic Asymmetric Addition of Nitroalkanes to Cyclic Enones with Trans-4,5-methano-l-proline," Organic Letters, Vol. 8, No. 21, 2006, pp. 4787-4786. doi:10.1021/ol0618407

59. R. G. Kostyanovsky, I. M. Gella, V. I. Markov and Z. E. Samojlova, "Asymmetrical Nonbridgehead Nitrogen—IV: Chiroptical properties of the Amines, N-Chloroamines and Cyanamides," Tetrahedron, Vol. 30, No. 1, 1974, pp. 39-45. doi:10.1016/S0040-4020(01)97214-X

60. T. Rosen, S. W. Fesik, D. T. W. Chu and A. G. Pernet, "Asymmetric Synthesis of 2-Substituted (4S)-4-Aminopyrrolidines. S_N2 Displacement at the 4-Position of the Pyrrolidine Moiety," Synthesis, No. 1, 1988, pp. 40-44. doi:10.1055/s-1988-27459

61. Y. N. Belokon', I. E. Zel'tzer, V. I. Bakhmutov, M. B. Saporovskaya, M. G. Ryzhov, A. I. Yanovsky, Y. T. Struchkov and V. M. Belikov, "Asymmetric Synthesis of Threonine and Partial Resolution and Retroracemization of .Alpha.-Amino Acids via Copper(II) Complexes of Their Schiff Bases

with (S)-2-N-(N'-Benzylprolyl)amino- benzaldehyde and (S)-2-N-(N'-Benzylprolyl)amino-acetophenone. Crystal and Molecular Structure of a Copper(II) Complex of Glycine Schiff Base with (S)-2-N-(N'-benzylprolyl)aminoacetophenone," Journal of the American Chemical Society, Vol. 105, No. 7, 1983, pp. 2010-2017.doi:10.1021/ja00345a057

62.   J. Long, J. Hu, X. Shen, B. Ji and K. Ding, "Discovery of Exceptionally Efficient Catalysts for Solvent-Free Enantioselective Hetero-Diels—Alder Reaction," Journal of the American Chemical Society, Vol. 124, No. 1, 2002, pp. 10-11. doi:10.1021/ja0172518

63.   H. Furuno, T. Hayano, T. Kambara, Y. Sugimoto, T. Hanamoto, Y. Tanaka, Y. Z. Jin, T. Kagawa and J. Inanaga, "Chiral Rare Earth Organophosphates as Homogeneous Lewis Acid Catalysts for the highly enantioselective Hetero-Diels—Alder Reactions," Tetrahedron, Vol. 59, No. 52, 2003, pp. 10509-10523. doi:10.1016/j.tet.2003.06.011

64.   K. Aikawa, R. Irie and T. Katsuki, "Asymmetric HeteroDiels—Alder reaction Using Chiral Cationic Metallosalen Complexes as Catalysts," Tetrahedron, Vol. 57, No. 5, 2001, pp. 845-851. doi:10.1016/S0040-4020(00)01048-6

Chapter 5

ITALIAN CHEMISTS' CONTRIBUTIONS TO NAMED REACTIONS IN ORGANIC SYNTHESIS: AN HISTORICAL PERSPECTIVE

Gianluca Papeo[1] and Maurizio Pulici[2]

[1]Department of Medicinal Chemistry, Nerviano Medical Sciences srl, Business Unit Oncology, Viale Pasteur 10, Nerviano 20014, MI, Italy
[2]Department of Chemical Core Technologies, Nerviano Medical Sciences srl, Business Unit Oncology, Viale Pasteur 10, Nerviano 20014, MI, Italy

ABSTRACT

From the second half of the 19th century up to modern times, the tremendous contribution of Italian chemists to the development of science resulted in the discovery of a number of innovative chemical transformations. These reactions were subsequently christened according to their inventors' name and so entered into the organic chemistry portfolio of "named organic reactions". As these discoveries were being conceived, massive social, political and geographical changes in these chemists' homeland were also occurring. In this review, a brief survey of known (and some lesser known) named organic reactions discovered by Italian chemists, along with their historical contextualization, is presented.

INTRODUCTION

At some point in the history of organic chemistry, someone arbitrarily decided to christen a certain chemical transformation after its discoverer's name. Arguably, this event might have occurred both to acknowledge the nominee's merits and to render the complicated chemical jargon more usable. From that point on, organic chemistry started to possess a new phraseology, still peacefully coexisting with the rigorous IUPAC-approved chemical language. Evidence of this can easily be found by glancing at any laboratory bookshelf: a textbook containing "named organic reactions" or one with a similar title will surely be present, frequently as a second (or higher) edition [1,2,3,4].

Eavesdropping on chemists' everyday working discussions is even more compelling: colleagues talking about a "Wittig reaction" surely outnumber those describing the "synthesis of an alkene, by reacting a given aldehyde with a suitable phosphorous ylide". Mentors can be heard proposing to their students the use of the "Meerwein salt" to alkylate a given oxygen atom much more frequently than those proposing "triethyloxonium tetrafluoborate" instead. Finally, medicinal chemists trying to improve the results of their "three-component reaction between acetoacetate, a given aldehyde and urea" will be in the minority with respect to those who are in the process of improving a "Biginelli reaction". Such examples are endless and so it is quite evident that the "named organic reactions" language encompasses a tremendous amount of embedded information. It represents a form of shorthand used to convey concepts that are otherwise much more complicated to exchange, with the additional advantage that it carries of "humanizing" chemistry, by recalling the original scientists standing behind the chemical transformations we are trying to exploit. However, the learning of this jargon, being two-step in nature, requires a double mnemonic effort. One must memorize the reaction and its discoverer's name (which ideally entails also learning the correct pronunciation!). Additionally, the subjectivity of the nomination process renders reaction-naming an error-prone process. The overlooking of a seminal paper, the misattribution of credits (due to poorly diffused journals and their language), and the misrecognition of contributions from co-authors (often left out from the reaction name for the sake of simplicity) are examples of potential shortcomings. This subjectivity also renders complex any attempt at defining fixed criteria for whether a given reaction deserves to enter the hall of fame, or "Olympus" of named organic reactions. The boundaries of this Olympus were also somewhat fuzzy, being more or less confined either geographically or temporally. Thus, another crucial question is not why a chemical transformation is propelled from anonymity, but rather for how long it can withstand the changes of the rapidly evolving science of synthesis. However, apart from any epistemological considerations, the "named organic reactions" jargon still persists in enthralling the chemical community. As long as science uses it vividly, this language continuously grows. This, on one hand, will increase the efforts required to master it whereas, on the other, it will presumably continue to serve chemists in rapidly (and elegantly) exchanging ideas and solutions to their problems. In one word: to communicate.

This special issue of the *Molecules* is dedicated to organic reactions discovered by Italian chemists. It is edited by Claudia Piutti, research chemist and grand granddaughter of Arnaldo Piutti, a renowned organic chemist of the early twentieth century, whose investigations on the relationships between stereochemistry and taste have been recently reviewed [5]. Considering the focus of this issue, we envisioned that a brief survey of some of the known and lesser known reactions named after their Italian discoverers may be useful for the reader interested in the topic. A short historical contextualization of these seminal contributions will hopefully render the reading more enjoyable.

FROM THE FIRST ITALIAN WAR OF INDEPENDENCE (1848–1849) TO THE PROCLAMATION OF THE KING-DOM OF ITALY (1861)

At the dawn of the second half of the 19th century, the Italian peninsula persisted to be fragmented into a number of independent States extremely variable in size. Furthermore, the majority of the more economically developed territories (the Lombardy-Venetia Kingdom) was still in the hands of the Austrian-Hungarian crown. The increasing discomfort generated by this politically obsolete situation, as well as the growing sense of a national consciousness, culminated in the uprising of Palermo against the Bourbon monarchy (January 1848) and, two months later, the almost simultaneous rising of Milan and Venice against the Austrians (March 1848). A number of volunteers arriving from every part of Italy gathered around Carlo Alberto, the King of Piedmont who, in the meantime, had declared war on Austria. The troops fought courageously, but they were defeated by the superior Austrian Army first in Custoza, near Verona (June 1848), then in Novara (March 1849). Carlo Alberto abdicated in favor of his son Vittorio Emanuele II. The unfortunate epic of the first Italian war of independence formally ended.

Like most of the intellectuals who lived during the pre-unification era, the chemists Raffaele Piria (1814–1865) and his student Cesare Bertagnini (1827–1857) were also patriots and were actively involved in the political life of their time. Working at the University of Pisa, they had joined that institutions' volunteer corps and took part in the battle of Curtatone and Montanara (near Mantova), one of the episodes of the first war of independence. Stanislao Cannizzaro (1826–1910), another of Piria's acolytes, who had at the time been on vacation in his native Sicily, took part instead in the revolution that broke out on the island against the Bourbons.

Professionally, the years spent at the University of Pisa were particularly fruitful for Piria, undoubtedly one of the most prominent scientists of his time who is remembered for his numerous contributions to organic chemistry [6]. In fact Piria published his seminal papers during his professorship at the local University.

Interestingly, what was referred to as the Piria reaction until a few years ago was the transformation of primary amines and/or amides to the corresponding alcohols and/or carboxylic acids by means of nitrous acid (Scheme 1). This reaction, which is nowadays known to proceed with retention of configuration, was discovered by Piria while working on asparagine and described in 1846 [7] and has retained a certain analytical value for the detection of amino- groups for quite a long time.

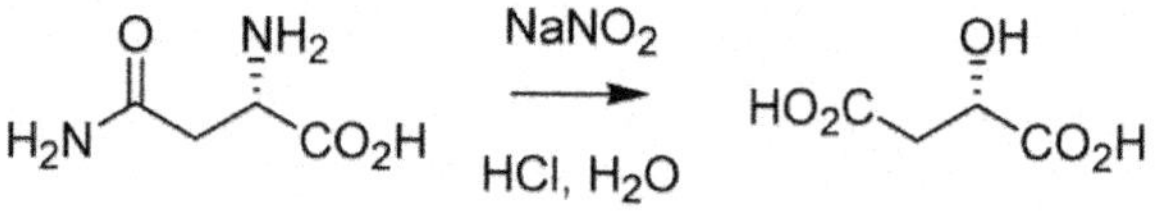

Scheme 1: The Piria diazotization reaction (1846).

However, what is generally acknowledged as the Piria reaction is the reduction of an aromatic nitro compound by means of a sulfite delivering a mixture of sulfamic- and aminosulfonic acid salts, the former being hydrolyzed to the corresponding amine upon acidic treatment [8,9]. The reaction was originally described on 1-nitronaphthalene (Scheme 2).

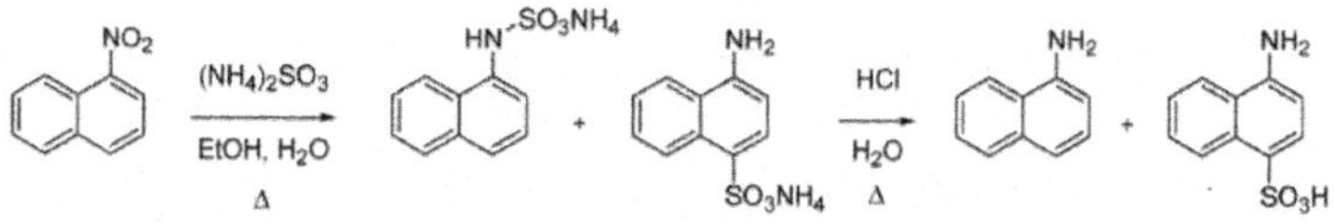

Scheme 2: The Piria aminosulfonic acid synthesis (1851).

In the context of this review, it is worth mentioning that both the reduction of carboxylic acids to aldehydes in the presence of calcium formate and the thermal decarboxylative coupling of carboxylic acid salts to form the corresponding symmetric ketones were formerly referred to as the Piria reaction(s) [10]. Cesare Bertagnini, one of Piria's favorite students inherited his Chair at the University of Pisa when his mentor moved to Turin. He is also remembered thanks to some outstanding work he published during his extremely short life. His name is associated with the formation of the so called Bertagnini's salt, the adduct between an aldehyde and an alkaline bisulfite (Scheme 3). This procedure was routinely used in the past to purify aldehydes and was first described by Bertagnini in 1851 [11,12,13].

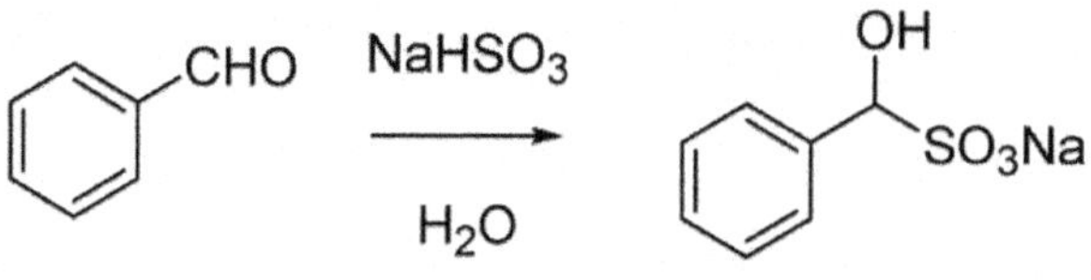

Scheme 3: "Bertagnini's salt" (1851).

Bertagnini's name is also linked to the Bertagnini-Perkin reaction. Although this carbon-carbon bond forming transformation is mainly reported as the Perkin reaction (the condensation between acetic anhydride and benzaldehyde delivering cinnamic acid), described by the English chemist as early as 1868 [14], the same conversion was anticipated by Bertagnini [15,16], who used acetyl chloride instead of acetic anhydride (Scheme 4). Additionally, the twosome Chiozza-Bertagnini is also found in references associated with the same type of transformation, thus recognizing, in a perhaps questionable form of parochialism, the contribution of Luigi Chiozza (1828–1889), who had conducted prior work on the synthesis of cinnamaldehyde [17].

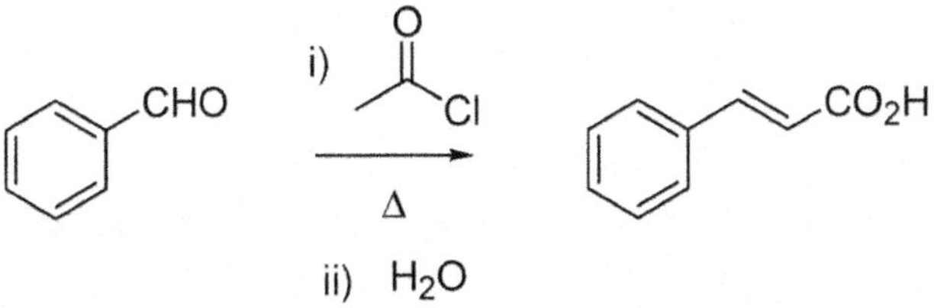

Scheme 4: The Bertagnini reaction (1856) later modified by Perkin (1868).

On the contrary, the name of Cannizzaro has come down to us more clearly in association with a specific organic reaction, the base-mediated disproportionation of a non-enolizable aldehyde into the corresponding alcohol and carboxylic acid. The original reaction that he described while he was at the Collegio Nazionale di Alessandria (Piedmont) in 1853 [18], was performed using bitter almond oil (benzaldehyde) and potash as the base (Scheme 5).

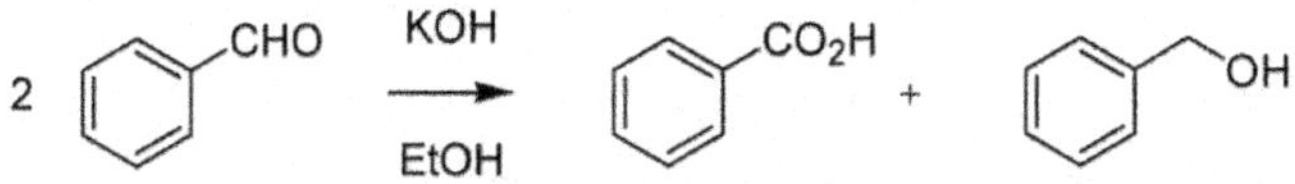

Scheme 5: The Cannizzaro reaction (1853).

In the meantime and despite the unhappy ending of the first Italian war of independence, the will to unify Italy was still smoldering under the ashes. At the beginning of 1859, Vittorio Emanuele II and his prime minister, the Count of Cavour, made a secret alliance with Napoleon III, the emperor of France, as they realized that there was no hope for Piedmont to defeat the Austrians alone. The key point of this alliance treaty was that France formally bound itself in helping Piedmont in case of aggression by the Austrians, as did in fact occur in April of that year. The war that followed (the second Italian war of independence) was rapidly won by the France-Piedmont allied troops and an armistice was signed in July. As a result, the vast majority of Lombardy was ceded to the King of Piedmont, while the Venetian region still remained part of the Austrian empire. In few months, with the consent of England and France, most of the population of central Italy was allowed to express their will to join the now enlarged Piedmont Kingdom. It was a complete success, but Unification of the country was still far from complete, with the Bourbon-controlled southern half of the Italian peninsula representing the next major obstacle to overcome. It took another year for the "red shirts", a volunteer army led by the warlord and politician Giuseppe Garibaldi, to occupy Sicily and turn towards North, finally defeating the Bourbons close to the Volturno River, north-west of Naples. It was October the 2nd, 1860.

While these dramatic events were taking place, another struggle, led by the Italian chemist Stanislao Cannizzaro, was occurring on scientific ground. In September 1860 the first international meeting of chemists, called by Kekulé, was held in Karlsruhe, Germany. Through his masterpiece "Sunto di un corso di filosofia chimica" (sketch of a course of chemical philosophy) [19] which he distributed at the very end of the conference, Cannizzaro successfully reiterated the hypothesis of Avogadro (Avogadro himself, was also, by the way, an Italian physical chemist), that certain gaseous elements possess diatomic molecules, thus requiring the doubling of their previously considered atomic weights.

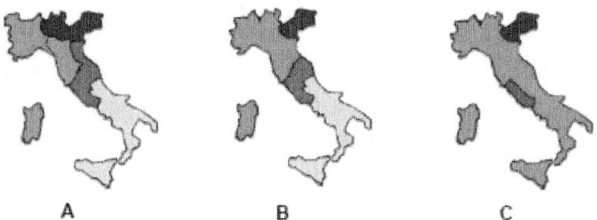

Figure 1: Maps of Italy in 1859 (**A**), 1860 (**B**) and 1861 (**C**).

On 17th March 1861 the Kingdom of Italy was proclaimed. However, the unification process was far from being truly accomplished. Apart from the Venetian region, there was still a significant chess-piece remaining on the board: Rome (Figure 1).

FROM FLORENCE AS THE NEW CAPITAL OF ITALY (1865) TO THE ASSASSINATION OF KING UMBERTO I (1900)

Following an agreement with Napoleon III, in June 1865 Italy moved its capital from Turin to Florence. One year later, an alliance with Prussia against the Austrians was signed, under the terms of which Italy would have gained rights to the Venetian region. War against Austria (known as the third war of independence) was shortly declared and quickly won, mainly due to the superiority of the Prussian army. The peace treaty was ratified in October 1866, and following a referendum, the Venetian region was annexed to the Kingdom of Italy. However, a large portion of the northeast territories still remained in the hands of the Austrians: the Trentino and Venezia-Giulia regions. The inhabitants of those lands would have to wait for the end of WWI before being recognized as Italian citizens.

The long-awaited opportunity for Rome's incorporation into the Italian Kingdom was presented by the Franco-Prussian war of 1870. Forced to withdraw its garrison from the city to reinforce its army and later defeated at Sedan by the Prussians, France was no longer in a position to protect Rome and thus Italian troops entered the city on 20th September of that year. Nine days later, "while the bells were ringing to celebrate the occupation of Rome", seven of the country's most prominent chemists, headed by Cannizzaro, met in Florence and founded the earliest Italian chemical journal, the "Gazzetta Chimica Italiana" (Italian Chemical Gazette) (Figure 2A), whose first issue was published on March 31st, 1871 (Figure 2B). In the meantime, a further referendum had ratified the annexation of Rome, which shortly became the capital of the Kingdom.

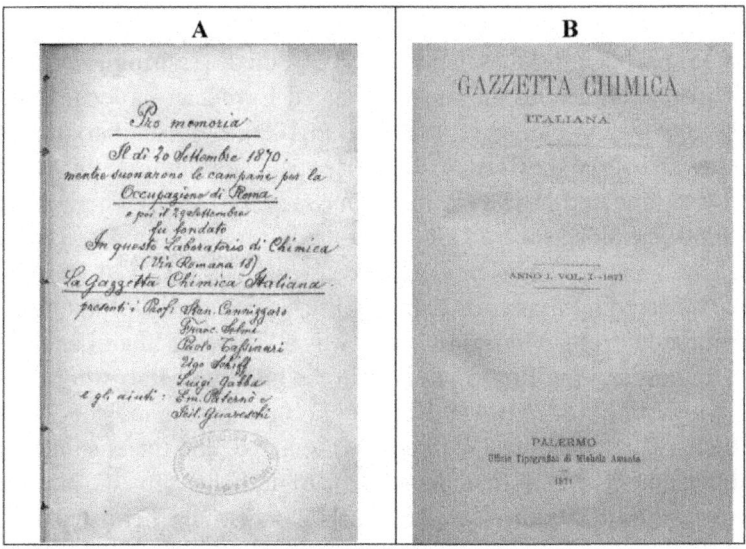

Figure 2: The founding act of the Gazzetta Chimica Italiana (Italian Chemical Gazette) (29 September 1870, **A**) and the frontispiece of its first issue (31 March 1871, **B**).

Italian chemistry of this period has another "giant" to whom recognition is due in the context of named organic reactions,*i.e.*, Hugo (often Italianized as Ugo) Schiff (1834–1915). Although German by birth, Schiff spent most of his life and career in Italy, and he was one of the "Magnificent Seven" subscribers of the founding act of the Gazzetta Chimica Italiana (Figure 2A). He was at the University of Pisa when he published his first paper on a "new type of organic base" [20], the compounds that still today are called the Schiff bases (Scheme 6). The work describing the Schiff fuchsin aldehyde test [21], of paramount importance for staining organic aldehydes in biological samples, was also published in that period.

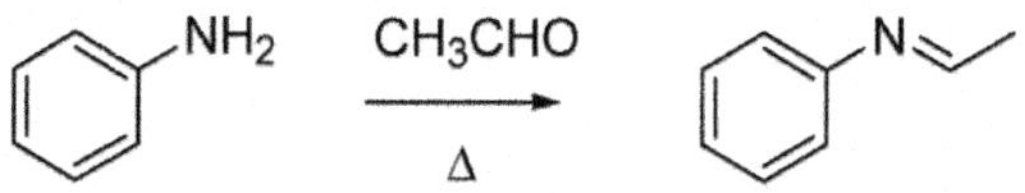

Scheme 6: The Schiff base synthesis (1864).

At the beginning of 1878, King Vittorio Emanuele II died. The throne was inherited by his son Umberto I who signed the Triple Alliance Treaty with Austria and Germany in 1882. This treaty forged an agreement of mutual military support in case of aggression by any other power, while a neutral behavior would be adopted in case of aggressive actions against others undertaken by any one of the signatories. The same year Giuseppe Garibaldi, the true military leader behind the unification of Italy, died. Desiring to be considered on an equal footing with the most advanced European nations, Italy started its own colonization campaign in Africa in 1885, by disembarking its troops in Massawa (Eritrea). Efforts to further expand its domination to Ethiopian territories culminated in 1887 with the massacre of Dogali where nearly all of the 500 Italian soldiers were killed after having fought against overwhelming forces. That same year the Triple Alliance Treaty was renewed. In 1892 one of the most outstanding figures of Italian politics, Giovanni Giolitti (1842–1928), was nominated Prime Minister. Giolitti was a left-wing liberal and is still the second-longest serving Prime Minister in the history of Italy after the dictator Benito Mussolini (1883–1945). During Giolitti's first mandate (1892–1893), the Socialist Party of Italian Workers was founded, and the ensuing years saw the establishment of the first association of Italian chemists, the Chemical Society of Milan (1895), followed by Turin's Industrial Chemistry Association (1899).

The Italian colonial adventure in Africa was reawakened in 1894. After a couple of modest military victories, the attempts to invade Ethiopia were finally frustrated first at Amba Alagi then at Adwa (1896) by the soldiers of

Menelik II, the Ethiopians' Emperor. A peace treaty was ratified at Addis Ababa at the end of the year. Accordingly, the Italian colonial Empire would be limited to Eritrea and Somalia. The cost of these war operations exacerbated an already difficult economic crisis. A number of meetings, demonstrations and strikes occurred throughout the Nation. In May 1898, a massive strike in Milan culminated in a massacre by the military forces present. It was during this highly agitated social climate that Gaetano Bresci, a Tuscan anarchist who had emigrated to the US, returned to Italy with the intention of assassinating the monarch to successfully accomplish his mission. On July 29, 1900 Bresci killed King Umberto I in Monza (near Milan), shooting him four times while he was attending a sports meeting.

On the scientific front during this period, in 1881 Giacomo Ciamician (1857–1922), at that time Cannizzaro's assistant at the University of Rome, co-authored together with Dennstedt a paper [22,23] that allowed also him to enter the Olympus of named reactions for the first time. The Ciamician-Dennstedt rearrangement entails expansion of the pyrrole ring to form a pyridine derivative by means of chloroform (or other halogeno- compounds) and a base through the *in situ* generation of dichlorocarbene (Scheme 7).

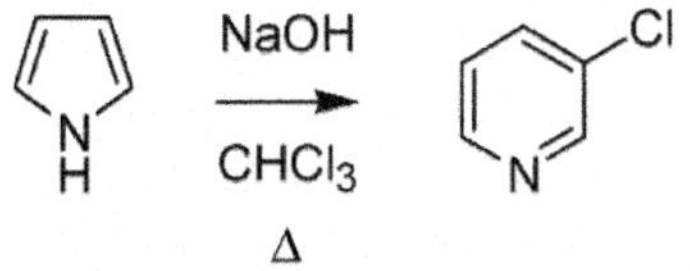

Scheme 7: The Ciamician-Dennstedt rearrangement (1881).

Ciamician's interest in pyrrole chemistry dated back to the years when he was a student, first in Vienna and then in Giessen, but persisted throughout his highly productive career. From pyrrole, through to plant chemistry, he gradually came upon the world of photochemistry, where he made a ground-breaking contribution through the discovery of several key reactions and is today unanimously recognized as a founding father of this branch of chemistry (Figure 3).

The remarkable scope and impact of Ciamician's work in the field of photochemistry can be perceived in the ambiguity with which the eponymous photochemical reaction has variously been described. According to most current textbooks for example [1], the Ciamician reaction consists in the reductive photocoupling of ketones leading to the formation of 1,2-diols [24] (Scheme 8).

Figure 3: The dawn of photochemistry. Ciamician surrounded by a number of sun-exposed flasks on his laboratory terrace at the University of Bologna.

However, during the golden age of photochemistry the Ciamician reaction was also proposed [25] as being the intramolecular [2 + 2] cycloaddition of alkenes, which Ciamician and his co-worker Paul (Paolo) Silber had described in reference to the formation of a camphor derivative during prolonged sunlight exposure of carvone [26]. Yet again, according to Giovanni Battista Bonino, who later wrote the entry for Giacomo Ciamician which appears in the authoritative Treccani Encyclopedia (considered in Italy to be the definitive source of reference for information) [27], the Ciamician reaction is the photochemical disproportionation of nitrobenzaldehyde to nitrosobenzoic acid, another significant transformation described by Ciamician and Silber [28].

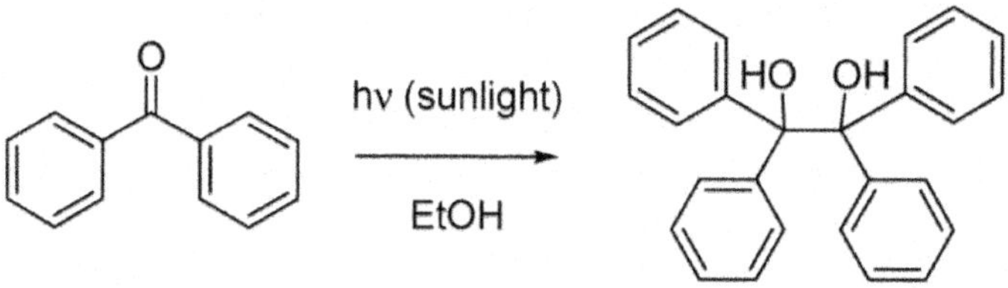

Scheme 8. The Ciamician photocoupling (1900).

These contributions reflect the profound interest and faith Ciamician had in photochemistry and its potential applications, which are highlighted in his famous dissertation on "the future of photochemistry" of 1912 [29], where he advocated the use of sunlight to produce energy and predicted the modern utilization of solar cells and can thus also be considered a pioneer of "green chemistry". In Bologna, the University where he spent most of his career, he was the founder of a school of Chemistry where many outstanding scientists were educated, and which still stands today. One of his earlier fellows, Giuseppe Plancher (1870–1929), is associated with a named reaction published before the end of the century. The Plancher rearrangement (sometimes referred to as the Plancher-Ciamician rearrangement) [30] was discovered by the reaction of 2-ethyl-3-methylindole with methyl iodide, leading to 1,2,3-trimethyl-3-ethyl indolenium iodide, which was later explained by a Wagner-Meerwein type migration of alkyl groups from position 3 to position 2 of an indoleninium species (Scheme 9). Since the migration reactions are in equilibrium, the process eventually leads to the most thermodynamically stable isomer.

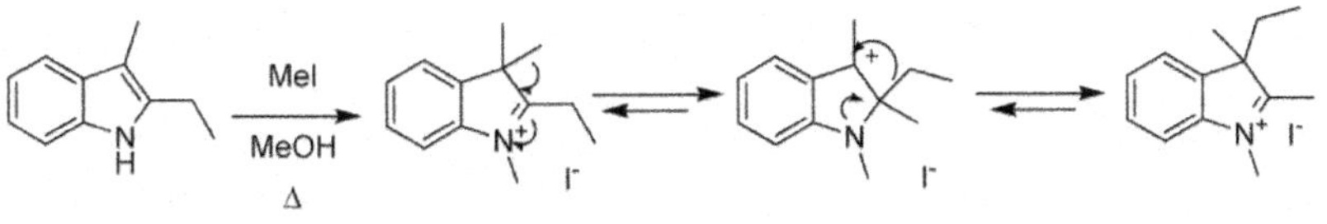

Scheme 9: The Plancher rearrangement (1898).

The last decade of the nineteenth century witnessed several important Italian contributions to organic chemistry. Pietro Biginelli (1860–1937) published the first accounts of his famous pyrimidine synthesis in 1891 (Scheme 10), soon after joining the laboratory of Hugo Schiff in Florence [31,32].

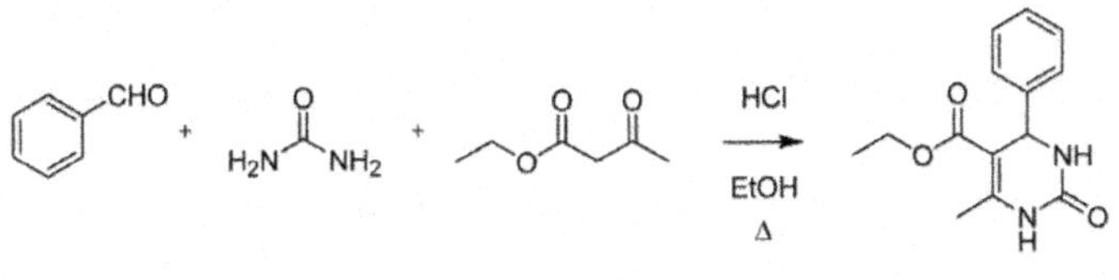

Scheme 10: The Biginelli pyrimidine synthesis (1891).

Biginelli, whose scientific activity culminated in Rome at the "Istituto Superiore di Sanità" (State Medicinal Institute) where he had the opportunity to develop an analytical method also known as the Biginelli test [33], soon decided to turn to a non-academic career [34]. His involvement in chemical research lasted a little more than a decade, but it was very intense, and left a well-defined mark. Arguably, this can be attributed to the education he received while he was a student in Turin, where his mentor was the founder of another important regional school of chemistry: Icilio Guareschi (1847–1918).

Icilio Guareschi stands out for being a complex figure of man, both scientific divulgator and chemist. In 1866, while still a high school student in

Parma, he participated as a volunteer in the third war of independence against the Austria-Hungary Empire. Later he became an active pacifist. He produced an impressive collection of works, encompassing dozens of publications devoted to chemical education, including a 13-volume encyclopedia dedicated to applied chemistry, unofficially known as "the Guareschi" [35], and to history of chemistry and science in general. The work that ensured him a place in the list of named organic reactions was published in 1896, while he already was a well-known professor at the University of Turin. The Guareschi reaction [36], which is also sometimes referred to as the Guareschi-Thorpe [37] reaction, deals with the synthesis of pyridines by condensation of cyanoacetic ester or primary amide with acetoacetic ester in the presence of ammonia (Scheme 11).

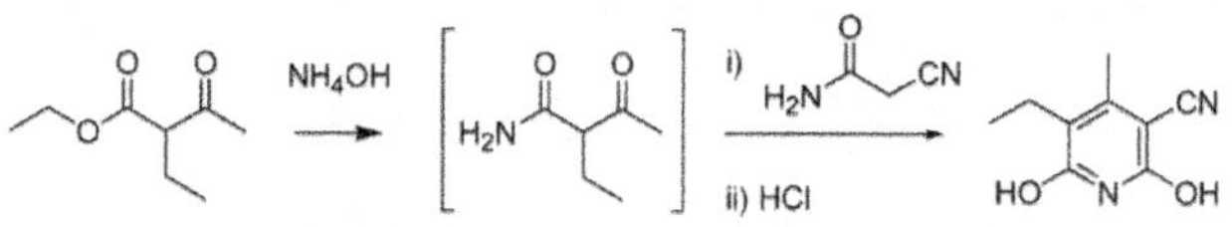

Scheme 11: The Guareschi reaction (1896).

Another chemist from the University of Turin who was later acknowledged thanks to the intense work he carried out on dioximes is Giacomo Ponzio (1870–1945). The reaction he published in 1897 on the oxidation of aldoxime by means of nitrogen dioxide in ether [38] is today known as the Ponzio reaction (Scheme 12).

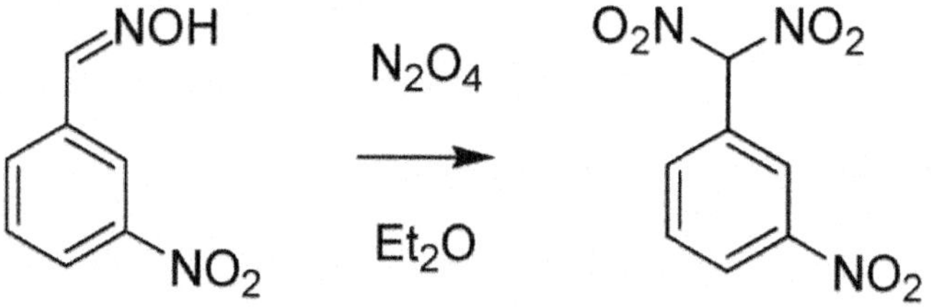

Scheme 12: The Ponzio reaction (1897).

A few years earlier, in 1894, another chemical transformation that would later to be referred to as a named reaction, had appeared in the literature. This is the Pellizzari reaction [39], after Guido Pellizzari (1858–1938) a professor of medicinal chemistry at the University of Florence and a former fellow of Hugo Schiff, who disclosed a new method for the synthesis of 1,2,4-triazoles (Scheme 13).

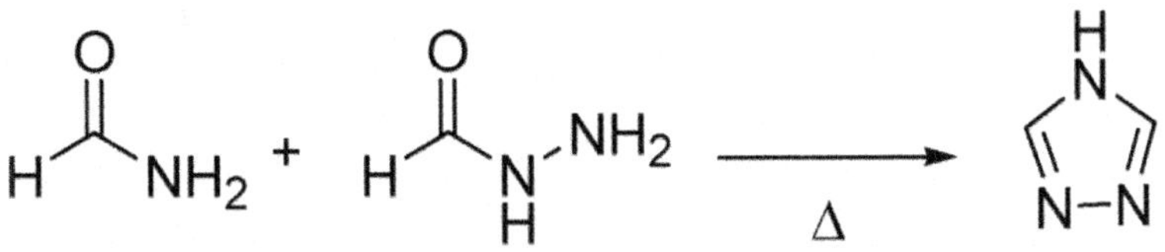

Scheme 13: The Pellizzari reaction (1894).

After his degrees in pharmacy and chemistry and before becoming an appreciated high-school teacher in Milan, Giovanni Ortoleva (1868–1939) worked as a pharmaceutical chemistry researcher at the University of Palermo. During this interlude he synthesized β-iodocinnamic acid by reacting cinnamic acid with elemental iodine in the presence of pyridine [40]. While further investigating the scope of this reaction, he then employed malonic acid as a substrate, which unexpectedly delivered a pyridinium betaine with loss of carbon dioxide [41]. This posed the basis for a novel and general transformation that encompasses the formation of N-alkyl pyridinium derivatives of activated methyl and methylene species, which in turn find a number of synthetic applications [2]. This reaction became later known as the Ortoleva-King reaction (Scheme 14) by acknowledging the contribution of L. Carroll King from Northwestern University, Illinois, who in 1944 further expanded the scope of the reaction [42].

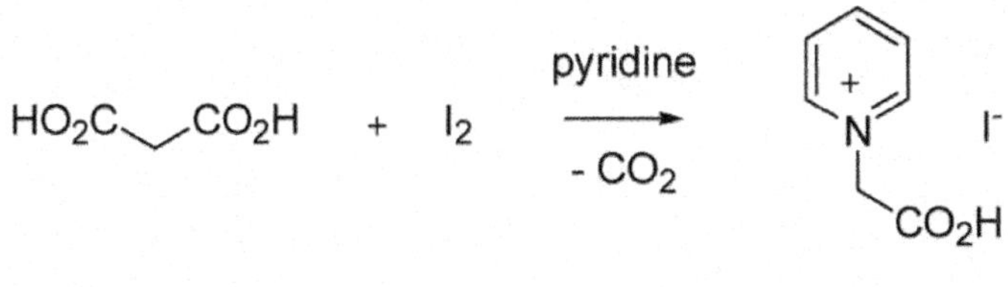

Scheme 14: The Ortoleva-King reaction (1900).

The Italian contribution to the chemistry of the last decade of the 19th century closed with another publication that appeared in 1900 by Mario Betti (1875–1942), a young graduate from the University of Pisa working in Schiff's laboratory in Florence. The Betti reaction [43] is nowadays considered as a multicomponent condensation between a phenol, an aldehyde and an aromatic amine that produces α-aminomethylphenols, though in the original paper the author described the use of a preformed imine (Scheme 15). This can be regarded as a special case of the more general (and subsequently discovered) Mannich reaction.

Scheme 15: The Betti reaction (1900).

FROM THE BEGINNING OF THE 20TH CENTURY TO THE OUTBREAK OF THE GREAT WAR

After the death of Umberto I, the crown was inherited by his son Vittorio Emanuele, who took the throne as Vittorio Emanuele III. He was the penultimate king of Italy. The beginning of the 20th century was one of the most economically rewarding periods for the Country. The national income doubled and savings increased four times. Italian foreign commerce was healthier than that of Germany and England.

Figure 4: The negative impact of war on scientific publications. Note the smaller size of the Gazzetta Chimica Italiana volumes between 1916 and 1919 (top) and between 1942 and 1945 (bottom) (from the Nerviano Medical Sciences library).

It was during these prosperous days, christened throughout Europe as the Belle Époque, that the Chemical Society of Rome was founded in 1902, once again thanks to the initiative of Cannizzaro, who since 1871 had been appointed Professor in Rome since, as well as one of his most talented disciples, the Marquis Emanuele Paternò. Seven years later, this Society merged with the that of Milan, giving birth to the "Società Chimica Italiana" (Italian Chemical Society), which is still in place today and which currently has more than 3,500 members. During Giolitti's fourth tenure as Prime Minister (1911–1914) a profound reform of education occurred, which raised the minimum age of compulsory education to twelve and which transferred management of the schools from city administrations to the State. A further crucial step towards the modernization of Italy was accomplished in June 1912 with extension of the right of vote to all 30-year-old male citizens, thus increasing the politically active population from 7% to more than 23%.

However, an economic recession was approaching. As a consequence, peace became as volatile as ether. To distract popular opinion from the incoming storm and after a secret agreement with France, the Italian government decided to enlarge its colonial dominions. In 1911 troops invaded the Ottoman territories of Tripolitania and Cyrenaica (now parts of Libya), forced the Dardanelles strait and obliged Turkey to recognize (in October 1912) Italian sovereignty over the conquered African lands. But this was just a skirmish. On May 24, 1915, after nearly one year of neutrality from the very beginning of the war that would stain Europe and the rest of the world with blood for more than four years, Italy joined the Triple Entente (United Kingdom, France and Russia) and entered into the conflict against the Central Powers (Austria-Hungary and Germany). It was the Great War, which would sadly be recorded in historical textbooks as the First World War (Figure 4, top). The first Italian named reaction of the 20th century has its roots back in 1896, when Angelo Angeli (1864–1931), an outstanding scientist who would later be nominated several times for the Nobel prize [44], published a paper describing the preparation of the sodium salt of nitro hydroxylamine ($Na_2N_2O_3$) [45]. This compound, known as Angeli's salt, is endowed with several pharmacological properties stemming from its instability, which leads to the formation of nitrous acid and nitroxyl (HNO) [44]. Subsequently, Angeli [46] noted that HNO could be "fixed" by aldehydes, forming hydroxamic acids. It was however Enrico Rimini (1874–1917), one of Ciamician's numerous students together with Angeli, who in 1901 disclosed a more practical version of this reaction, that he employed to detect aldehydes [47]. The reaction, using N-hydroxybenzensulfonamide (Piloty's acid) rather than the Angeli's salt as a donor of nitroxyl, is devoid of the interference of nitrous acid and was later christened with the name of both scientists. The Angeli-Rimini reaction (Scheme 16) has both analytical (for the identification of aldehydes) and preparative value (for the synthesis of hydroxamic acids).

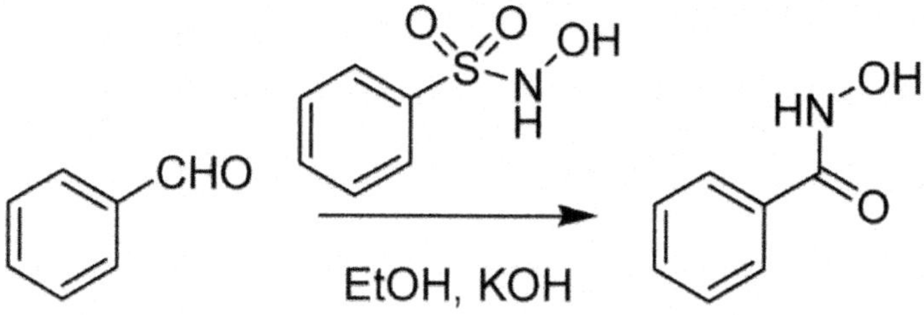

Scheme 16: The Angeli-Rimini reaction (1901).

It is worth mentioning that the name of Enrico Rimini is often associated with another analytical method, the so-called Rimini-Schryver reaction [48]. This test is used to quantitatively estimate allantoin in biological fluids, and is based on a procedure developed by Rimini for the detection of formaldehyde [49] later implemented by Schryver [50].

Guido Bargellini (1879–1963) was a brilliant chemist who, after a post-doctoral experience in Emil Fischer's laboratory, was appointed at the University of Rome, where he spent most of his career. His interests in coumarins prompted Bargellini to investigate a multicomponent reaction between phenol, chloroform and acetone in the presence of alkali previously described in a German patent [51], thus finding that the structure attributed to the product was misassigned. He discovered that the reaction produced a carboxylic acid (instead of a phenol as originally proposed) [52] (Scheme 17), thus paving the way for a new transformation, the Bargellini reaction, whose number of variations contributed to its wide application both in organic and medicinal chemistry.

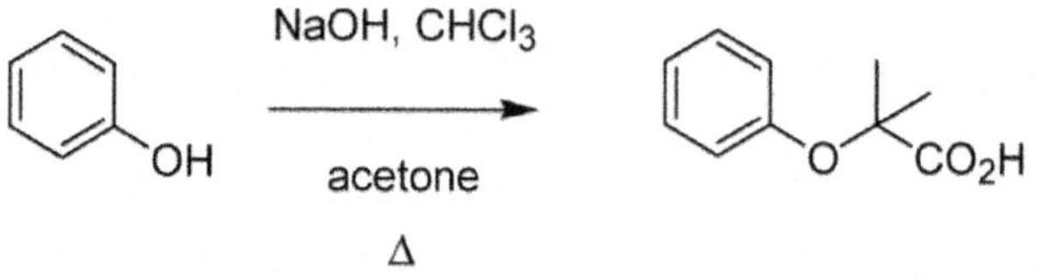

Scheme 17: The Bargellini reaction (1906).

Bargellini represents a sort of joining link between the classical era of Italian chemistry and the modern times. While from an anagraphical point of view he is projected towards the latter, he is however bound to the former by his direct link with two great masters: Cannizzaro, who first called him to Rome, and Cannizzaro's successor at the chair of general chemistry of the same University: Emanuele Paternò.

Emanuele Paternò (1847–1935), born Marquis of Sessa, was another prominent scientific personality who was also actively involved in the political life of his native homeland. As already mentioned (see Section 2), he was one of the founders and the first director (from 1870 to 1919) of the Gazzetta Chimica Italiana. He served on many other fronts as well, being for instance nominated Dean at the University of Palermo and Senator of the Kingdom. As a chemist, he is often associated with the pioneering studies on tetrahedral carbon and on cryoscopy. In organic synthesis, he studied the photochemical cycloaddition of olefins to carbonyl compounds leading to oxetanes [53]. Paternò published his results about this reaction in 1909, but he had no means to establish its regiochemical outcome. It was only 45 years later that Büchi, by capitalizing on spectroscopic experiments, discovered that the reaction can be regio- and stereoselective depending upon the nature of the reactants [54], thus providing a fundamental piece of information of what is nowadays called the Paternò-Büchi reaction (Scheme 18).

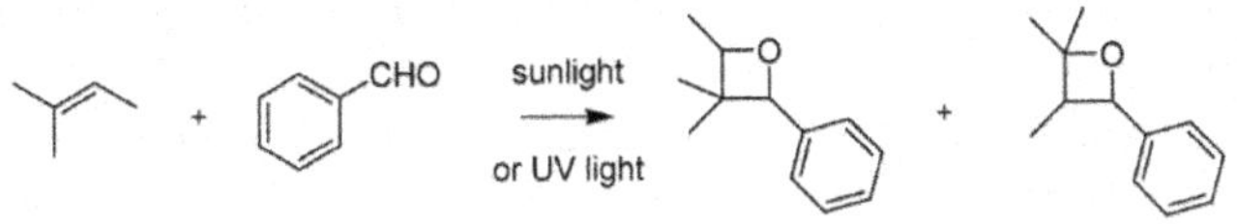

Scheme 18: The Paternò-Büchi [2 + 2] cycloaddition (1909).

ITALY BETWEEN THE WARS

Italy fought WWI chiefly on the "Italian front" (the northeast of the Country) against the Austrian-Hungarian troops. It was essentially a war of position, with a number of tactically useless assaults and counterassaults on a difficult mountainous terrain, which meaninglessly moved the front at a high cost in human lives. After nearly 2 years and a half of stagnation, the Italian army was defeated at Kobarid (Caporetto in Italian, now in Slovenia) in October 1917. However, the Austrians did not capitalize on this success. The centrifugal forces that were acting within the Austrian-Hungarian empire due to claims for independence of its constituting nations contributed to the dissolution of its multiethnic army. The decisive battles close to the river Piave and the town of Vittorio Veneto (near Treviso, Venetian region) allowed the Italians to victoriously conclude the war. The armistice between Italy and Austria was formally signed on November 3, 1918 at Villa Giusti (Padua). As a consequence, Trentino and Venezia-Giulia territories became part of the Kingdom. However, nationalists were still not satisfied. The annexation of the largely Italian-speaking Dalmatian region and the city of Rijeka (Fiume in Italian, now in the State of Croatia) to Italy was denied by the other WWI co-winners gathered at Versailles (France). The already tense political and economical situation arising from the post-war crisis was further exacerbated by the socialists who systematically and often violently, fomented public opinion. During the following year (1919) several significant events occurred. Benito Mussolini, a prior left-wing journalist, founded the "Fasci Italiani di Combattimento" (Italian Groups of Combat), an organization which recruited a significant number of followers, wary of the rise of the socialism (with which Mussolini himself had previously flirted). In addition, the Popular Party (the future Christian Democrats) had its first meeting, and the right of vote was granted to all 21-year-old (and older) male citizens. In September the poet, novelist and swashbuckler Gabriele D'Annunzio and his legionnaires occupied the Croatian city of Rijeka, which they held until they were forced to leave by the Italian troops in December 1920. Six months earlier Giolitti had become prime minister for his fifth (and last) time. In January 1921 the left wing of the socialistic movement founded its own party: the Italian Communist Party. During these agitated years, Italy was poisoned by a number of violent strives between the different political factions. None of these was able to build up a coalition in parliament strong enough to withstand the ascent of Mussolini's fascists, who formed a fully-fledged political party in November 1921 and nearly one year later its members (the so-called "black shirts") occupied Rome. Vittorio Emanuele III had little choice. He asked Benito Mussolini to form the new government, leaving the stage to the darkest period in the history of the Country.

One of the most talented chemists during the interlude between the wars was Carlo Gastaldi (1884–1962), a knowledgeable professor of chemistry who spent his earlier and later career in Sassari (Sardinia). In 1916 he had just joined the University of Turin, when he was drafted into army and sent to the front. After spending three years at the army's "pyrotechnic laboratory", he went back to the laboratory directed by Ponzio, and continued the research that culminated with the publication of his pyrazine synthesis [55] (Scheme 19).

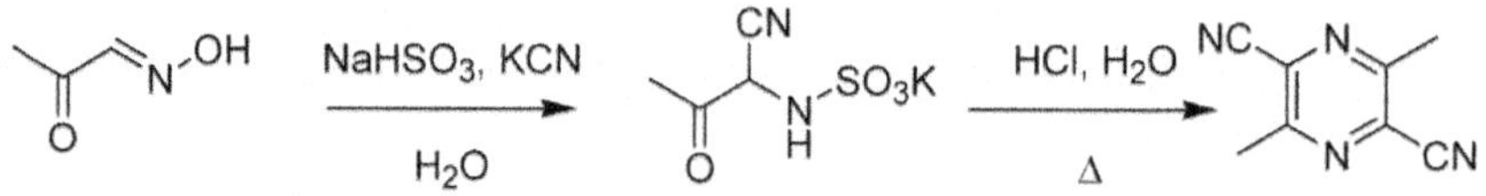

Scheme 19: The Gastaldi pyrazine synthesis (1921).

During the course of 1921 another publication appeared in the Gazzetta Chimica Italiana for which his author would later be remembered for the eponymous reaction [56,57]. Its author, Mario (Torquato) Passerini (1891–1962), had also taken part in the war while still a student. Passerini had graduated from the University of Florence in 1916 during army leave and after completing military service he worked at the same University for several years. The Passerini reaction is a multicomponent reaction where an isonitrile, a carboxylic acid and a carbonyl compound are reacted in an apolar solvent leading to a α-acyloxyamide (Scheme 20). Its publication, fruit of the scientist's Florentine years, provided the first important account on the chemistry of isonitriles, and is undoubtedly one of the most cited Italian named reactions.

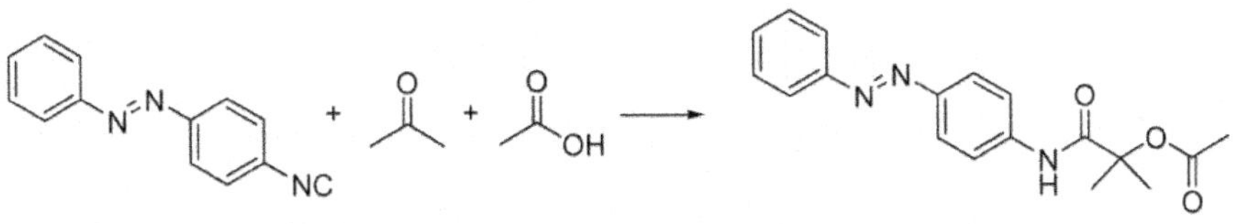

Scheme 20: The Passerini reaction (1921).

The treaty of Lausanne (July 1923), which definitively recognized Italian sovereignty over the Dodecanese islands (now part of Greece) and the re-annexation of Rijeka, formalized in January 1924, increased the government's consensus, up to the point at which a few months later (June 1924), it brazenly commissioned the murder of Giacomo Matteotti, a young socialist leader and a dictatorial regime soon followed in January 1925. Mussolini was subsequently the target of several assassination attempts, giving him the pretext to render illegal all opposition parties and newspapers, and to reintroduce the death penalty that had been banned since 1889. Aiming at a positive resolution of the conflict between the Italian Kingdom and the Catholic Church dating back to the annexation of Rome, Mussolini signed the Lateran Treaty in 1929, a formal reconciliation with the Vatican. Later the same year, the Wall Street crash sparked the deepest global economic crisis ever experienced.

During this politically and economically turmoiled period, Mario Amadori (1886–1941), originally at the University of Padua and subsequently at the University of Modena, performed his research on sugars [58] discovering what nowadays is referred to as the Amadori rearrangement. This transformation involves the reaction of an aldose with a suitable amine leading to the corresponding glycosylamine which in turn rearranges to the corresponding ketoseamine (Scheme 21).

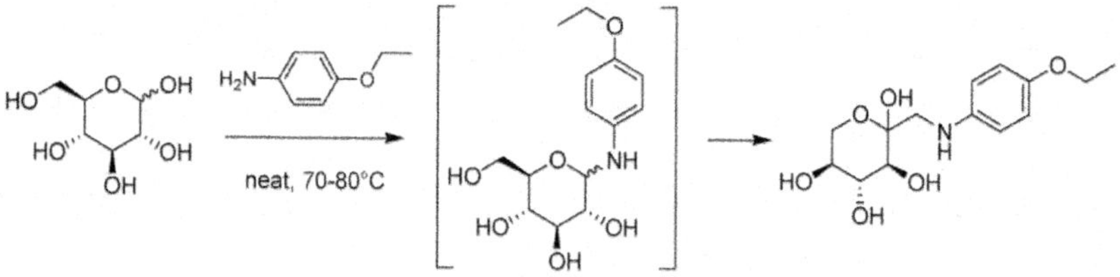

Scheme 21: The Amadori rearrangement (1925).

In June 1934 Mussolini and Hitler, the leader of the Nazi party and Chancellor of Germany, met for the first time. In pursuing his dream of an Italian "Empire" and despite the economical and financial sanctions inflicted by the League of Nations, Mussolini invaded Ethiopia in 1935. After having re-conquered Adwa, the Italian troops finally entered the capital Addis-Ababa in May 1936. It was the "magic moment" of the fascist era.

In the same year, a novel method for the synthesis of fluorene via the formation of an intermediate diazonium ion was published [59] (Scheme 22). The method is known as the Mascarelli fluorene synthesis after its discoverer, Luigi Mascarelli (1877–1941), who grew up at the school of Ciamician in Bologna and later inherited the chair of Guareschi at the University of Turin.

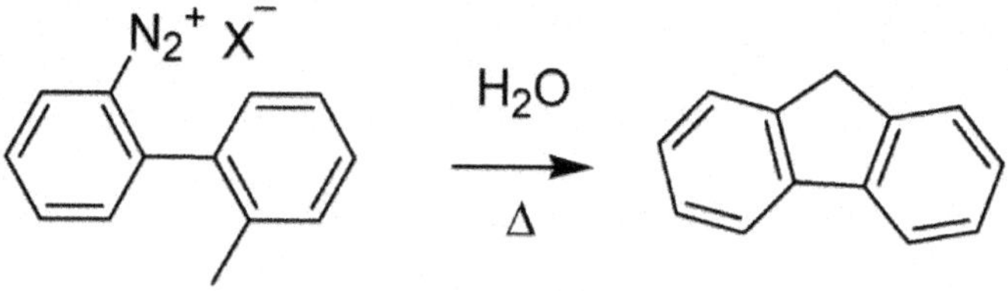

Scheme 22: The Mascarelli fluorene synthesis (1936).

1936 also saw outbreak of the Spanish civil war, in which Italy supported the Nationalist front headed by the general Francisco Franco against the pro-government Republicans by providing Franco with a military contingent. This volunteer army heavily contributed to Franco's seizure of power (April 1939) who established a fascist-like dictatorship. Aiming at being a protagonist of the political situation in Europe and not just a background actor, Mussolini aped Hitler by introducing in 1938 the racial laws against Jews. In April 1939 the Italian troops occupied Albania. With the invasion of Poland by the Nazis (1 September 1939) and the consequent declaration of war by France and England against Germany, the Second World War officially commenced. Erroneously convinced that the war would end swiftly with a German triumph, Mussolini entered the conflict at Hitler's side in June 1940. At the time, although many Italians realized how inopportune this act was, few imagined the immense catastrophe the country would undergo.

WWII AFTERMATHS AND THE CONTEMPORARY TIMES

Italy entered WWII with a technically and organizationally unprepared army. After some useless skirmishes with the French, occurring just few days before France was forced to sign the armistice with Germany, and a handful of modest successes in northern and eastern Africa, the Italian troops headed towards a misfortunate fate on every front they fought. The futile invasion of Greece (started in October 1940) was disastrously conducted and only the arrival of the Germans allowed the Italians to not to be completely defeated. Joint German and Italian forces were subsequently driven out of Africa (May 1943) and their advance inside the Russian territories ended with the dramatic full retreat of the ARMIR ("Armata Italiana in Russia", Italian Army in Russia) and the German defeat at Stalingrad (January 1943, the bloodiest battle in history). In July 1943 the Anglo-American troops (US entered the conflict in December 1941) disembarked in Sicily. They encountered a feeble resistance by the Italians. While the Allied army relentlessly continued its march up North, everybody realized that the Fascist age was close to the end. On 25 July 1943 the "Gran Consiglio del Fascismo" (Grand Council of Fascism), the main body of the fascist government, voted the deposition of Mussolini, who was suddenly arrested and replaced by Marshal Badoglio, the signer of the unconditional armistice between Italy and the Allied forces that would occur on 8 September 1943. Immediately thereafter, Badoglio and the King Vittorio Emanuele III left the Capital to settle the new government in the South of Italy. Few days later, the now enemy German troops occupied Rome and then liberated Mussolini from his confinement. Supported by the Nazis, the Italian dictator founded the "Repubblica Sociale Italiana" (Italian Social Republic) with jurisdiction over the North of the Country up to Rome (the part of Italy still in Germans' hands). It was a hopeless, pathetic attempt to reinstall the fascist regime. While most of the opponents entered the partisan brigades to fight the fascists in what was a true civil war, the Nazis were defeated by the Allied army both in France and in Italy, where an armistice was signed on 29 April 1945. In February the same year, the Allied heads of state had met in Yalta (Crimea) to define the political and geographical asset of post-war Europe. They had also ratified the project concerning the creation of the United Nations. The insurrection of the partisans in Milan (25 April), and the execution of Mussolini (28 April) two days before the suicide of Hitler in his bunker in an already Russian-occupied Berlin, led to the end of the WWII in Europe. The unconditional surrender was signed by the Germans on 8 May 1945. It would take four extra months of war and two atomic bombs on Hiroshima and Nagasaki to force also Japan (that had entered the conflict in December 1941) to surrender (2 September 1945). The biggest human tragedy of the modern era was finally ended. Back from Auschwitz, the sadly famous Nazi concentration camp, Primo Levi, an Italian Jewish chemist and writer, collected his memoires of lager survivor in the book "Se questo è un uomo" (If this is a man), a literary masterpiece that should be present in everyone's library.

In May 1946, the King Vittorio Emanuele III abdicated in favor of his son Umberto, who took the throne as Umberto II. However his Kingdom lasted roughly a month: on 2 June 1946 a constitutional referendum set forth the birth of the Italian Republic. Umberto II went into exile in Portugal, where he died in 1983.

The first president of the newborn Republic (Enrico de Nicola, a member of the Italian Liberal Party) was elected at the end of June 1946. One and a half years later, in December 1947, the Constitution of the Italian Republic, the chief legislative act ruling the general principles on which the Republic itself is founded, was enacted. In the late '40s Italy started its economic resurgence, heavily supported by the US-sponsored European Recovery Program (Marshall Plan). The consciousness of the importance of these American monetary aids, along with the fear of becoming a satellite of Soviet Union in case of victory of the joined Italian Communist and Socialist Parties, are some of the reasons that potentially explain the massive triumph of the Christian Democracy during April 1948 elections (women's suffrage was granted in 1945). This was not an occasional event: from then on this party would rule the Nation for more than thirty years. In 1949 Italy joined the North Atlantic Treaty. With the restitution of the northeast-located town of Trieste and some of its neighboring territories, which were held in trust for the Allied since the end of WWII, Italy definitely acquired its actual geographical asset in 1954. The Country was admitted to join the United Nations organization at the end of 1955, and in 1957 it hosted the European Economic Community constitution summit. During this period, Italy witnessed a spectacular economic growth, characterized by an unprecedented raise of both salaries and occupational levels due to a dramatic increase in production and investments. In order to fulfill the manpower request by the industries, a massive migratory flow of work-seeking people from the south of the Country to the more developed northern regions occurred.

After the predictable depression in the scientific production witnessed by the chemical community during the wartime (Figure 4, bottom), it took some time before seeing another Italian ascending the "hall of fame". This is literally what happened to Giulio Natta (1903–1979) whose discoveries made during the '50s in the field of polymerization were awarded the Nobel Prize together with Karl Ziegler in 1963. Natta graduated from the Polytechnic of Milan in 1924 and spent several years as a chemistry professor in prestigious Universities. In 1938, he was appointed Professor of General Chemistry at his*alma mater* as the application of the racial laws against Jews forced the chair's former holder, Professor Mario Giacomo Levi, to leave. This event generated some out-of-context controversies about a potential adhesion of Natta to the Fascism [60,61], which however recently resulted in a more equilibrated judgment on the facts [62].

Natta's work, first published in 1955 [63], led to the improvement of the organometallic-based catalysts for the low-pressure polymerization of lower alkenes earlier discovered by Ziegler [64].

The so called Ziegler-Natta catalyst allowed the stereospecific polymerization of propylene (Scheme 23) for the first time thus revolutionizing the world of macromolecular chemistry. As the Nobel Prize committee member Arne Fredga said during the prize assignment ceremony with reference to Nature's ability in synthesizing stereoregular polymers, "the work of Professor Natta has broken this monopoly".

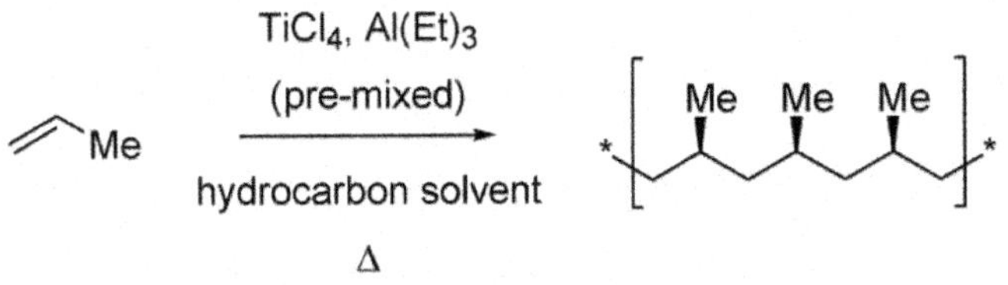

$$TiCl_4, Al(Et)_3$$

(pre-mixed)
$$\longrightarrow$$
hydrocarbon solvent

$$\Delta$$

Scheme 23: The Ziegler-Natta isotactic polypropylene synthesis (1955).

Roughly in the same years, Giancarlo Berti (1924–2005), a young graduate from the University of Pisa, was doing his PhD at the University of Notre Dame, Indiana. He defended his dissertation entitled "The Pyrolysis of Organic Sulfites" in 1953, and gave an account of the work he had performed in a paper that was published the following year [65]. This later became known as the Berti olefination (Scheme 24), a reaction in which methyl sulfites are used to prepare unsaturated hydrocarbons. Methyl sulfites are more easily accessible than the corresponding xanthates, which are exploited in the classical Chugaev reaction [66]. In addition, their pyrolysis, although requiring slightly higher temperature than for xanthates, proceeds with higher yields thus making the Berti olefination more advantageous.

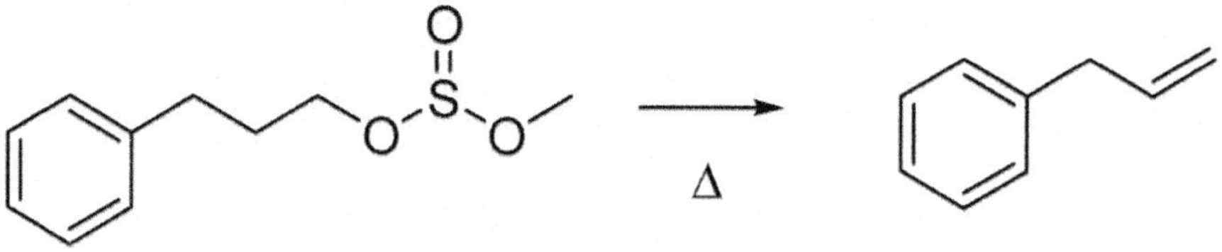

Scheme 24: The Berti Olefination (1954).

The publication of Berti's paper in English was harbinger of an epochal change that was about to occur. During the subsequent decade, the habit of using English as the official language for scientific communication caught on in the Italian chemical community. Work performed at the Polytechnic of Milan represented a landmark in this respect thanks to Luciano Caglioti, who in 1964 published his first paper [67] on the reduction of tosylhdrazones to the corresponding saturated derivatives by means of NaBH$_4$ (Scheme 25). The method, which is very mild and tolerates a number of functional groups, proceeds through the formation of a substituted tosylhydrazide that yields an unstable diimide, in turn decomposing to the saturated final product.

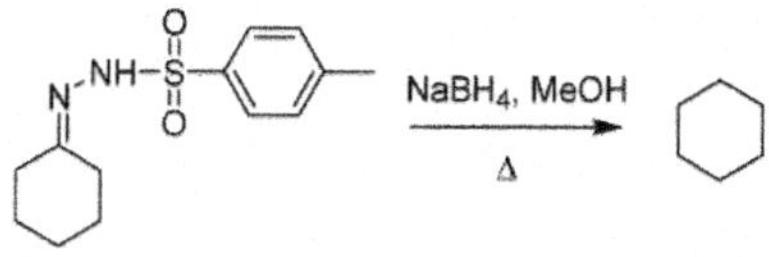

Scheme 25: The Caglioti reduction (1964).

In the late '60s, massive movements of protest, originally mounted in US, flooded throughout Europe. Characterized by a deep socialistic connotation, these movements were highly patchy, being constituted by students, workers and poor people fighting against the inappropriateness of the political, social and educational structures of the western Nations. Originally peaceful in nature, these ideologically-driven protests rapidly became violent culminating, in Italy, with the exacerbation of the conflicts among left wing extremists, the far right and the government apparatus. Terrorists from both sides committed a number of attacks and murders aiming at destabilizing the Country. To increase the already dramatic situation, Italy experienced in the early '70s a profound economical crisis due to the decision of the Organization of the Arab Petroleum Exporting Countries (OAPEC) to proclaim an oil embargo.

While political radicals were acting insanely, chemical radicals were managed to react in a fruitful fashion. Thus, in 1971, a group at the Polytechnic of Milan led by Francesco Minisci, at that time still at the University of Parma, published what can be considered the first account of the so called Minisci reaction [68]. This article deals with the alkylation of heteroaromatic bases by a carbon-centered radical (Scheme 26).

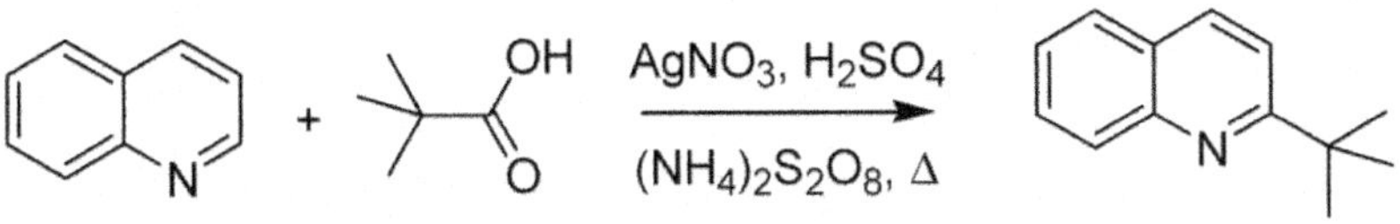

Scheme 26: The Minisci reaction (1971).

An important cultural event occurred one year after the publication of Minisci's seminal paper. After a tough ideological struggle inside the Italian Chemical Society, a decision was finally taken that all the contributions to the Gazzetta Chimica Italiana had to be written in English.

The seventies of last century witnessed the publication of another reaction that subsequently attracted a lot of attention. This is the Piancatelli rearrangement, which was published by Giovanni Piancatelli and co-workers from the University of Rome in 1976 [69]. The reaction, which entails the synthesis of 4-hydroxycyclopent-2-enones starting from the corresponding 2-furyl carbinols, proceeds with high stereocontrol furnishing a relative *trans* orientation of the substituent at position 5 and the 4-hydroxy-group (Scheme 27). The rearrangement found some popularity as it allows accessing advanced intermediates in the synthesis of prostaglandins and other natural products.

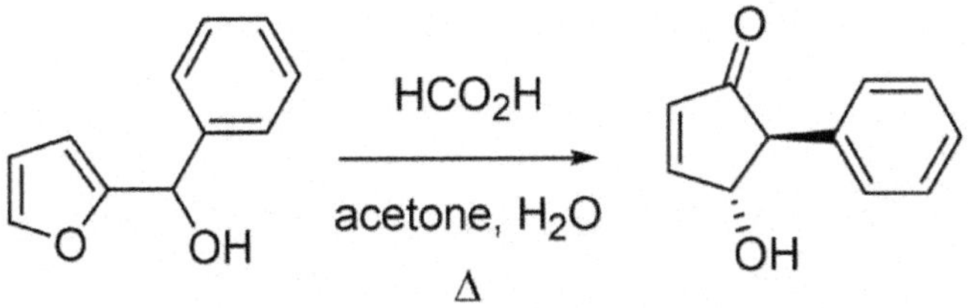

Scheme 27: The Piancatelli rearrangement (1976).

In the meantime, the terrorist violence that was shocking Italy culminated, in March 1978, with the kidnapping of Aldo Moro, the president of the Christian Democracy party (at the time still the major party), and the assassination of his five bodyguards by a commando of the "Brigate Rosse" (Red Brigades), a left wing paramilitary organization. Despite a number of efforts by the government to safely solve the issue, Moro was killed by his kidnappers nearly two months later. However, contrary to the extremists' expectations, this tragic event, and all the others to come up to early '80s, did not increase the popular adhesions to their ideologies. Furthermore, a number of important arrests beheaded the terroristic organizations which slightly dissolved alongside their utopian, and by then outdated, ideologies.

During the '80s Italy witnessed a tremendous economic renaissance. The Nation ranked fifth in the World's most developed countries due to the impressive growth of the gross domestic product and to the decrease of the inflation. For the first time in its history, Italy had a socialist prime minister, Bettino Craxi, who was elected in 1983. Two years later, Michail Gorbachev became president of the Soviet Union. He was the protagonist of a number of political and social reformations that shook his country and the entire communist bloc from their very roots. One of this process most touchy outcomes was the fall of the Berlin wall (November 1989) which, aside from physically separating the eastern from the western side of the German city, was the main icon of two opposite ideologies.

From a chemical standpoint, the '80s opened with a couple of olefination reactions. The first one was developed by Luciano Lombardo, an Italian citizen resident in Canberra, Australia, that in 1982 published the first account on the preparation and reactivity of what would later be known as the "Lombardo's reagent" [70]. This is a highly electrophilic reagent consisting of a $TiCl_4$/Zn/ CH_2Br_2 complex, which in its original version had been disclosed by Takai's group [71]. However, only when prepared according to the procedure reported by Lombardo this organometallic species is obtained in a very active form, enabling the mild methylenation of ketones. Noteworthy, these are not enolized by the reagent thus preventing isomerization issues (Scheme 28).

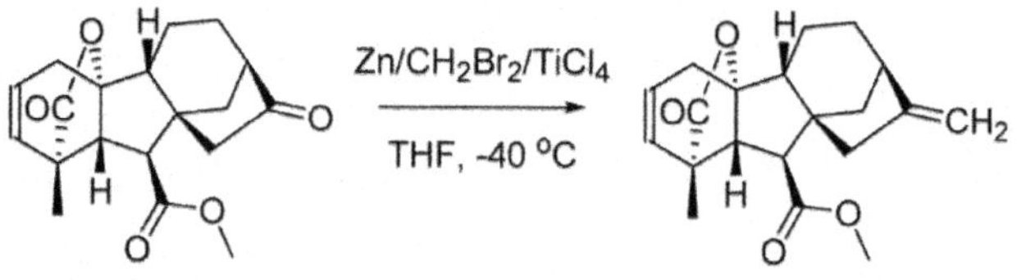

Scheme 28: The Lombardo olefination (1982).

A valuable modification of the Horner-Emmons olefination reaction leading to the selective synthesis of Z-unsaturated esters saw the light at the Columbia University (New York, NY, USA) in 1983. Authors of the work [72] were Clark Still and Cesare Gennari, who was later to become professor at the University of Milan. The Still-Gennari reaction (Scheme 29), as this transformation is referred to nowadays, uses electrophilic bis(trifluoroethyl) phosphonoesters and a strongly dissociated base system like KN(TMS)$_2$ and crown ether. The reaction is quite general in scope and it is one of the most cited methods for the synthesis of Z-olefins.

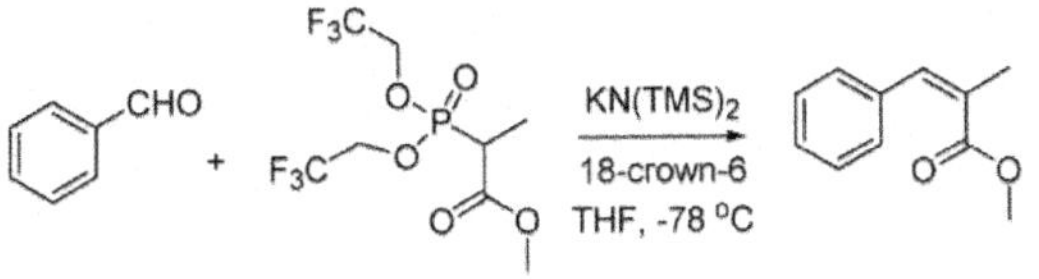

Scheme 29: The Still-Gennari reaction (1983).

In 1984 two groups independently published their results on the enantioselective oxidation of sulfides to sulfoxides. The first team was headed by Henri Kagan at the University of Orsay, France [73], and the second was based at the University of Padova and coordinated by Giorgio Modena [74]. The two methods are very similar and based on the Sharpless asymmetric epoxidation reaction [75], using titanium isopropoxide as the catalyst, diethyl tartrate (DET) as the chiral auxiliary, *tert*-butyl hydrogen peroxide as the oxidant. In both cases, the reaction is carried out below 0 °C. They basically differ for the solvent system employed that is dichloromethane and water in the Kagan's method and dichloroethane in the Modena's one. Accordingly, this oxidation procedure is generally referred to as the Kagan-Modena reaction (Scheme 30).

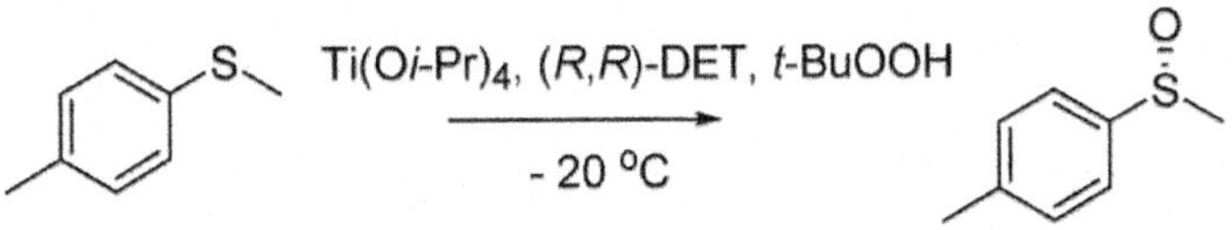

Scheme 30: The Kagan-Modena reaction (1984).

The Guarna-Brandi, sometimes called the Brandi reaction, consists in the formation of dihydro- or tetrahydropyridone derivatives in a one-pot two-step process involving the preparation of isoxazolines or isoxazolidines respectively, followed by their thermal rearrangement (Scheme 31). The reaction was disclosed by Antonio Guarna, Alberto Brandi and their co-workers at the University of Florence in two different moments. In its first version of 1985 [76], they exploited the cycloaddition of alkyl- or aryl-nitrile oxides with methylenecyclopropane to regioselectively synthesize isoxazolines that were thermally rearranged to 2-substituted dihydropyridine-4-ones. One year later [77] they further expanded the scope of this reaction by employing nitrones instead of nitrile oxides. The cycloaddition reaction is slightly less regioselective, but only the 5-spirocyclopropaneisoxazolidine isomer rearranges to the corresponding tetrahydropyridone.

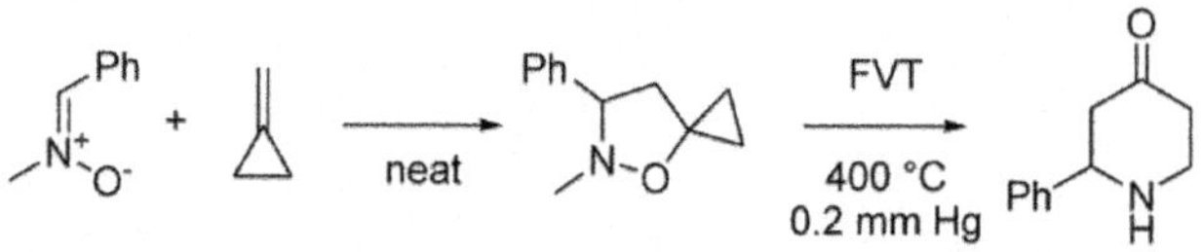

Scheme 31: The Guarna-Brandi reaction (1986).

In 1986 Alessandro Dondoni and co-workers at the University of Ferrara published an improved protocol on the use of 2-trimethylsilylthiazole in the homologation reaction of aldehydes [78], thus disclosing what is nowadays referred to as the Dondoni homologation reaction (Scheme 32). Previously, the team in Ferrara had already unveiled both the general potentialities of this reagent [79] and the aldehyde homologation sequence [80].

The reaction of 2-trimethylsilylthiazole (Dondoni's reagent) with chiral aldehydes proceeds with high diasteroselectivity, and, together with the subsequent thiazole degradation sequence, delivers the homologated aldehydes in high yield.

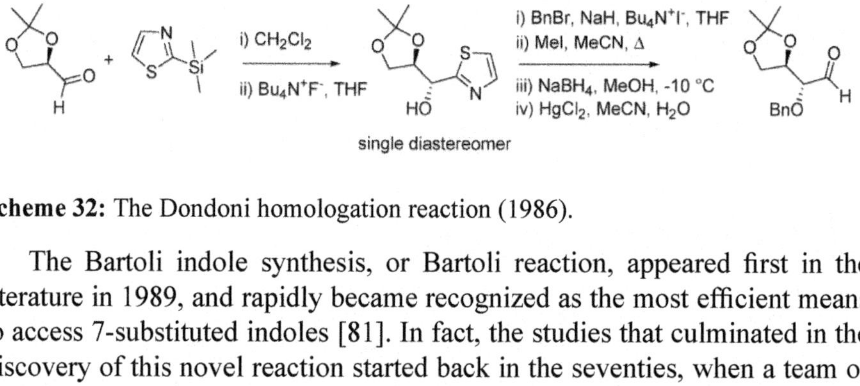

single diastereomer

Scheme 32: The Dondoni homologation reaction (1986).

The Bartoli indole synthesis, or Bartoli reaction, appeared first in the literature in 1989, and rapidly became recognized as the most efficient means to access 7-substituted indoles [81]. In fact, the studies that culminated in the discovery of this novel reaction started back in the seventies, when a team of the University of Bologna systematically investigated nitroarenes reactivity towards organometallic compounds. This work had led to the discovery of another reaction that sometimes is referred to as the Rosini or Rosini-Bartoli reductive nitroarene alkylation, where some benzo-fused nitro derivatives are reductively alkylated in a selective manner by means of Grignard reagents. In its first version [82] this reaction allowed to prepare alkyl nitroso derivatives, however, when performed in the presence of copper (I) iodide, the reaction led to the attainment of the more useful anilines [83] (Scheme 33).

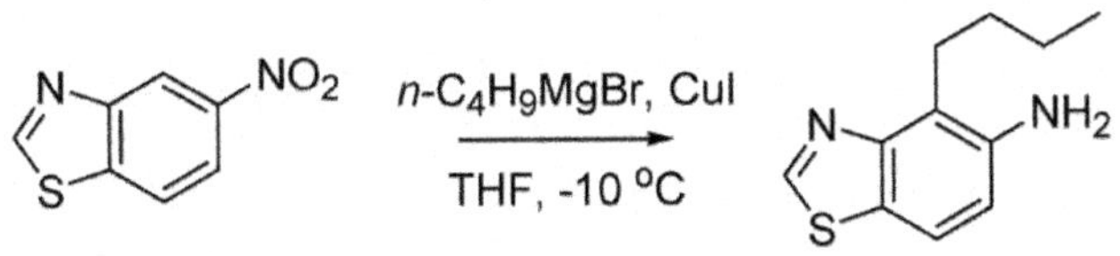

Scheme 33: The Rosini-Bartoli reductive nitroarene alkylation (1978).

A more popular variant of this reaction, the Bartoli indole synthesis, was discovered when an *ortho*-substituted nitroarene was reacted with a vinyl Grignard reagent. In this case the use of three moles of the organometallic species per mole of nitroarene smoothly leads to the formation of the corresponding 7-substituted indole in good yields [81] (Scheme 34).

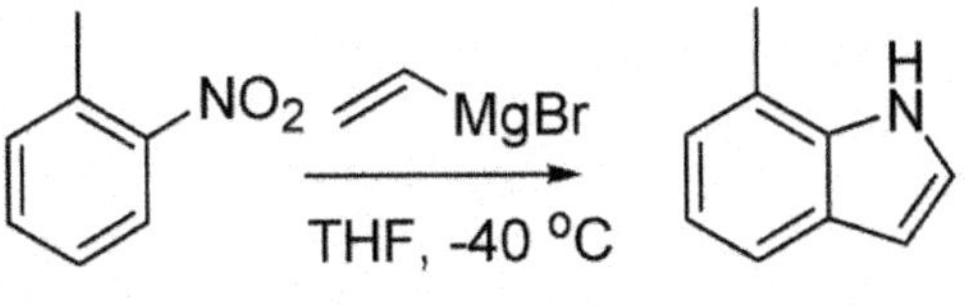

Scheme 34: The Bartoli indole synthesis (1989).

Indoles were also the target of the first Italian named reaction of the nineties, the Cacchi reaction (sometimes Arcadi-Cacchi reaction) [84], developed by Sandro Cacchi and co-workers at the University of Rome, providing an efficient entry into 2,3-disubstituted indoles. This is achieved starting from *o*-alkynyltrifluoroacetanilides that are reacted with vinyl triflates or aryl halides by means of a palladium-catalyzed reaction (Scheme 35).

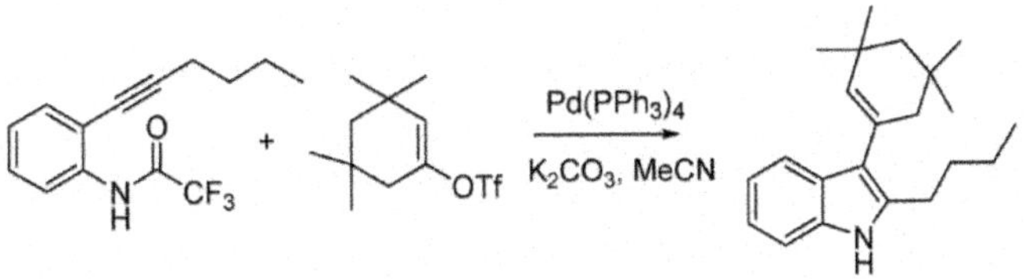

Scheme 35: The Cacchi reaction (1992).

Palladium chemistry was one of the workhorses of organic synthesis during the nineties, and besides the Cacchi reaction a further important contribution from Italian scientists would appear during the decade (*vide infra*). Before that, however, another hot topic of chemical research of the time witnessed the publication of a seminal paper by an Italian group. This is the chemistry of fullerenes, in which Maurizio Prato of the University of Trieste and his colleagues at Padua left their mark [85] thanks to the engagement of C_{60} in a 1,3-dipolar cycloaddition reaction, using azomethine ylides generated *in situ*, resulting in the corresponding fulleropyrrolidines (Scheme 36).

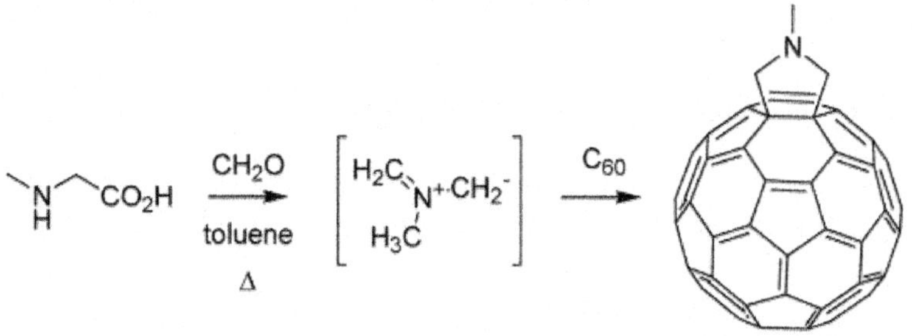

Scheme 36: The Prato reaction (1993).

In 1992, Italy was shocked by an unprecedented scandal that, starting from Milan, rapidly ran over the whole country. A number of kickbacks were discovered to be cashed in by the politicians in exchange for contracts. The politicians, in their turn, deposited part of the illegal incomes in their parties' accounts. The scandal that initially involved only minor characters within the parties, finally reached their summits, with a number of parties' secretaries recognized to be involved. The party which paid the highest price was the Socialist one whose leader and former Prime Minister Craxi resigned and left Italy for Tunisia, where he would die in 2000.

None of the economical sectors was exempted from the corruption: the pharmaceutical industry, by illegally financing the health minister and his managers for having the price of their drugs increased or falsely recognized as indispensable; the chemical industry, for the enormous amount of kickbacks linked to certain public and private companies' joint ventures maneuvers.

This scandal, however, had the merit to unearth the corrupted connections between the unscrupulous component of the private sector and part of the public institutions. The moralization wind which followed originated what was colloquially called the "Second Republic". By the nomination of high-profile rulers (e.g., the Prime Minister Carlo Azeglio Ciampi, elected in 1993, who would later become the 10 President of the Republic), the government tried to distance from the maladministration of the past and to regain the trust of the citizens. Between 1994 and 1997, a center-right alliance, a technical government and finally a center-left coalition succeeded in ruling the Nation [86].

In the twilight of the century, when the Italian chemical and pharmaceutical industry was still suffering the repercussions of the scandal that had gripped it a few years before, a highly significant reaction belonging to the universe of palladium-mediated organic synthesis was published by Marta Catellani and co-workers [87] at the University of Parma. The Catellani reaction provides a very elegant entry into the synthesis of *o,o*-disubstituted vinylarenes starting from aryl iodides. The reaction exploits a multicomponent protocol where, together with the reactants and catalyst, norbornene or another strained olefin is used. The latter is essential as it enters the complex catalytic cycle by activating three adjacent positions of the arene, being recycled at the end of the process (Scheme 37).

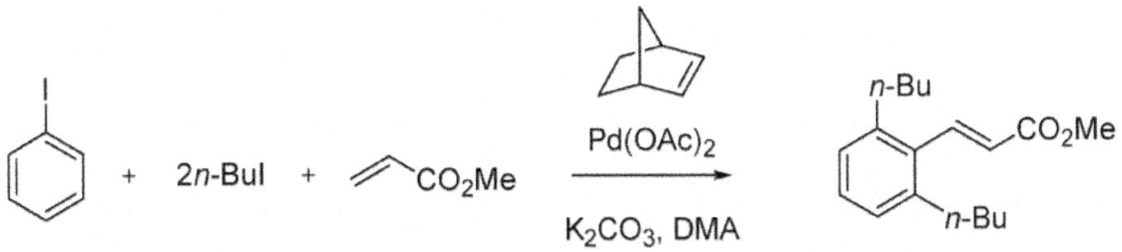

Scheme 37: The Catellani reaction (1997).

In December 1997, the Gazzetta Chimica Italiana published its last issue, thus ceasing its activity after 127 years of glorious history. Since 1998, in fact, most of the European Societies' journals have flowed into the two newly-founded European Journal of Organic Chemistry and European Journal of Inorganic Chemistry [88]. The last article of the Gazzetta, authored by Giorgio Montaudo, dealt with the contribution of Paternò and Cannizzaro (two of the founding members of the journal) to the discovery of the tetrahedral carbon (Figure 5). It was an elegant way of closing the circle.

Traveling back and forth through the animated landscape of about 150 years of Italian history, we have finally reached the twenty-first century, where our tale finishes. The present day is too close to be recounted with adequate objectivity, just as for current politics, as for newly discovered organic reactions. They are either still in the need to pass the test of time or simply awaiting for someone to christen them according to their inventors' names.

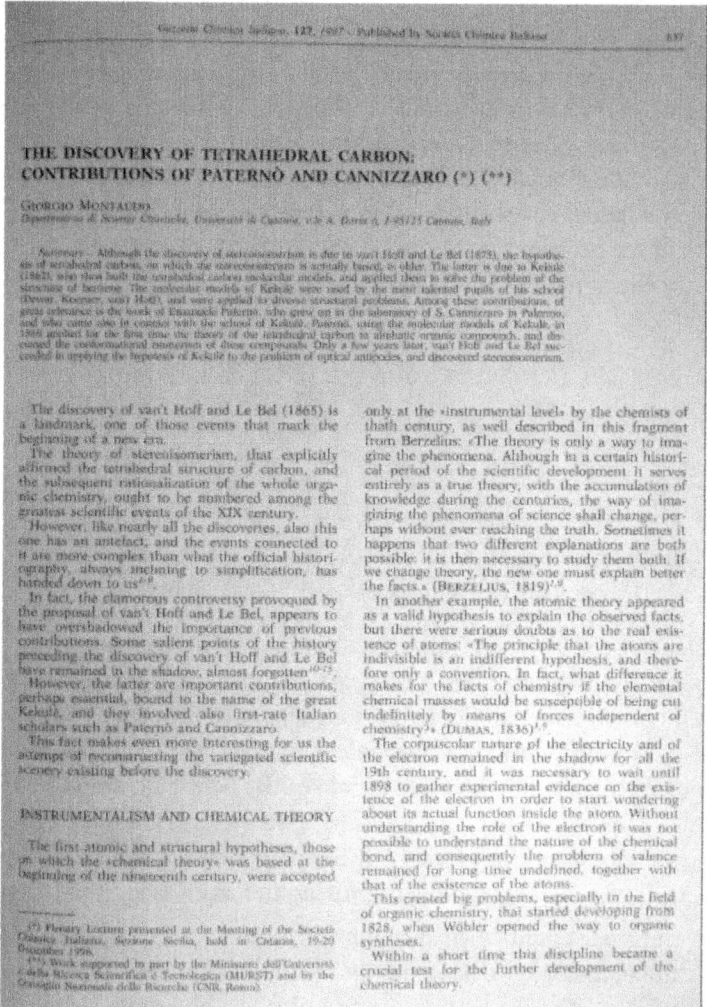

Figure 5. The first page of the last article published in the Gazzetta Chimica Italiana (**1997**, *127*, 837–842) (from the Nerviano Medical Sciences library).

CONCLUSIONS

We have discussed the contribution to the discovery of new chemical reactions by Italian scientists, whose names have been associated to these reactions. By walking through almost 150 years of Italian history, we have also aimed at contextualizing these discoveries within the brighter and darker moments of the country's history. Indeed, because of the historical in nature of this review, reactants and reaction conditions shown in the schemes were, whenever possible, taken from the original articles. Readers interested in the topic are encouraged to refer to recent literature for the developments, scopes and limitations of each chemical transformation reported. Finally, as stated in the introductory section, the "named organic reactions" chemical jargon is far from being undisputedly objective. Thus, we apologize to the chemists whose contributions may have involuntarily been omitted from this review. Oversights, misrecognitions and misattributions in science are fortunately infrequent but, being science a human act, they cannot be completely eradicated.

ACKNOWLEDGMENTS

Several people have constructively contributed to the ultimate version of the present review: Alessandra Badari, Daniele Donati, Eduard Felder, Arturo Galvani, Chiara Marchionni, Achille Panzeri and Francesca Quartieri. Many others have provided us with their encouragement. To all of them our deepest gratitude.

REFERENCES AND NOTES

1.   Hassner, A.; Namboothiri, I. *Organic Synthesis Based on Name Reactions*, 3rd ed.; Elsevier: Amsterdam, The Netherlands, 2011.

2.   Li, J.J. *Name Reactions: A Collection of Detailed Reaction Mechanisms*, 4th ed.; Springer-Verlag: Berlin, Germany, 2009.

3.   Kürti, L.; Czakó, B. *Strategic Applications of Named Reactions in Organic Synthesis*; Elsevier: Amsterdam, The Netherlands, 2005.

4.   Wang, Z. *Comprehensive Organic Name Reactions and Reagents*; Wiley: Hoboken, NJ, USA, 2009.

5.   Gal, J. The discovery of stereoselectivity at biological receptors: Arnaldo Piutti and the taste of the asparagine enantiomers—History and analysis on the 125th anniversary. *Chirality* 2012, *24*, 959–976.

6.   Cannizzaro, S. *Sulla vita e sulle opere di Raffaele*; Piria, E., Ed.; Loescher: Turin, Italy, 1883; Available online: http://books. google.it/books/about/Sulla_vita_e_sulle_opere_di_Raffaele_Pir. html?id=LloFRQAACAAJ&redir_esc=y (accessed on 5 August 2013).

7.   Piria, R. Studi sulla costituzione chimica dell'asparagina e dell'acido aspartico. *Il Cimento* 1846, *4*, 55–73.

8.   Piria, R. *Sull'azione del solfito d'ammoniaca sulla nitronaftalina e i suoi prodotti che da quella derivano*; Nistri: Pisa, Italy, 1850.

9.   Piria, R. Ueber einige Produkte Einwirkung des schwefligsauren

Ammoniaks auf Nitrophtalin. *Liebigs Ann.* 1851, *78*, 31–68.

10. Some Italian organic chemistry textbooks (*i.e.*, Fusco, R.; Bianchetti, G.; Rosnati, V. *Chimica Organica Volume 1* (Organic Chemistry Volume 1), 2nd ed.; CEA: Milano, 1974) refer to the "Piria reaction" to indicate the synthesis of aldehydes from the corresponding carboxylic acids by heating the latter in the presence of calcium formate and, by extension, the thermal decarboxylative coupling of calcium salts of carboxylic acid to form the corresponding simmetric ketones. These reactions were indeed first performed by Piria, who described the method in a letter to S. Cannizzaro dated 26 June 1854, where he acknowledge "some experience on ketones" by Williamson, who had predicted the transformation but never performed it (Williamson, A. Ueber aetherbildung (On ethers formation). *Liebigs Ann.* 1852,*81*, 73–78.

11. Bertagnini, C. Sulle combinazioni di alcuni oli essenziali con i solfiti alcalini. *Annali delle università toscane* 1851, 31–54.

12. Bertagnini, C. Ueber die verbindungen einiger flüchtigen oele mit den zweifach-schwefligsauren alkaline. *Liebigs Ann.*1853, *85*, 179–196.

13. Bertagnini, C. Ueber die verbindungen einiger flüchtigen oele mit den zweifach-schwefligsauren alkaline. *Liebigs Ann.*1853, *85*, 268–288.

14. Perkin, W.H. On the hydride of aceto-salicyl. *J. Chem. Soc.* 1868, *21*, 181–186.

15. Bertagnini, C. Produzione artificiale dell'acido cinnamico, ottenuto per lungo riscaldamento a 120°–130° di aldeide benzoica e cloruro di acetile anidri. *Nuovo Cimento* 1856, 46–48.

16. Bertagnini, C. Ueber die künstliche darstellung der zimmtsäure. *Liebigs Ann.* 1856, *100*, 125–127.

17. Chiozza, L. Ueber die künstliche bildung des cinnamylwasserstoffs. *Liebigs Ann.* 1856, *97*, 350–352.

18. Cannizzaro, S. Ueber den der benzoësäure entsprechenden alkohol. *Liebigs Ann.* 1853, *88*, 129–130.

19. Cannizzaro, S. *Sunto di un corso di filosofia chimica*; Sellerio: Palermo, Italy, 1991.

20. Schiff, H. Eine neue reihe organischer basen. *Liebigs Ann.* 1864, *131*, 118–119.

21. Schiff, H. Eine neue reihe organischer diamine. *Liebigs Ann.* 1866, *140*, 92–137.

22. Ciamician, G.L.; Dennstedt, M. Sull'azione del cloroformio sul composto potassico del pirolo. *Gazz. Chim. Ital.* 1881,*11*, 300–314.

23. Ciamician, G.L.; Dennstedt, M. Ueber die einwirkung des chloroforms auf die kaliumverbindung pyrrols. *Berichte der deutschen chemischen Gesellschaft* 1881, *14*, 1153–1168.

24. Ciamician, G.; Silber, P. Chemische Lichtwirkungen. *Berichte der deutschen chemischen Gesellschaft* 1900, *33*, 2911–2913.

25. Schönberg, A. *Preparative Organic Photochemistry*, 2nd ed.; Springer-Verlag: Berlin, Germany, 1968.

26. Ciamician, G.; Silber, P. Chemische Lichtwirkungen. *Berichte der deutschen chemischen Gesellschaft* 1908, *41*, 1928–1935.

27. Enciclopedia italiana Treccani. Available online: http://www.treccani.it/ enciclopedia/giacomo-ciamician_(Dizionario-Biografico)/ (accessed on 5 August 2013).

28. Ciamician, G.; Silber, P. Chemische Lichtwirkungen (Chemical effects of light). *Berichte der deutschen chemischen Gesellschaft* 1901, *34*, 2040–2046.

29. *La Fotochimica Dell'avvenire*; Available online: http://books.google. it/books?id=gQ_nAAAAMAAJ&q=ciamician+la+fotochimica+del l%E2%80%99avvenire&dq=ciamician+la+fotochimica+dell%E2% 80%99avvenire&hl=it&sa=X&ei=0bnSUbCPBa304QSjwICwAw- &ved=0CDQQ6AEwAA (accessed on 5 August 2013).

30. Plancher, G. II. Ricerche sull'azione dei ioduri alcolici sugli indoli. Sulla β-etil-β-N-dimetil-α-metilindolina. *Gazz. Chim. Ital.* 1898, *28*, 374–391.

31. Biginelli, P. Intorno ad uramidi aldeidiche dell'etere acetilacetico. *Gazz. Chim. Ital.* 1891, *21*, 497–500.

32. Biginelli, P. Intorno ad uramidi aldeidiche dell'etere acetilacetico. II. *Gazz. Chim. Ital.* 1891, *21*, 455–461.

33. Biginelli, P. Used for the detection of picric acid in biological liquids. *Ann. Chim. Appl.* 1924, *14*, 209–222.

34. Tron, G.C.; Minassi, A.; Appendino, G. Pietro Biginelli: The man behind the reaction. *Eur. J. Org. Chem.* 2011, *28*, 5541–5550.

35. Originally, he collaborated with professor Selmi to the compilation of the "Enciclopedia di Chimica scientifica e industriale" (Scientific and Industrial Chemistry Encyclopedia), whose publication commenced in 1868. Later he edited his own encyclopedia: "Nuova enciclopedia di chimica scientifica, tecnologica e industriale diretta dal prof. Icilio Guareschi con la collaborazione di distinti chimici italiani" (New Scientific, Technological and Industrial Chemistry Encyclopedia edited by Professor Icilio Guareschi with the collaboration of distinguished

Italian chemists), Torino: Unione Tipografico Editrice, 1915.

36. Guareschi, I. Sintesi di composti piridinici dagli eteri chetonici coll'etere cianacetico in presenza dell'ammoniaca e delle amine. *Mem. Reale Accad. Sci. Torino* 1896, *46*, 1–30.

37. Thole, F.B.; Thorpe, J.F. The formation and reactions of iminocompounds. Part XV. The production of imino-derivatives of piperidine leading to the formation of the ββ-disubstituted glutaric acids. *J. Chem. Soc. Trans.* 1911, *99*, 422–448.

38. Ponzio, G. Azione del tetrossido d'azoto sugli isonitrosochetoni. *Gazz. Chim. Ital.* 1897, *27*, 271–279.

39. Pellizzari, G. Nuova sintesi del triazolo e dei suoi derivati. *Gazz. Chim. Ital.* 1894, *24*, 222–229.

40. Ortoleva, G. Azione del jodio sull'acido cinnamico in soluzione piritica. *Gazz. Chim. Ital.* 1899, *29*, 503–509.

41. Ortoleva, G. Azione del jodio sull'acido malonico in soluzione piridica. *Gazz. Chim. Ital.* 1900, *30*, 509–514.

42. King, L.C. The reaction of iodine with some ketones in the presence of pyridine. *J. Am. Chem. Soc.* 1944, *66*, 894–895.

43. Betti, M. Sull'addizione di basi aldeido-aminiche ai naftoli. *Gazz. Chim. Ital.* 1900, *30*, 301–309.

44. Paolocci, N.; Wink, D.A. The shy Angeli and his elusive creature: The HNO route to vasodilatation. *Am. J. Physiol. Heart Circ. Physiol.* 2009, *296*, H1217–H1220.

45. Angeli, A. Sopra la nitroidrossilammina. *Gazz. Chim. Ital.* 1896, *26*, 17–28.

46. Angeli, A.; Angelico, F. Sopra l'acido nitroidrossilamminico. *Gazz. Chim. Ital.* 1900, *30*, 593–595.

47. Rimini, E. Sopra una nuova reazione delle aldeidi. *Gazz. Chim. Ital.* 1901, *31*, 84–93.

48. Young, E.G.; Conway, C.F. On the estimation of allantoin by the Rimini-Schryver reaction. *J. Biol. Chem.* 1942, *142*, 839–853.

49. Rimini, E. Sul riconoscimento della formaldeide negli alimenti. *Ann. di Farmacot. e Chim.* 1898, 97–101.

50. Schryver, S.B. The photochemical formation of formaldehyde in green plants. *Proc. R. Soc. Lond. Ser. B* 1910, *82*, 226–232.

51. Link, G. Verfahren zur Darstellung von Oxy-i-butyryl-Phenolen. German Patent 80,986, 1894.

Italian Chemists' Contributions to Named Reactions in Organic....   159

52. Bargellini, G. Azione del cloroformio e idrato sodico sui fenoli in soluzione nell'acetone. *Gazz. Chim. Ital.* 1906, *36*, 329–338.

53. Paternò, E.; Chieffi, G. Sintesi in chimica organica per mezzo della luce. Nota II. Composti degli idrocarburi non saturi con aldeidi e chetoni. *Gazz. Chim. Ital.* 1909, *39*, 341–361.

54. Büchi, G.; Inman, C.G.; Lipinsky, E.S. Light-catalyzed Organic Reactions. I. The reaction of carbonyl compounds with 2-methyl-2-butene in the presence of ultraviolet light. *J. Am. Chem. Soc.* 1954, *76*, 4327–4331.

55. Gastaldi, C. Sulle pirazine (On pyrazines). *Gazz. Chim. Ital.* 1921, *51*, 233–255.

56. Passerini, M. Sopra gli isonitrili (I). Composto del *p*-isonitril-azobenzolo con acetone ed acido acetico. *Gazz. Chim. Ital.* 1921, *51*, 126–129.

57. Passerini, M. Sopra gli isonitrili (II). Composti con aldeidi o con chetoni ad acidi organici monobasici. *Gazz. Chim. Ital.* 1921, *51*, 181–189.

58. Amadori, M. Prodotti di condensazione tra il glucosio e la para fenetidina. Parte I. *Atti della Accademia Nazionale dei Lincei, Classe di Scienze Fisiche, Matematiche e Naturali, Rendiconti* 1925, *2*, 337–342.

59. Mascarelli, L. Contributo alla conoscenza del bifenile e dei suoi derivati.— Nota XV. Passaggio dal sistema bifenilico a quello fluorenico. *Gazz. Chim. Ital.* 1936, *66*, 843–850.

60. Hargittai, I.; Comotti, A.; Hargittai, M. Giulio Natta. *Chem. Eng. News* 2003, *81*, 26–28.

61. Hargittai, I.; Hargittai, M. Giulio Natta: A complex portrait. *Chem. Eng. News* 2003, *81*, 8–10.

62. Girelli, A. I chimici e il fascismo. *La Chimica e l'Industria* 2003, *85*, 26–27.

63. Natta, G. Una nuova classe di polimeri di alfa-olefine aventi una eccezionale regolarità di strutture. *Atti Acc. Naz. Lincei Mem.* 1955, *4*, 61–71.

64. Ziegler, K. Neuartige katalytische Umwandlungen von Olefinen. *Brennst. Chem.* 1952, *33*, 193–200.

65. Berti, G. The Pyrolysis of Sulfites. III. Methyl Alkyl Sulfites. A new method for the preparation of olefins. *J. Am. Chem. Soc.* 1954, *76*, 1213–1219.

66. Chugaev, L. Uber eine neue methode aur darstellung ungesättigter kohlenwasserstoffe. *Berichte der deutschen chemischen Gesellschaft* 1899, *32*, 3332–3335.

67. Caglioti, L.; Grasselli, P. A new method for the reduction of aldehydes

and ketones to CH_2 under mild conditions.*Chem. Ind. Lond.* 1964, 153.

68. Minisci, F.; Bernardi, R.; Bestini, F.; Galli, R.; Perchinummo, M. Nucleophilic character of alkyl radicals-VI: A new convenient selective alkylation of heteroaromatic bases. *Tetrahedron* 1971, *27*, 3575–3579.

69. Piancatelli, G.; Scettri, A.; Barbadoro, S. A useful preparation of 4-substituted 5-hydroxy-3-oxocyclopentene.*Tetrahedron Lett.* 1976, *17*, 3555–3558.

70. Lombardo, L. Methylenation of carbonyl compounds with $Zn-CH_2Br_2$-$TiCl_4$. Applications to gibberellins. *Tetrahedron Lett.* 1982, *23*, 4293–4296.

71. Takai, K.; Hotta, Y.; Oshima, K.; Nozaki, H. Effective methods of carbonyl methylenation using CH_2I_2-Zn-Me_3Al and CH_2Br_2-Zn-$TiCl_4$ system. *Tetrahedron Lett.* 1978, *19*, 2417–2420.

72. Still, W.C.; Gennari, C. Direct synthesis of Z-unsaturated esters. A useful modification of the Horner-Emmons olefination. *Tetrahedron Lett.* 1983, *24*, 4405–4408.

73. Pitchen, P.; Kagan, H.B. An efficient asymmetric oxidation of sulfides to sulfoxides. *Tetrahedron Lett.* 1984, *25*, 1049–1052.

74. Di Furia, F.; Modena, G.; Seraglia, R. Synthesis of chiral sulfoxides by metal-catalyzed oxidation with *t*-butyl hydroperoxide. *Synthesis* 1984, 325–326.

75. Katsuki, T.; Sharpless, K.B. The first practical method for asymmetric epoxidation.*J. Am. Chem. Soc.* 1980, *102*, 5974–5976.

76. Guarna, A.; Brandi, A.; Goti, A.; de Sarlo, F. 4,5-dihydroisoxazole-5-spirocyclopropanes. Synthesis and thermolytic rearrangement to 5,6-dihydro-4-pyridones. *J. Chem. Soc. Chem. Commun.* 1985, 1518–1519.

77. Brandi, A.; Guarna, A.; Goti, A.; de Sarlo, F. Rearrangement of nitrose cycloadducts to methylene cyclopropane. Synthesis of indolizidine and quinolizidine derivatives. *Tetrahedron Lett.* 1986, *27*, 1727–1730.

78. Dondoni, A.; Fantin, G.; Fogagnolo, M.; Medici, A. Synthesis of long-chain sugars by iterative, diastereoselective homologation of 2,3-*o*-isopropylidene-d-glyceraldehyde with 2-trimethylsilylthiazole. *Angew. Chem. Int. Ed.* 1986, *25*, 835–837.

79. Medici, A.; Fantin, G.; Fogagnolo, M.; Dondoni, A. Reactions of 2-trimethylsilylthiazole with acyl chlorides and aldehydes synthesis of new thiazol-2-yl derivatives. *Tetrahedron Lett.* 1983, *24*, 2901–2904.

80. Dondoni, A.; Fogagnolo, M.; Medici, A. Pedrini, P. Diastereoselectivity

in the 1,2-addition of silylazoles to chiral aldehydes. Stereocontrolled homologation of α-hydroxyaldehydes. *Tetrahedron Lett.* 1985, *26*, 5477–5480.

81. Bartoli, G.; Palmieri, G.; Bosco, M.; Dalpozzo, R. The reaction of vinyl Grignard reagents with 2-substituted nitroarenes: A new approach to the synthesis of 7-substituted indoles. *Tetrahedron Lett.* 1989, *30*, 2129–2132.

82. Bartoli, G.; Rosini, G. Reductive alkylation of 6-nitrobenzothiazoles with Grignard reagents: Synthesis of 7-alkyl-6-nitrosobenzothiazoles. *Synthesis* 1976, 270–271.

83. Bartoli, G.; Medici, A.; Rosini, G.; Tavernari, D. Reductive C-Alkylation of nitroarenes with Grignard reagents; Synthesis of alkyl-amino-arenes. *Synthesis* 1978, 436–437.

84. Arcadi, A.; Cacchi, S.; Marinelli, F. A versatile approach to 2,3-disubstituted indoles through the palladium-catalysed cyclization of *o*-alkynyltrifluoroacetanilides with vinyl triflates and aryl halides. *Tetrahedron Lett.* 1992, *33*, 3915–3918.

85. Maggini, M.; Scorrano, G.; Prato, M. Addition of azomethine ylides to C_{60}: Synthesis, characterization, and functionalization of fullerene pyrrolidines. *J. Am. Chem. Soc.* 1993, *115*, 9798–9799.

86. The vast majority of the historical information reported in this review has been taken from: Montanelli, I. In *Storia d'Italia*; RCS Libri S.p.A.: Milano, 2004; This monumental twelve-volume work is considered one of the most popular and authoritative divulgation books about Italian history.

87. Catellani, M.; Frignani, F.; Rangoni, A. A complex catalytic cycle leading to a regioselective synthesis of *o,o* -disubstituted vinylarenes. *Angew. Chem. Int. Ed.* 1997, *36*, 119–122.

88. Gennari, C.; Peruzzini, M. The Italian chemical society is 100 years old. *Eur. J. Org. Chem.* 2009, *19*, 3095–3097.

Chapter 6

SYNTHESIS AND REACTIONS OF ACENAPH-THENEQUINONES-PART-2. THE REACTIONS OF ACENAPHTHENEQUINONES

El Sayed H. El Ashry, Hamida Abdel Hamid, Ahmed A. Kassem and Mahmoud Shoukry

Department of Chemistry, Faculty of Science, Alexandria University, Alexandria, Egypt

ABSTRACT

The reactions of acenaphthenequinone and its derivatives with different nucleophiles, organic and inorganic reagents are reviewed. This survey also covers their oxidation and reduction reactions, in addition to many known reactions such as Friedel Crafts, Diels-Alder, bromination and thiolation.

INTRODUCTION

The broad spectrum of applications of acenaphthenequinone and its derivatives as biologically active compounds, dyes, etc has prompted us to review their chemistry and uses. The syntheses of acenaphthenequinone (1) and its derivatives, which are based mainly on the use of starting materials having the carbon skeleton of 1 and their reactivity towards nitrogen nucleophiles have been discussed in the first part of this series [1]. In this part, the reactions of acenaphthenequinones are reviewed.

REACTIONS OF ACENAPHTHENEQUINONES

Ring opening and Enlargement

Ring cleavage of acenaphthenequinone (1) with an aqueous potassium hydroxide solution in dimethylsulphoxide at room temperature led to formation of 1, 8-naphthaldehydic acid (2) which exists in equilibrium with the corresponding cyclic structure [2,3,4]. The reaction of 1 in aqueous alkali gave 2,3-dimethylbenzoic acid (3) [5]. The alkaline permanganate oxidation of 1 gave 2,6-dicarboxy-phenylglyoxylic acid [6]. On the other hand, oxidation of 1 with molecular oxygen in propionic acid containing a homogeneous catalyst such as cobalt (II) acetate or manganese (II) acetate gave 1,8-naphthalic anhydride (6) [7]. Addition of potassium bromide to this reaction mixture increased the rate of reaction. A similar transformation was effected by oxygen in the presence of copper (I) chloride and pyridine [8]. On the other hand, the oxidative cleavage of 1 by the oxygen adduct of cobaltocene gave cobaltocinium carboxylate (4) that upon reaction with hydrogen chloride in ether, acid halides or dialkyl sulphates gave naphthalene-1,8-dicarboxylic acid (5), or its anhydride or ester, respectively [9]. When ozonolyses of vinyl ethers were conducted in presence of 1, it afforded 6 in addition to unreacted 1 [10]. This was attributed to the transfer of an oxygen atom from the carbonyl oxides, generated from the vinyl ethers, to 1 to give a Baeyer-Villiger type product. Heating of 1 with sodamide and treating with water gave naphthalene and oxalamide [11].

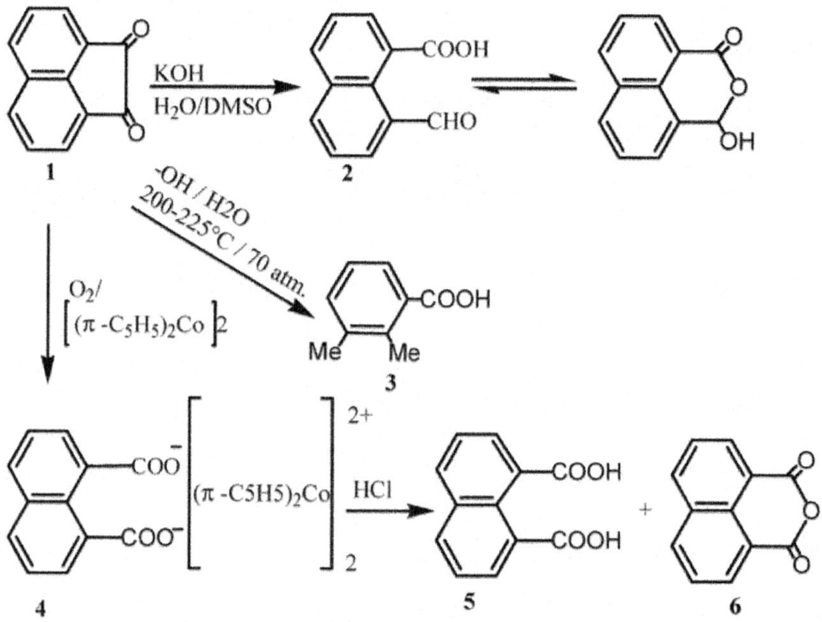

Scheme 1:

Photolysis of acenaphthenequinone in methylene chloride saturated with oxygen [12,13] gave 1,8-naphthalic anhydride (6). When an olefin such as cyclohexene was included in the reaction, it was converted to a mixture of oxidized products consisting, mainly the allylic hydroperoxide (7), epoxide

(8), and adipaldehyde (9) in addition to 6. The quantum efficiency for quinone oxidation was independent of quinone and olefin concentrations. A mechanism was suggested in which an initial reaction between excited quinone and oxygen resulted in covalent bond formation whose subsequent rearrangement accounted for the formation of the products.

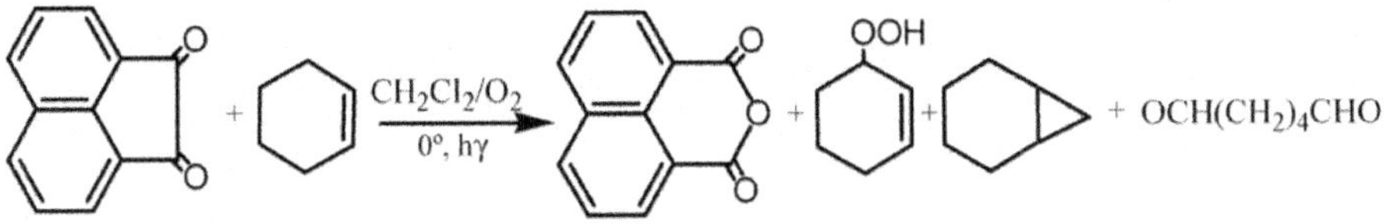

Scheme 2:

The acenaphthenequinone was cleaved electrochemically in presence of oxygen to give, after methylation, the corresponding ester of 1,8-naphthalene dicarboxylic acid [14]. Schmidt rearrangement of 1 with sodium azide gave naphthalic anhydride [15]. Reaction of 1 with the diazoalkanes 10 or 11 yielded 3-substituted-2-perinaphthen-2-o1-1-one 12or 13, respectively (Scheme 3) [16]. The products could be extracted from the reaction mixture with dilute aqueous alkali in order to prevent undesirable side reactions [17] which led to difficulties in isolating the products.

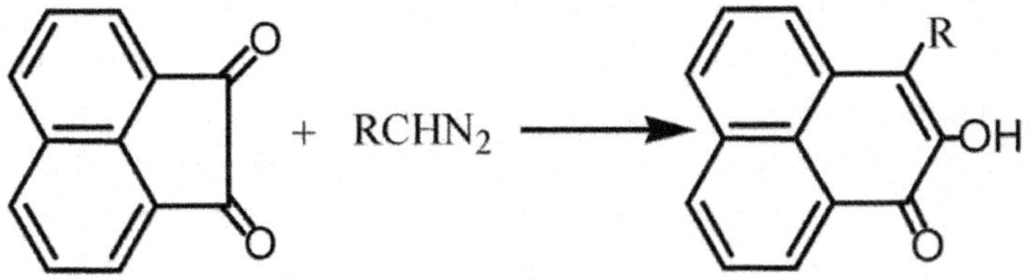

Scheme 3:

The cyanohydrin 14 undergoes facile base-catalyzed carbon-to-oxygen acy1 rearrangement to peri ring-expanded naphthalides 15 [18]. The proposed mechanism (Scheme 4) involved base-catalyzed formation of an intermediate α-oxanol followed by bridgehead carbon-carbon bond cleavage to an aromatic carbanion isoelectronic with the 14 π-electron phenalenyl carbanion. The reaction could also be extended to other analogues of 14 where the CN is replaced by other substituents.

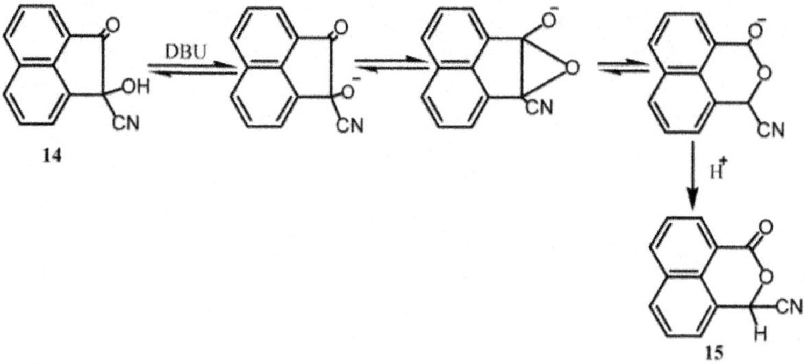

Scheme 4:

Reduction

Acenaphthenequinone is easily reduced as a consequence of the involvement of its carbonyl groups in the conjugated system [19]. Treatment of 1 with iron in acetic acid, until a water soluble colorless compound is formed, yields easily soluble alkali salts, which are of a violet-blue color, in the presence of an excess of caustic alkali. Condensing the reduced products with 3-hydroxy-1-thionaphthene or indoxyl derivatives gave vat-dyeing materials [20]. Reduction products of acenaphthenequinone were obtained by confining the reduction to the formation of compound 16 which is poorly soluble in water and yields with alkalies deep blue salts which are also poorly soluble in water (Scheme 5). On the other hand reduction of 1 could be carried until the formation of 17, which is soluble in water. It forms with excess of alkalies readily soluble violet-blue salts [21,22]. When acenaphthenequinone absorbed five moles of hydrogen, in the presence of platinum in aqueous ammonium hydroxide or dilute alkali [23], it yielded exclusively the bimolecular substance 18. Its catalytic hydrogenation in presence of nickel salts was also studied [24].

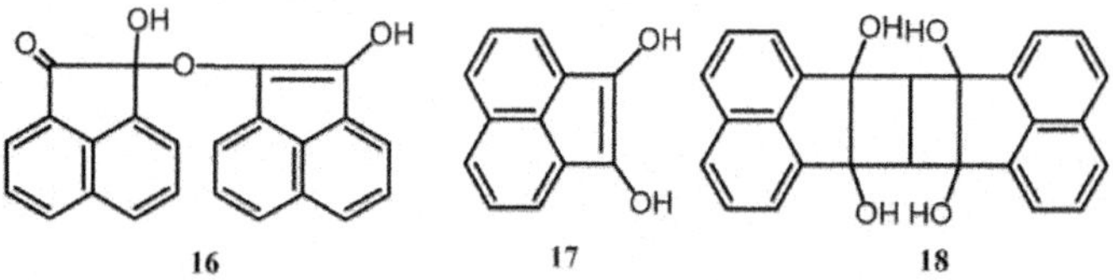

Scheme 5:

Clemmensen reduction of 1 with amalgamated zinc in hydrochloric acid gave acenaphthene 19 (Scheme 6) [25]. When the reduction was carried out with amalgamated sodium in ethanol, in an atmosphere of nitrogen, it gave 38% of the transglycol21 [26]. The product did not give a condensation product with acetone and it did not decolorize bromine in warm chloroform. Catalytic reduction of 1 in presence of platinum in ethanol gave a mixture of *cis-* and *trans* acenaphthylene glycols (20 and 21). Reduction with LiAlH$_4$ gave also the trans diol accompanied by the *cis* diol whose derivatives were prepared [27]. The *cis*-diol could be prepared by selenium dioxide oxidation of acenaphthene [25].

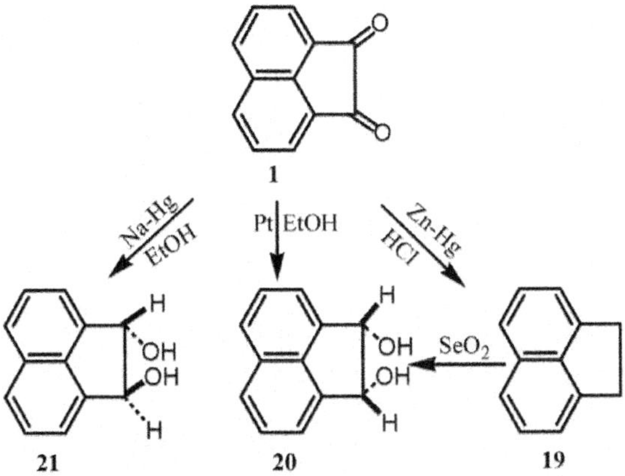

Scheme 6:

Treatment of 1 with alkali metals e.g. sodium and potassium in tetrahydrofuran gave three reduced forms [28], which behave like a monovalent, a divalent, and a trivalent base, respectively. It was found that tris(triphenylphosphine) chlororhodium is an effective catalyst for the homogeneous reductive hydrosilylation of quinones [29] which offer an easy procedure for protecting the highly reactive quinonic moiety. Thus, reductive silylation of 1 with Et$_3$SiH over tris(triphenylphosphine)-chlororhodium as a catalyst gave the bis(silyl) ethers of the hydroquinone which could be oxidatively desilylated with PhI(OAc)$_2$ [30]. The respective 1,2-bis(trimethylsiloxy)ethene analogue was prepared from reaction of 1 with hexamethyldisilane in presence of Pd or Pt catalyst [31].

Electrochemical reductions of 1 at a mercury cathode were carried out under a constant potential, in presence of nonelectroactive aroyl chlorides to give the 1,2-diaroyloxyacenaphthylene derivatives 22, in good yields (Scheme 7) [31,33]. Their formation corresponds to the transfer of an overall two-electron process. However, when acetic anhydride was used, a one-electron transfer process had taken place to give meso-bis (1-acetoxy-2-oxoacenaphthen-1-yl) (23). The structure of the last compound was determined by x-ray crystallography [32]. The effect of metal ions and solvents on the polarographic reduction of 1 was studied [34]. The dependence of limiting currents and half-wave potentials were determined [35]. The mechanism and kinetics of the polarographic reduction of 1 in DMF and in the presence of phenol as proton donor was found to involve 4-electrons in successive 1-electron steps [36]. Reductive methylation of 1 had taken place electrochemically in presence of methylhalides via coupling of the radical anion of 1 with the methyl radical [37]. Electrochemical reduction in DMF-Bu$_4$NI gave a binucleophile 24, which underwent cyclization with 25 to give heterocyclic macrocycles 26 [38].

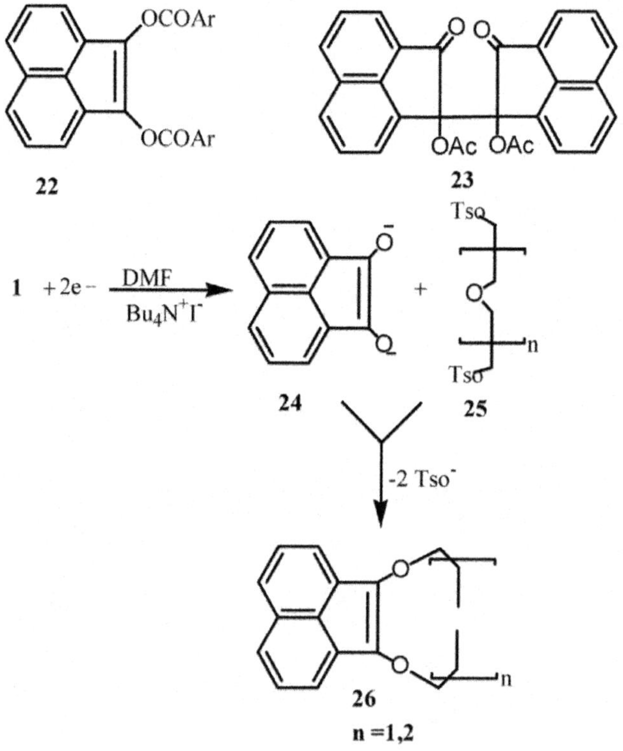

Scheme 7:

The three stereoisomers 27-29 of the six possible dodecahydroacenaphthylene were prepared (Scheme 8)[39]. The configurations were confirmed *inter alia* by X-ray analysis of the precursors.

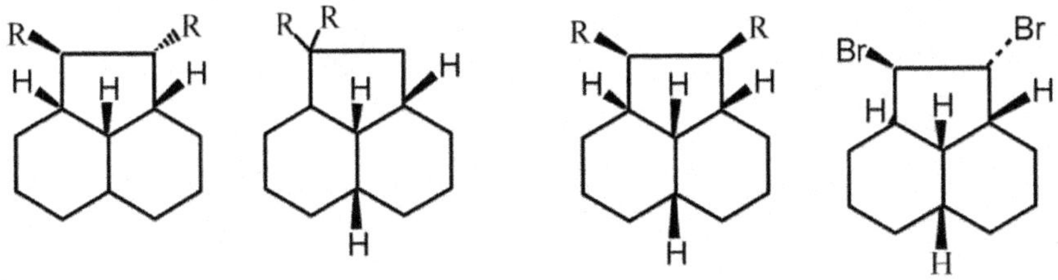

27 R = H   28 R, R = H, H   29 R = H   30
27a R = OAc   28a R, R = Me—⟨ ⟩—S-NHN   29a R = OH

Scheme 8:

Protonation

Protonation of acenaphthenequinone (1) gave a diprotonated species [40], the structure of which was determined as 31 by[1]H- and ^{13}C-NMR. The relative photochemical reactivity of acenaphthenequinone as an α-diketone was investigated in hydrogen donating solvents [41].

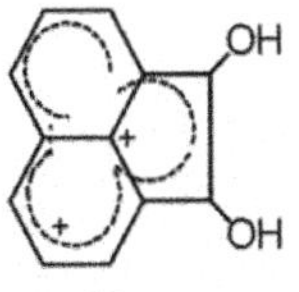

31

Reaction with active Methylene Compounds

Reaction of 1 with malononitrile gave 1-(dicyanomethylene) acenaphthen-2-one (32) whose reaction with hydrazine gave33 that hydrolyzed with sulfuric acid to give 35 (Scheme 9) [42]. Reaction of 32 with substituted hydrazines gave deeply colored hydrazones 34, which are classified as azacyanine types polymethine dyes [43]. Reaction of 1 with malononitrile in presence of bases yielded a blue compound namely 6β-hydroxy-8-imino-7, 8-dihydro-6β-*H*-cyclopenta[a] acenaphthylene-7,7,9-tricarbonitrile [44].

Reaction of 1 with o-phenylene diacetonitrile in presence of piperidine at room temperature [45,46] gave the dinaphthylenenitrile amide (36) and not the expected dinitrile 37 (Scheme 10). The product could not be hydrolyzed to the dicarboxylic acid; hydrolysis ceases at the diamide stage [47]. However, the cyclocondensation of o-phenylene diacetonitrile with 1 was reported in a more recent publication to give 37 that followed by decyanation to give benzo[k] fluoranthene (38) [48].

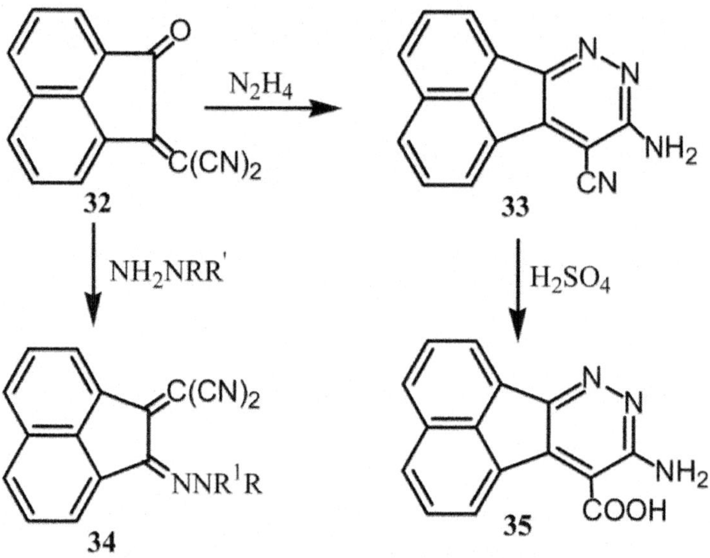

Scheme 9:

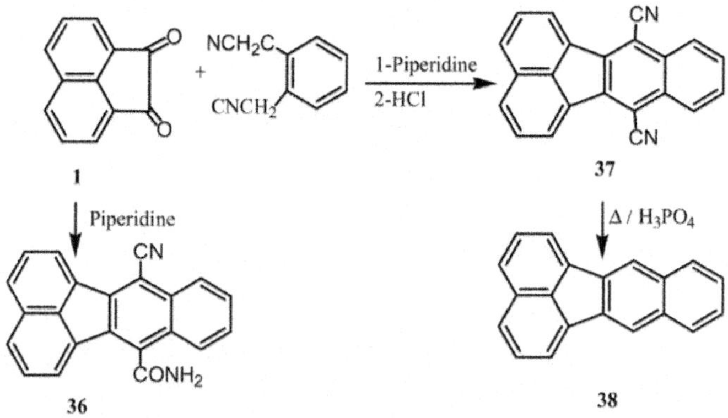

Scheme 10:

Condensation of ethyl cyanoacetate and malonic acid with 1 gave the acid 39a and the ester 39b, respectively (Scheme 11) [49]. Esterification of 40a gave the corresponding ester 40b. Dehydration of 40a gave 39a.

Hydrogenation of 39b in presence of Adams' catalyst gave 41. Reduction of 41 gave either the lactone 42 (R=CN) or the hemiacetal 43 depending upon the conditions employed. The lactone (42, R=H) was prepared by reaction of hemiacetal 43 with hot alkali followed by acidification. Claisen-Stobbe condensation of 1 with phenylacetic esters afforded the benzylidene derivatives 44 [50].

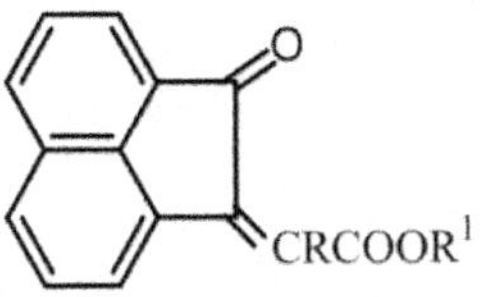

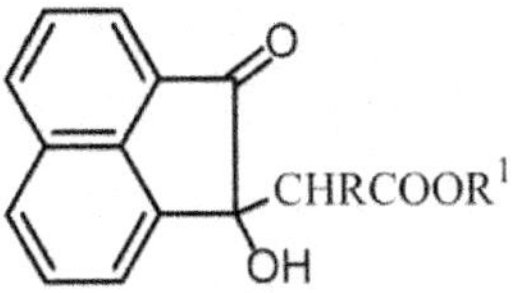

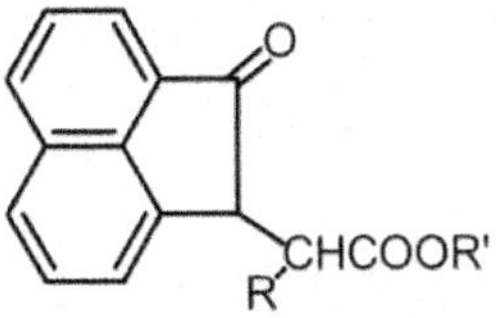

39a R = R^1 = H
39b R = CN, R^1 = Et

40a R = R^1 = H
40b R = H, R^1 = Et

41 R = CN, R^1 = Et

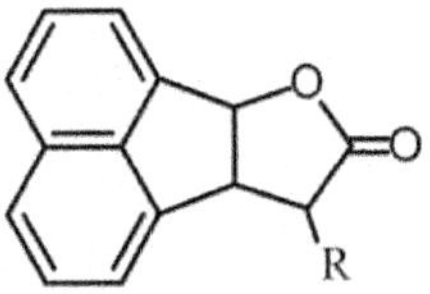

42

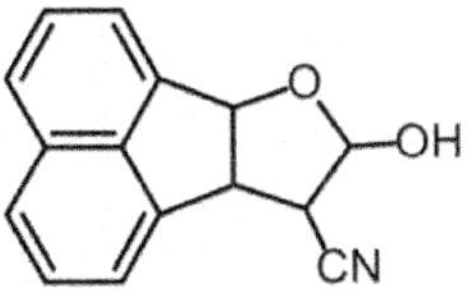

43

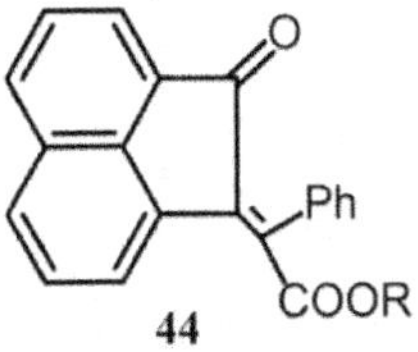

44

Scheme 11:

Reaction of 1 with diethylacetone dicarboxylate gave the substituted cyclopentadienone (45) (Scheme 12) [51]. Reduction of 45 by zinc and acetic acid followed by hydrolysis and decarboxylation afforded the ketones 46, 47 and 48.

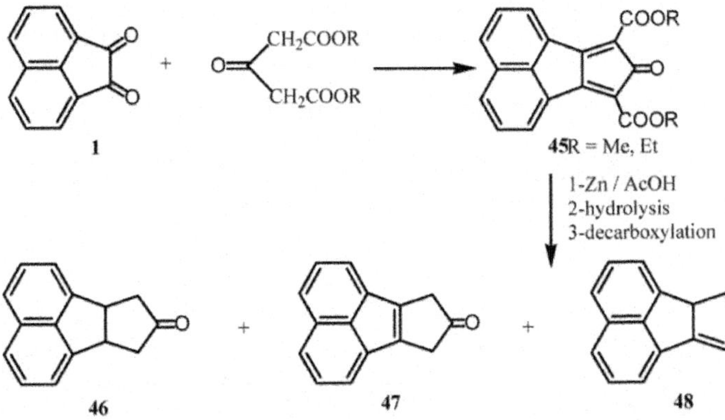

1

45 R = Me, Et

1-Zn / AcOH
2-hydrolysis
3-decarboxylation

46   **47**   **48**

Scheme 12:

The ketone 46 is a valuable a precursor for the synthesis of peri-diketone 52 via the conversions to 49-52 (Scheme 13).

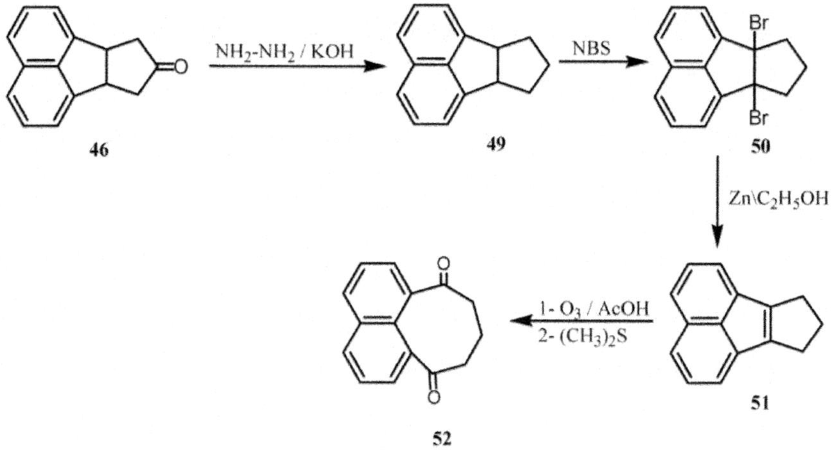

Scheme 13:

Condensation of acenaphthenequinone (1) with dimethyl pentenedioate (dimethyl glucatonate) [52] gave the epimeric diesters 53 and 54 (Scheme 14). The diester 53 was oxidized by lead tetraacetate to 55 whose reduction was effected by magnesium in methanol in order to reduce the conjugated 8,9double bond while preserving the keto-ester groups, whereby compounds 56, 57, and 58 were obtained.

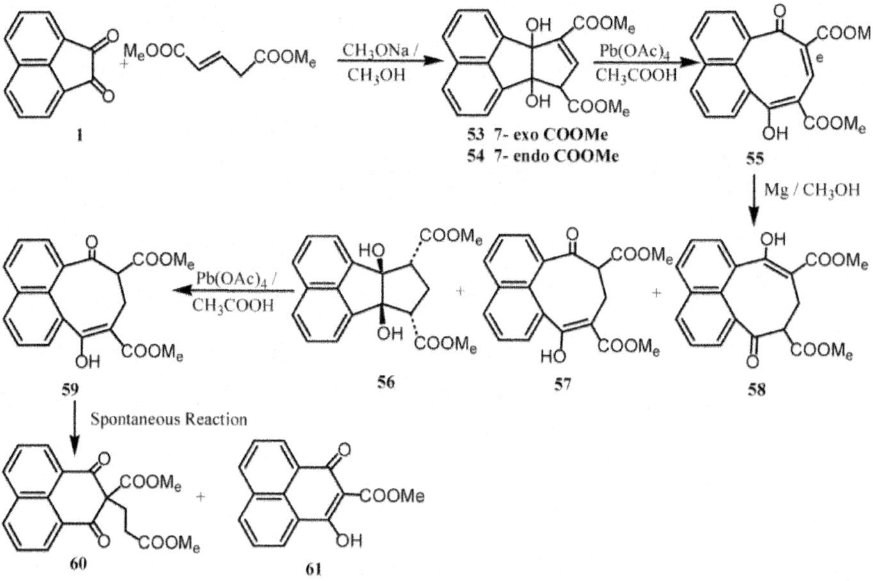

Scheme 14:

Reactionof56withleadtetraacetategave59,whichcouldbeaprecursorforthe peri-diketone 52, however, 59 was spontaneously transformed to 60 and 61 and consequently this approach for 52 was precluded. Acenaphthenequinone (1) was condensed with $S(CH_2CO_2Et)_2$ or $p\text{-}NO_2BnSCH_2CO_2Et$ in the presence of base to give tetrahydroacenaphthothiophenes 62a and 62b (Scheme 15). Dehydration of 62a in sulfuric acid or acetic anhydride gave acenaphtho[1,2-c] thiophene (64a) [53]. Heating of 64a with copper and quinoline gave 65 [54]. On the other hand, when 1was treated with $O(CH_2CO_2Et)_2$, it gave 63, which upon reaction with acetic anhydride gave 66 [55].

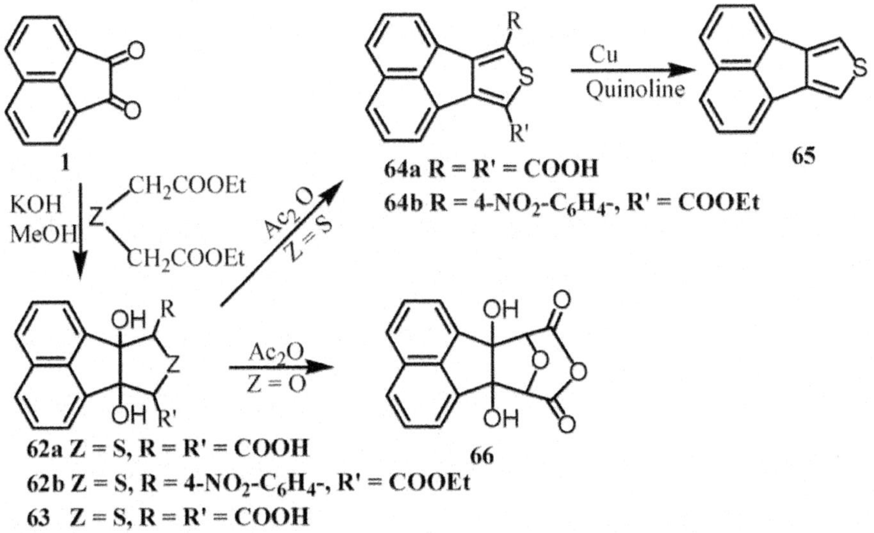
1
KOH
MeOH
Z
CH₂COOEt
CH₂COOEt
Ac₂O
Z = S
R
S
R'
Cu
Quinoline
S
64a R = R' = COOH
64b R = 4-NO₂-C₆H₄-, R' = COOEt
65
OH
R
Z
OH
R'
Ac₂O
Z = O
OH
O
O
O
OH
O
62a Z = S, R = R' = COOH
66
62b Z = S, R = 4-NO₂-C₆H₄-, R' = COOEt
63 Z = S, R = R' = COOH

Scheme 15:

Condensation of 1 with 1,3-indandione (Scheme 16) gave 2,2-bis(1,3-indandion-2-yl) acenaphthene-1-one (67) [56,57,58]. The reaction of 1 and 2,2-dihydroxy-1,3-phenylenedione gave 1,8-naphthalic anhydride in high yield [59]. Condensation of 1with the appropriate 2-propanones gave 68 [60,61].

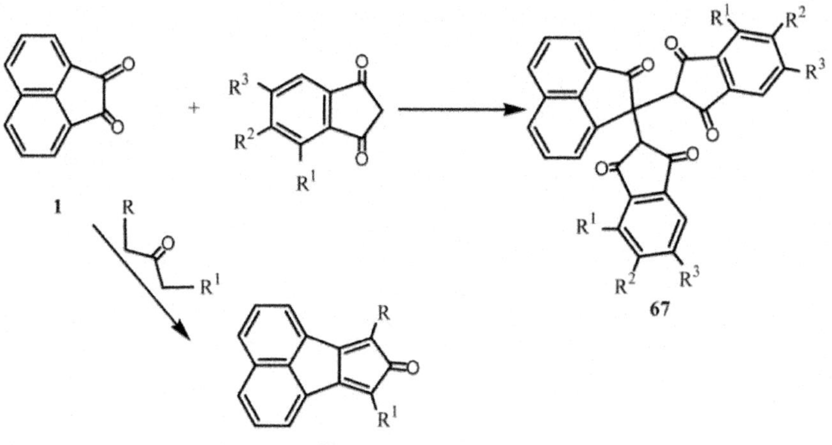

Scheme 16:

Reaction of acenaphthenequinone and 6-chloro-3-hydroxy-thionaphthene gave 2-(6-chloro-thionaphthene)acenaphthylene indigo [6-chloro-2-(2-oxo-1-acenphthylidene)-3(2H) thionaphthenone] which was used as a dye [62]. Similarly, condensation of substituted acenaphthenequinone with 3-hydroxythionaphthenes or indoxyl in the presence of a catalytic amount of hydrochloric acid gave 69 or its isomer which was not specifically identified (Scheme 17) [63,64,65,66]. Reaction of 1 or its halogen derivatives with thiohydantion [67], pseudothiohydantion [68] or rhodanine gave products of the type 70 or 71 [69].

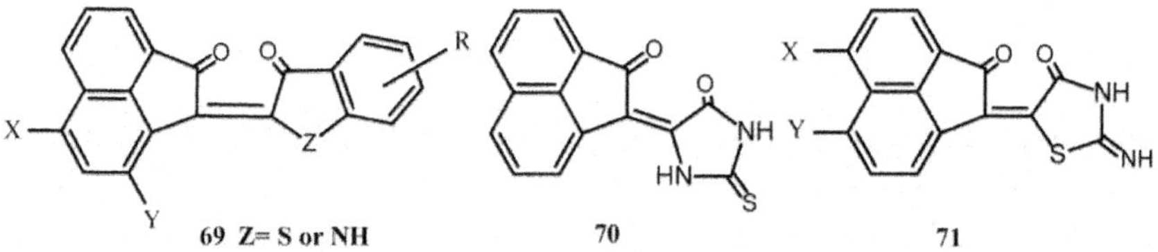

Scheme 17:

Reaction of 1 with 3,4–dehydro-DL-proline gave a product in which the pyrroline ring is converted into an N-substituted pyrrole [70]. Cyclocondensation of 2,3-dimethylquinoxaline-1,4-dioxide with 1 gave phenazine dioxide 72 (Scheme 18) [71]. When dialkyl-N-aminoazinium salts 73 or the quinazoline derivative 74 have been condensed with 1, they gave the new heterocycles 75 and 76, respectively, with a quaternary N in the bridgehead position (Scheme 19) [72,73].

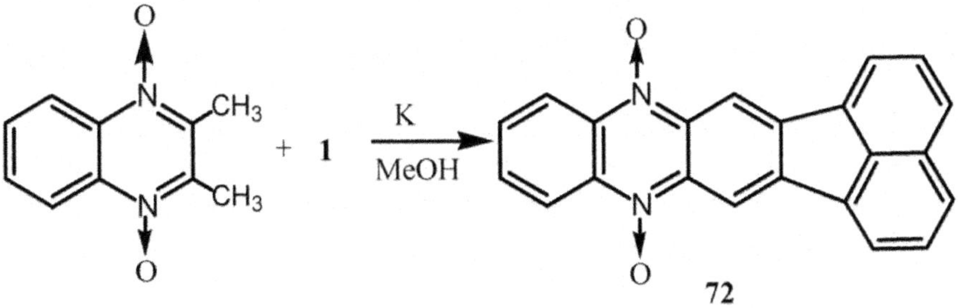

Scheme 18:

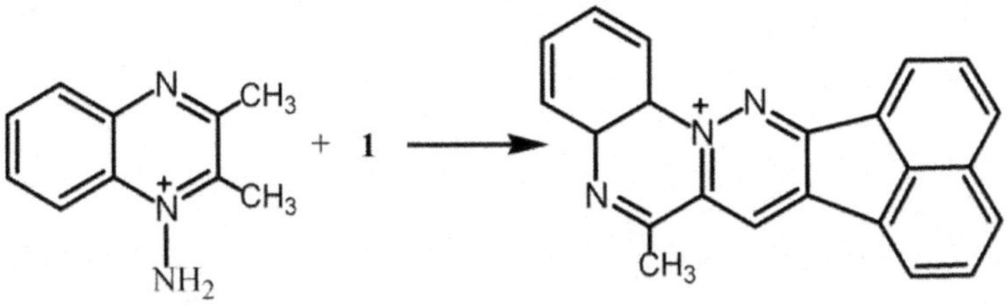

73   75

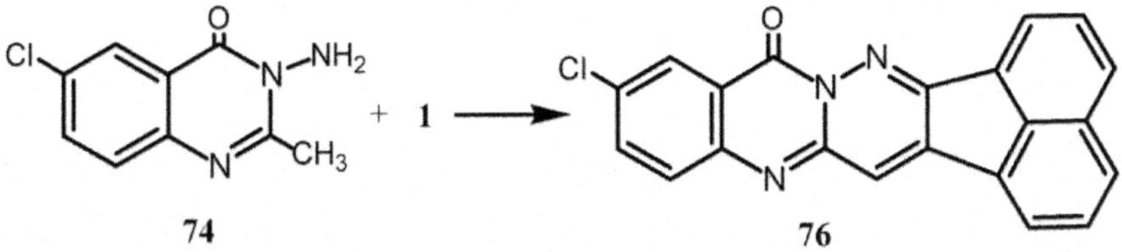

74   76

Scheme 19:

When 1 was treated with the furoxan derivative 77, an addition product was formed whose heating gave the diisocyanate 78 (Scheme 20) [74].

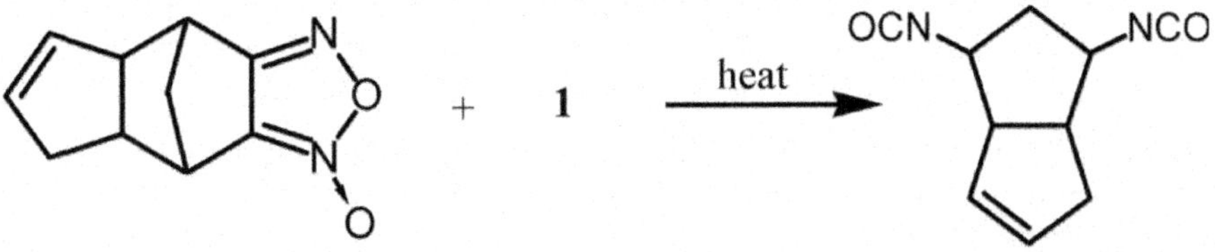

77   78

Scheme 20:

The Westphal condensation was used for the synthesis of different types of heterocycles from 1. Thus, condensation of 2-methylpyridinium, quinolinium or isoquinolinium salts with 1 in presence of a sodium acetate yielded the quinolizium salts79 [75,76]. When 1-ethoxycarbonylmethyl-2, 6-dimethylpyridinium salt was heated with acenaphthenequinone in the presence of di-n-butylamine, deep purple precipitate of [2,3,3]cyclizin -1-one derivatives 80 were formed whose formation was rationalized as presented in Scheme 21. All cyclazinone derivatives 80 were isolated as hydrobromides 81 [77].

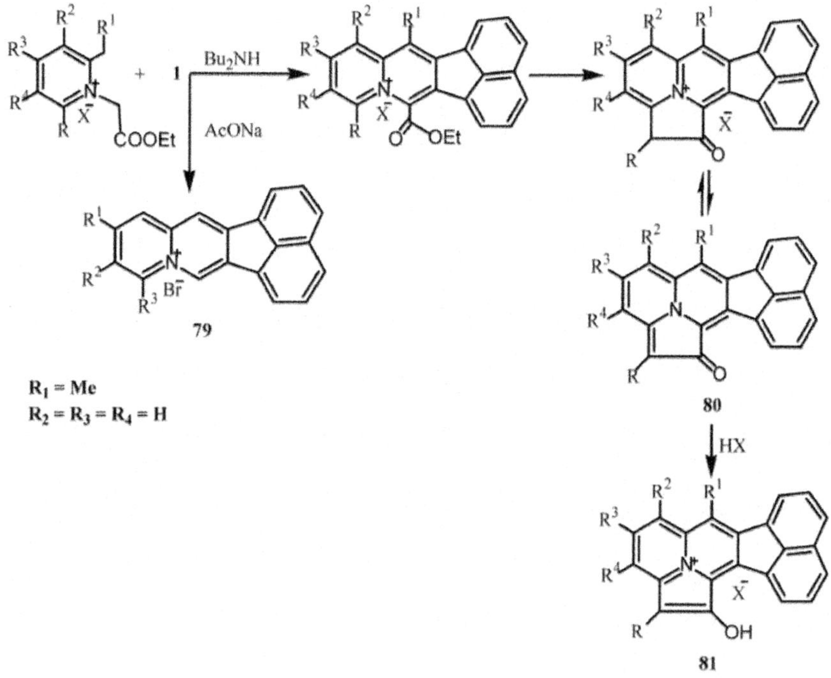

R_1 = Me
R_2 = R_3 = R_4 = H

Scheme 21:

The [2,3,3] cyclazin-6-one was prepared by condensation of acetoxymethylpyridinium bromide and 1 in presence of sodium acetate to yield the 4-ethoxycarbonylquinolizinium-1-olate (82) which upon reaction with hydrobromic acid produced 83 (Scheme 22). Treatment of the latter with sodium carbonate followed by DMAD gave [2,3,3] cyclazin-6-one derivative 84.

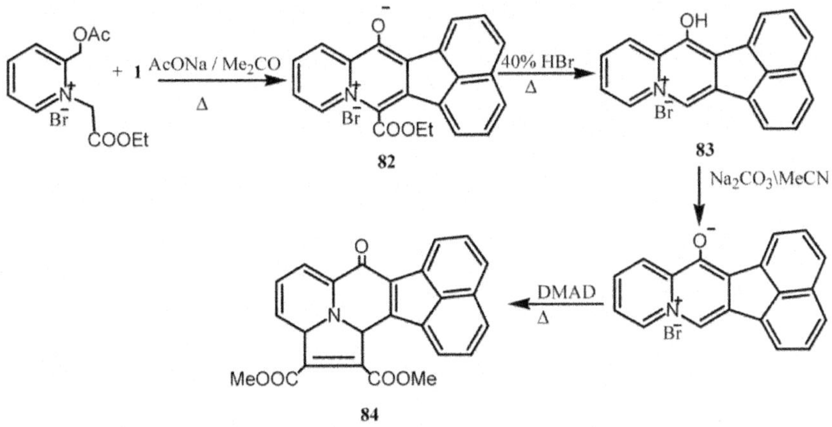

Scheme 22:

The Westphal condensation was also used for the synthesis of π-donor-π-acceptor heterocycles such as pyridopyrrolopyrazinium 86 by the condensation of 1 with pyrrolopyrazinium compounds 85 (Scheme 23) [78]. Similarly, the 2-methylthiazolium salts (87) were used as 1,4-dinucleophiles for the synthesis of thiazolo [3,2-a] pyridinium salts 88[79].

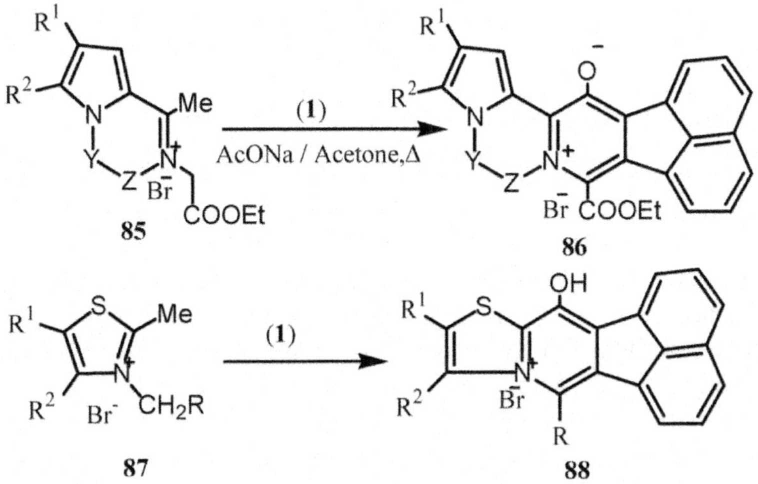

Y-Z	R_1	R_2
-CH$_2$-CH$_2$-	H	H
-CH=CH-	H	H
-CH$_2$-CH$_2$-	-(CH=CH)$_2$-	

Scheme 23:

Condensation of 1 with 2-alkyl-1-aminopyridinium, quinolinium or 1-alkyl-2- aminoisoquinolinium salts gave in presence of base, pyrido[1,2-b] pyridazinium salts 89 in good yield [80]. Similarly, the pyridazinopyrrolopyrazinium derivatives 91were prepared from 90 (Scheme 24) [80].

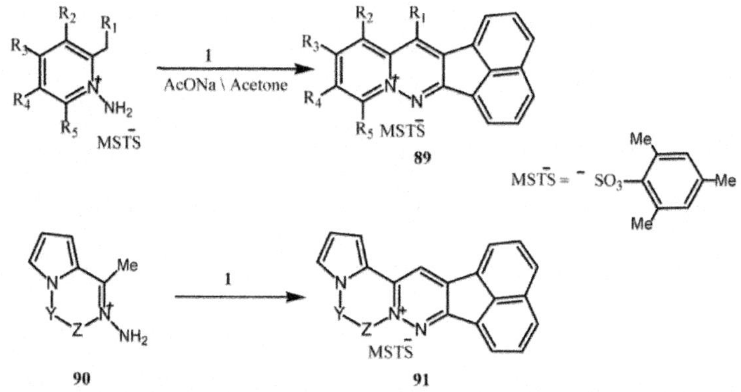

Scheme 24:

Reaction of 1 with nitromethane in alkali followed by acidification gave an adduct $C_{13}H_9NO_4$ which was formed by a 1,2-addition on one of the carbonyl groups [81].

Reaction with Aldehydes and Ketones

It has been shown that 1 could be condensed with aldehydes in a general manner to afford 92 that gave a violet-red color with concentrated sulfuric acid (Scheme 25) [82]. Reaction of acenaphthenequinone with acetone in presence of potassium hydroxide gave mono-acetoneacenaphthenequinone [83]. Reaction of 1 with p-chlorobenzaldehyde, p-acetamidobenzaldehyde, o-nitrobenzaldehyde in presence of ammonia gave oxazoles 99 at 0 °C and imidazoles 100 at higher temperatures. On the other hand, o-chloro-benzaldehyde, o-hydroxybenzaldehyde and m-hydroxybenzaldehyde under similar condition gave a mixture of oxazoles and imidazoles, which cannot be separated [83], and at higher temperatures, only imidazoles were obtained. p-nitrobenzaldehyde, p-hydroxybenzaldehyde and p-methoxybenzaldehyde gave only imidazoles. Vanillin and p-bromosalicyladehyde react very slightly at 0°C, but at higher temperature imidazoles 98 were formed. Heating the corresponding oxazole with ammonia in a sealed tube caused a partial conversion into the imidazole. When the reaction of 1 with o-hydroxybenzaldehyde was exposed to light for one month, it gave the monosalicylyl derivative of acenaphthenequinol [83]. When a suspension of 1 in isoamyl alcohol or anhydrous ethanol was treated with benzaldehyde in presence of ammonia, a variety of products 93-97 were obtained depending upon the condition of the reaction [84]. The structure of 97 was though to be either 97a or 97b. Its hydrolysis with dilute hydrochloric acid gave 1, naphthylimide and an unidentified compound $C_{12}H_8N_2O$.Substituted benzaldehydes were also used. Reaction of 1 with various aldehydes in boiling ammonium hydroxide and in dry ammonia gave aryl acenaphthimidazoles [83]. When dry ammonia was passed through a hot solution of 1 and p-acetamidobenzaldehyde in ammonium hydroxide, 4-acetyl-amino-2-phenyl-acenaphthoxazole (99), and the iminazole (100, R=NHAc) were obtained [85]. Treatment of 1 with ammonium acetate in presence of p-nitrobenzaldehyde gave (101,R=NO$_2$) [86]. It may be supposed that oxazoles are first formed which by subsequent replacement of the ring oxygen atom by NH forms the iminazoles.

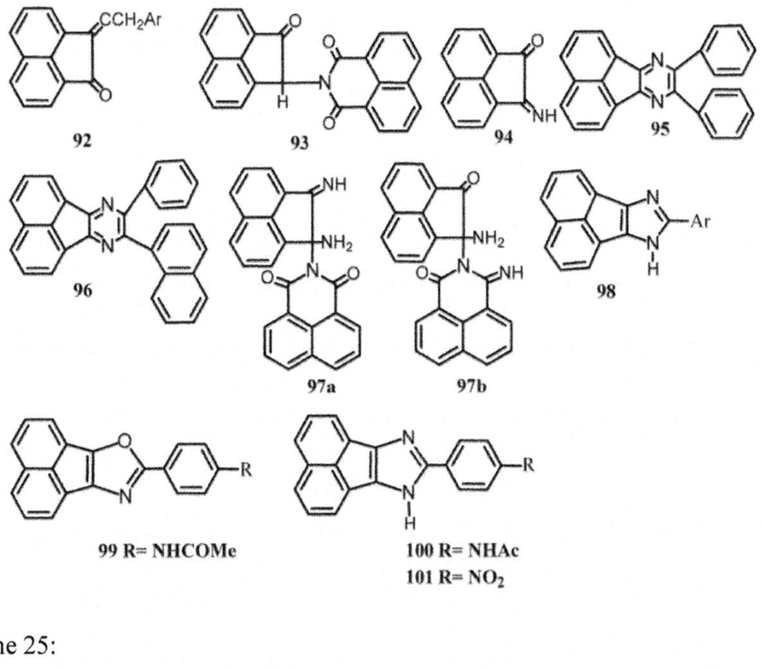

99 R= NHCOMe

100 R= NHAc
101 R= NO$_2$

Scheme 25:

Reaction with Wittig Reagents

Reaction of 1 with several Wittig reagents has been studied [87]. Thus, its reaction with equimolar amount of benzylidenetriphenylphosphorane at room temperature gave the corresponding benzylidene-acenaphthenones in fairly good yields. When the reaction of 1 was done with two molar equivalents of benzylidenetriphenylphosphorane under severe conditions it afforded also the benzylidene-acenaphthenone and no dibenzylidene derivative could be obtained. When 1 was reacted with 3(methoxyphenethyl)triphenylphosphonium bromide followed by cyclization-dehydration of the intermediate 102 gave 10-methoxybenzo[j]fluoranthene (103) exclusively (Scheme 26) [88].

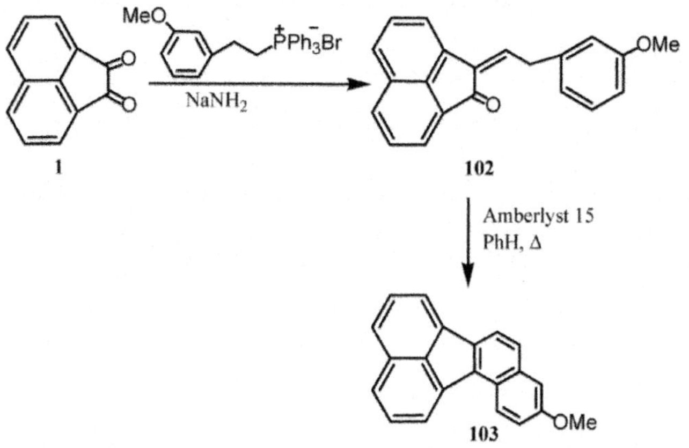

Scheme 26:

The adducts 104 were isolated from the reaction of 1 with triethyl phosphonoacetate (Scheme 27) [89]. The reaction of 1 with a resonance-stabilized phosphorane [90], such as acetonylidene-, phenacylidene-and p-chlorophenacylidene-phosphoranes afforded the expected α,β-unsaturated ketones 105. The ethoxycarbonyl-methylene acenaphthenone was obtained from the reaction of 1 with diethyl ethoxycarbonyl methyl phosphonate. Methylene-phosphorane was reacted with 1 to give methyleneacenaphthenone in poor yield. On the contrary, the reaction with ethylidenephosphorane gave 2,2'-methylenebisacenaphthenone, which was also formed by the reaction of acenaphthenone with glyoxal, in good yield.

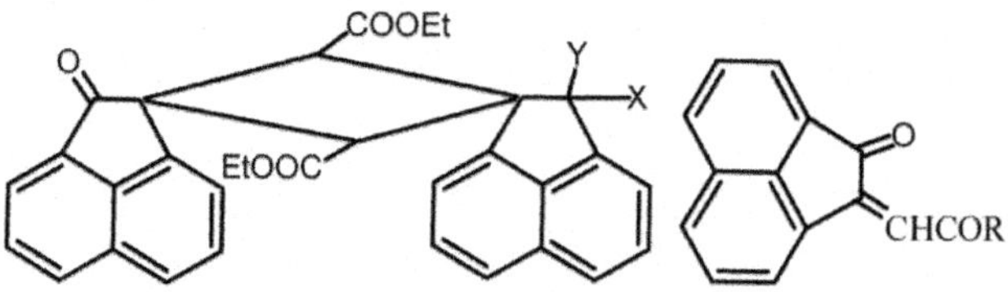

104 X, Y = O; X = OH, Y = CH₂COOEt   105 R = Me, Aryl

Scheme 27:

When equimolar amounts of the bisphosphonium salt 106 and 1 were treated with aqueous 5 M-lithium hydroxide (Scheme 28),

8,10-dimethylfluorantheno[8,9-c]thiophene (110) and 2-(2,4,5-trimethyl-3-thienyl-idene)acenaphthylen-2-one (111) were obtained [91]. Compound 111 was isolated in only one configuration and its formation may arise either *via* the o-quinomethaneylide intermediate 109 or the monoylide 108. Intramolecular Witting reaction of 109 gave 110.

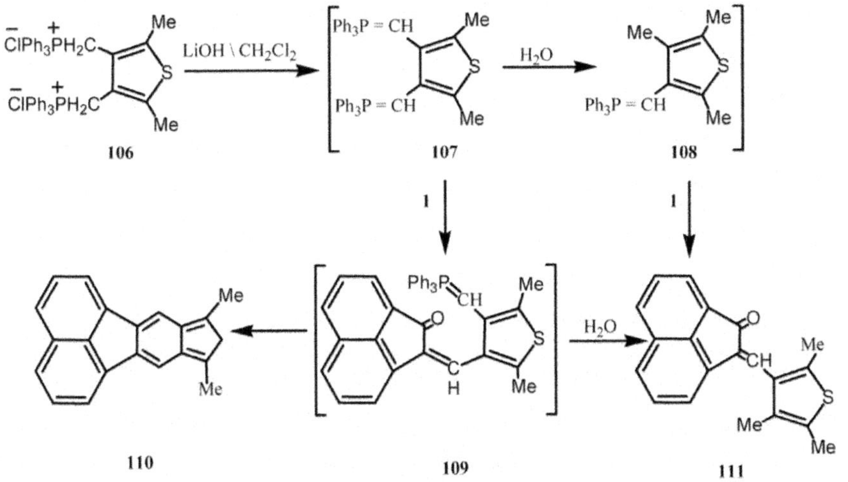

Scheme 28:

Wittig reaction of bis(triphenylphosphonium) dibromides with 1 under phase-transfer conditions gave 112 (Scheme 29) [92].

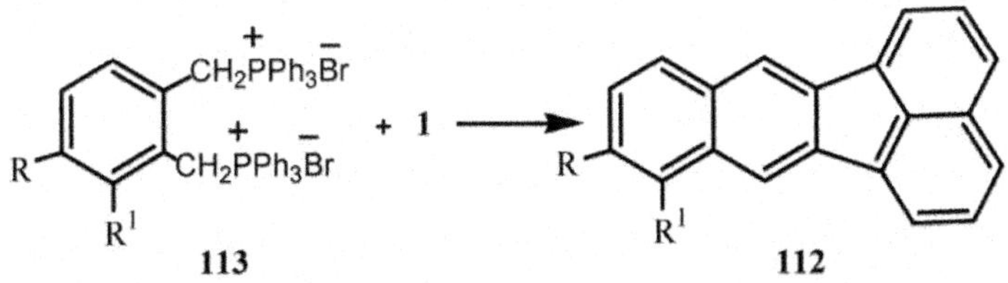

Scheme 29:

Reaction of 113 with hydroxylamine gave the oxime 114 and the pyrrole derivative 115, whereas its reaction with hydrazines afforded the pyridazinones 116 (Scheme 30) [93]. Reaction of the monoxime of 1 with $Ph_3P=CHCO_2Me$ afforded stereoisomeric products of 114, whereas the reaction with ylide $Ph_3P=CHCOMe$ gave the pyridine derivatives 118. Hydrogenation of 114 and thermal cyclization gave the polycyclic compounds 117 [94].

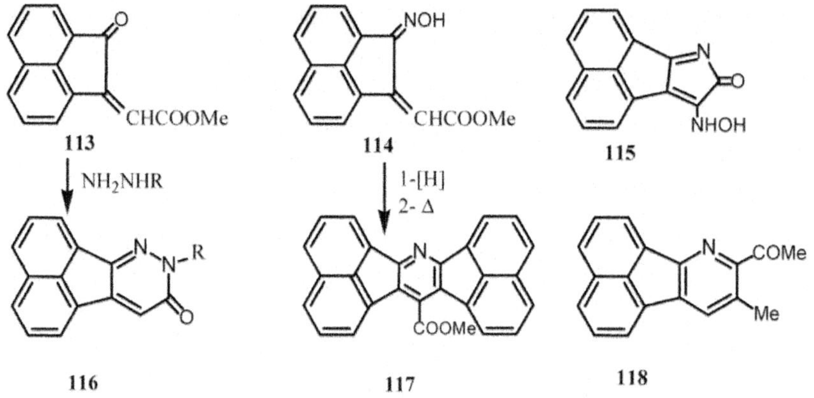
CHCOOMe
113
NH2NHR
NOH
CHCOOMe
114
1-[H]
2- Δ
NHOH
115
R
116
COOMe
117
COMe
Me
118

Scheme 30:

Spiro acenaphthyleneisoxazoles 119 were prepared by the regioselective 1,3-dipolar cycloaddition reaction of 113 with nitrile oxides (Scheme 31) [95]. The reaction of 1 with nitrile oxide gave acenaphthylenedioxazoles 120, which underwent a Wittig reaction with $Ph_3P{=}CHCO_2Et$ to give 121. The cycloaddition reaction of 120 with PhCNO gave a mixture of the dispiro compounds 122 and 123. Catalytic hydrogenation of 121-123 over Pd/C cleaved the dioxazole rings, whereas the isoxazole ring of 119 was cleaved by reduction over Raney nickel or by treatment with sodium ethoxide.

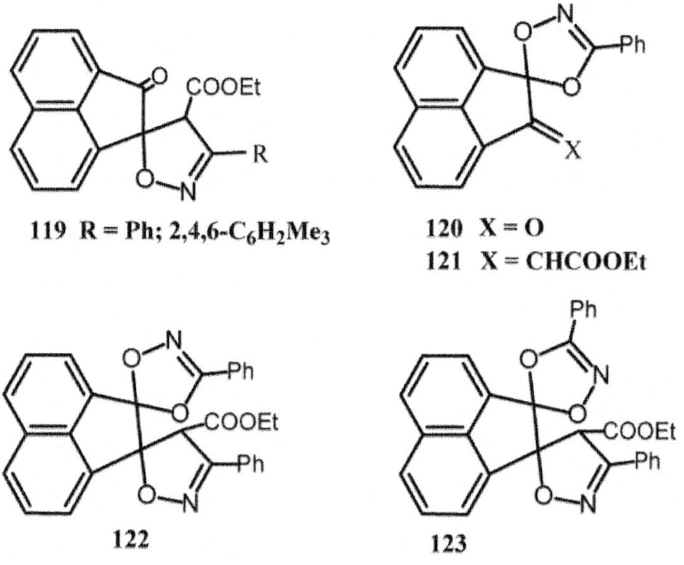
COOEt
R
119 R = Ph; 2,4,6-C6H2Me3
Ph
X
120 X = O
121 X = CHCOOEt
Ph
COOEt
Ph
122
Ph
COOEt
Ph
123

Scheme 31:

Reaction with Magnesium and Lithium Reagents

Various reports on the reaction of Grignard reagents with 1 have been published [96,97,98,99,100,101,102]. Thus, reaction of 1 with EtMgBr gave 1,2-diethylacenaphthoglycol 124 [96] whose dehydration with acid gave 1,2-diethylideneacenaphthene 125 (Scheme 32). The latter could be oxidized back to 1 with sodium dichromate in acetic acid.

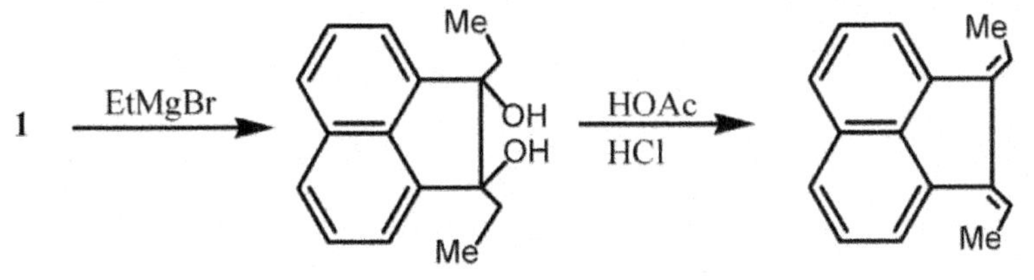

124   **125**

Scheme 32:

The dehydration behavior of 124 is unlike that of the diphenyl derivative, which gave a pinacoline under this treatment [96]. Reaction of 1 with arylmagnesium bromides gave 7,8-diarylnaphthenediols (126) (Scheme 33) [97,98] in which the simple aryl group migrates exclusively to give the 7,7-diaryl-acenaphthenones (127) [99]. The latter could be cleaved into the naphthoic acid derivatives 128 by alkali. Some derivatives of 126 gave low yields of 127 due to the formation of 131. The latter were prepared from 126 *via* 130. Ring opening of 126 was achieved by chromic acid to give the aroyl naphthalene 129[99,100,101,102,103].

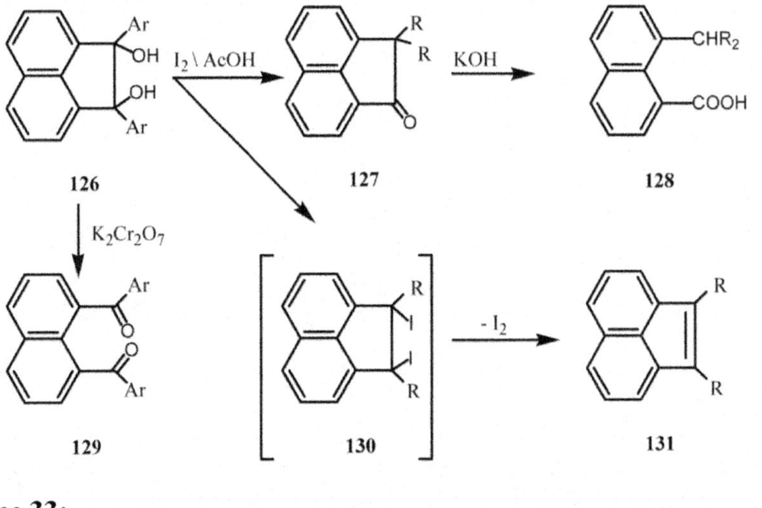

126   **127**   **128**

129   **130**   **131**

Scheme 33:

Reaction of acenaphthenequinone with several organolithium and organomagnesium reagents gave 132 (Scheme 34) [104]. This and its derived pinacol rearrangement product failed to cyclize under a variety of the acidic conditions, perhaps due to the poor stereoelectronic alignment for cyclization. On the other hand, when a less rigid analogue such as 1-lithio-3,4-dihydronaphthalene was used the tricyclo[4.3.0.0]nonane 133 was formed by di-oxy-Cope rearrangement followed by the unprecedented criss-cross 2π+2π cycloaddition of the two enolate ions formed [104]. The structure was confirmed by X-ray crystallography.

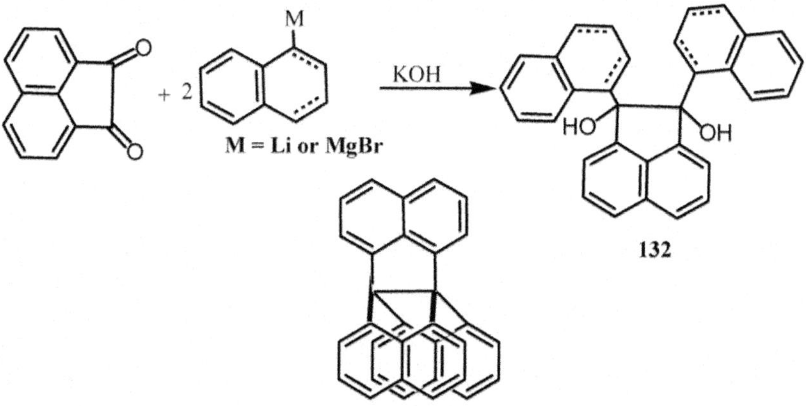

Scheme 34:

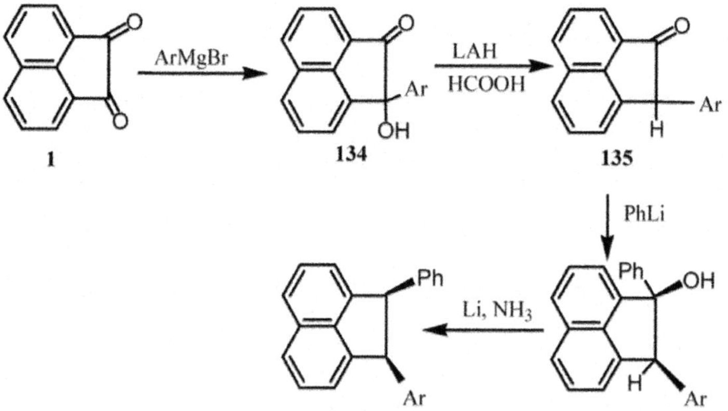

Scheme 35:

The addition of mesityl or triisopropyl magnesium bromide to 1 led (*via* a single electron transfer) to the formation of the corresponding semiquinone

[105] to give 134 that upon reduction with lithium aluminum hydride gave 135 (Scheme 35). Reaction of the latter with phenyl lithium gave 136 that upon Birch reduction gave the diaryl derivative 137 [106]. Reaction of vinyl chloride with 134 gave the corresponding vinyl ethers [107].

When 1,8-diiodonaphthalene and 5, 6-dibromoacenaphthene were lithiatied and treated with acenaphthenequinone ,and then the diol cycloaddition products were treated with hydrofluoric acid, acenaphth[1,2-a]acenaphthylenes, 138-140 were obtained (Scheme 36) [108].

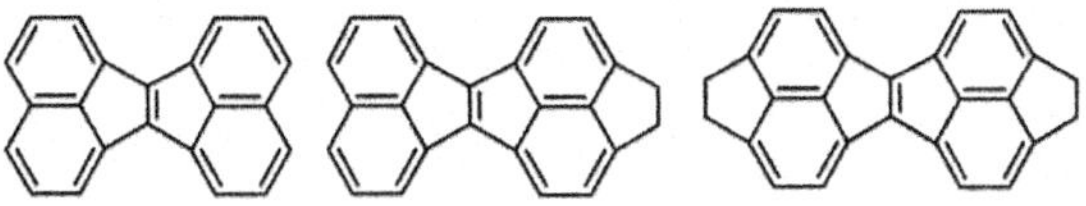

Scheme 36:

Friedel Crafts reaction

Condensation of acenaphthenequinone with benzene in presence of aluminum chloride gave 1,1-diphenyl-2-acenaphthenone (141) (Scheme 37) [109]. Boiling alcoholic potassium hydroxide converts 141 into 8-diphenylmethyl-1-naphthoic acid (142). Its distillation with barium hydroxide gave diphenyl naphthylmethane (143) whereas its treatment with chromium trioxide in acetic aid gave the lactone derivative 144.

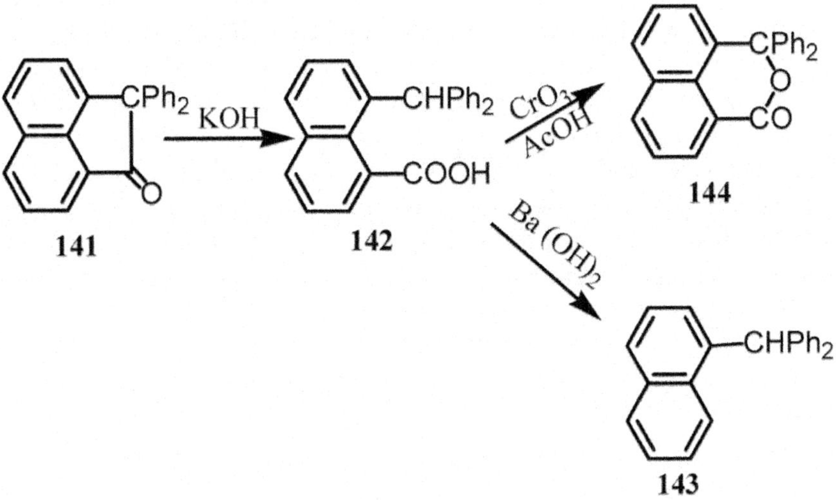

Scheme 37:

Diels-Alder reaction

Photochemical Diel's-Alder reaction of acenaphthenequinone with olefin led to the formation of **145** and **146** (Scheme 38) [110].

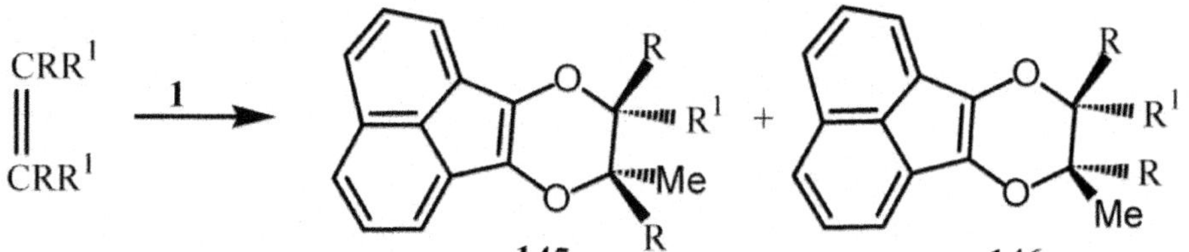

Scheme 38:

Reaction with Phenolic Compounds

The reaction of phenol with 1 was successful in presence of concentrated sulfuric acid [111,112]. Since the para-position is the most reactive in phenol, the product may probably be given the structure 1,1-bis [4'-hydroxyphenyl]-2-acenaphthenone (147) (Scheme 39). Reaction of 1 with resorcinol in presence of zinc chloride gave 148 [109], and the reaction with hydroquinone gave 149 [111]. To explain the formation of such anhydro derivatives it must be assumed that the oxygen of the quinone reacts with a hydrogen atom in the o-position to one of the hydroxyl groups in the dihydroxy compound [110]. The compound from catechol formed no anhydride and as the hydrogen atom in the p-position to one of the hydroxy groups in the catechol is the most reactive, the product must be 1,1-bis[3',4'-dihydroxyphenyl]-2-acenaphthenone. Condensation of acenaphthenequinone with asymmetric oxylenol, or gave 150. Thymol yields dithymol acenaphthenone. Nitrophenols reacted more sluggishly with 1 and a large excess of the phenol and the condensing agent was required [114]. This was explained on the basis of theoretical electrostatic considerations. Reaction with o-nitrophenol gave 1,1-bis(3-nitro-4-hydroxy-phenyl)-2-acenaphthenone, and with p-nitrophenol gave anhydro-1,1-bis(5-nitro-2-hydroxyphenyl)-2-acenaphthenone.

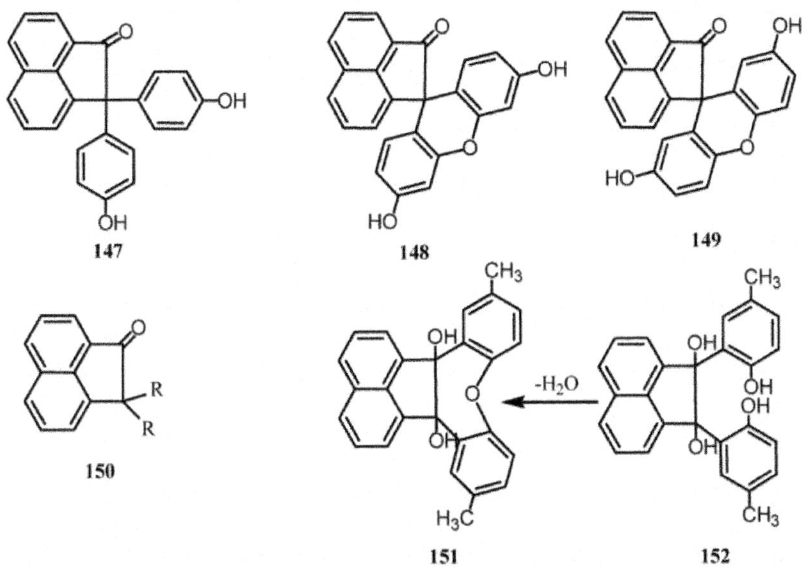

Scheme 39:

It was found that condensation of acenaphthenequinone with cresols and naphthols does not always give, like mono-and dihydric phenols, compounds of the type 150. Thus, reaction of 1 with p-cresol, gave 151 and 152; the relative amounts of which depend on the amount of sulfuric acid used as condensing agent [115].

Halogenation Reactions

Treatment of 1 with N-bromosuccinimide in polar solvents gave 5-bromoacenaphthenequinone [116]. Reaction of 1 with bromine gave the 3-bromo derivative [114]. No further bromination could be effected with bromine alone but in presence of iron filings, acenaphthenequinone gave the 2,3,5-tribromo derivative. Prolonged bromination of 1 in presence of iron gave the 2,3,4,5-tetrabromo derivative. The position of the bromine atoms has not been definitely determined [117]. On the other hand, treatment of 1 with bromine in chlorobenzene as a solvent gave 1,8-naphthalic anhydride whereas the reaction did not take place in nitrobenzene [118].

Haloacenaphthenes or haloacenaphthylenes were successively brominated, dehydrobrominated, chlorinated and hydrolyzed with sulfuric acid to give haloacenaphthenequinones [119]. Reaction of 1 with PCl$_5$ gave dichloroacenaphthenone whose reduction with powdered Fe in glacial acetic acid gave acenaphthenone (Scheme 40) [120,121].

Haloacenaphthenequinones 153 were prepared from the corresponding tetrachlorides 154 by successive reaction with sulfuric acid and sodium sulfite and hydrolysis (Scheme 40). The tetrachloro derivatives 154 were prepared by chlorination [122] of dibromo derivatives 155, which were prepared by bromination of 156.

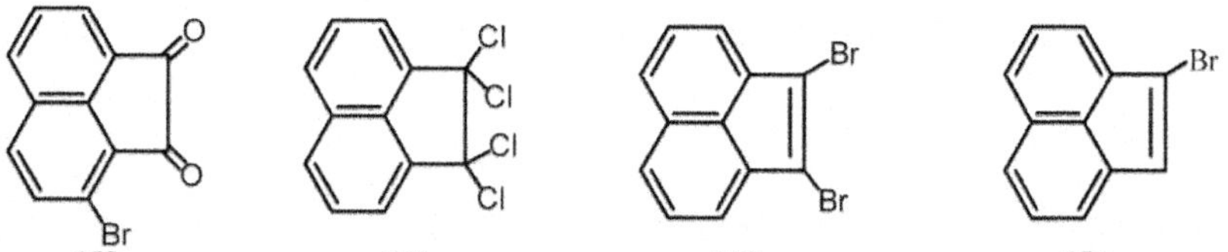

Scheme 40:

3-Bromo- and 3-iodoacenaphthenequinone were obtained from the 2–bromo- and 2-iodo-acenaphthenes, respectively by oxidation with sodium dichromate [63]. Reaction of 3-iodo-acenaphthenequinone and sodium dichromate in acetic acid gave 2-iodo-1,8-naphthalic anhydride.

Thiation Reactions

Acenaphthenequinone underwent mono thiation [123] on treatment with dithiadiphosphetane disulfide **157**.

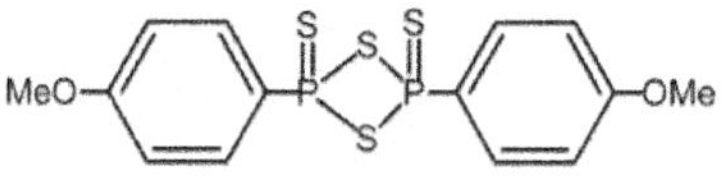

Alkylation Reactions

When acenaphthenequinone was treated with sodium in dry tetrahydrofuran, followed by 1,4-dichlorobutane], the product was 1,4-dioxacine derivative 158 (Scheme 41) [124]. Mercuration of acenaphthene-quinone under various conditions was failed [125]. Irradiation of a solution of 1 in acetonitrile in presence of allylic stannanes afforded homoallylic alcohols in good yields. When unsymmetrical allylstannens were used, the allylic groups were introduced predominantly at the α-positions. Complete regioselective introduction could be achieved by irradiation in presence of sodium hydroxide or cobalt chloride [126]. Reaction of 1 with trialkylallyltin gave the corresponding allylhydroquinone 159 [127] which was catalyzed by the Lewis acid [127].

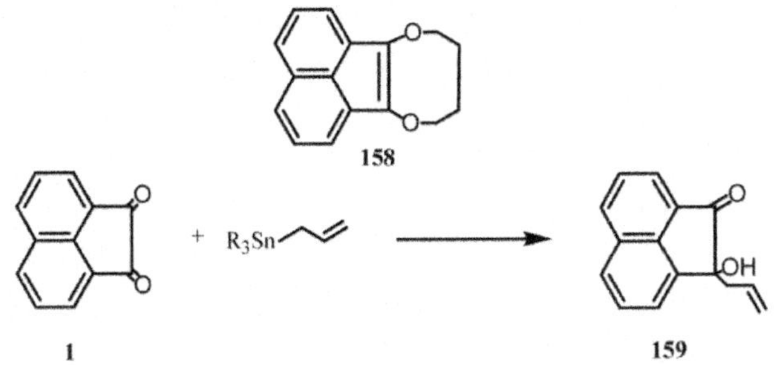

158

$+$ R_3Sn

1   **159**

Scheme 41:

The photoaddition of **1** to cycloheptatriene gave various cycloadducts, $(2+2)\pi$-, and $(2+6)\pi$- cycloadducts together with ene product [128]. Irradiation of **1** in benzene in the presence of 2,3-dimethyl-2-butene led to a facile formation of a single photoproduct (Scheme 42) [129].

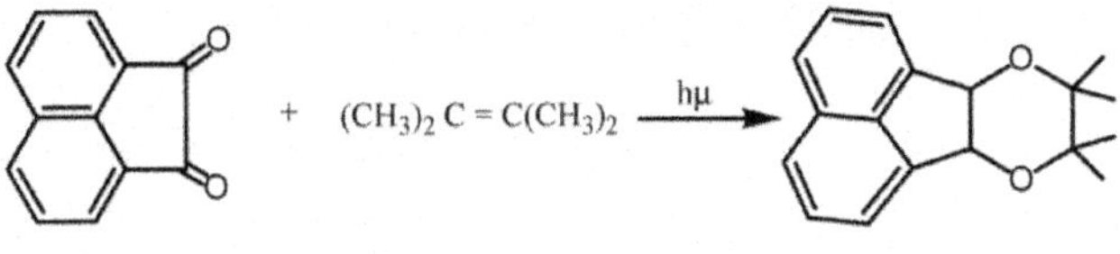

$+$ $(CH_3)_2 C = C(CH_3)_2$ $\xrightarrow{h\mu}$

1   **160**

Scheme 42:

The photoaddition reaction of 1 with α-silyl n-electron donors *via* triplet single electron transfer desilylation and triplet hydrogens abstraction pathways was explored [129]. Thus, photoaddition of $Et_2NCH_2SiMe_3$ to **1** produced 2-hydroxy-2-[(diethylamino)   methyl]   acenaphthylen-1-one   (161), whereas the photoaddition of n-$PrSCH_2SiMe_3$ to **1** generates two photoproducts 162 and 163 along with a photoreduction dimer of **1** (Scheme 43).

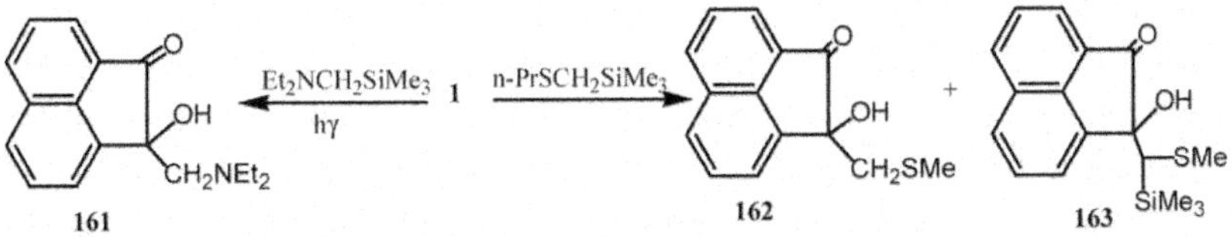

161   **162**   **163**

Scheme 43.

Ketal Derivatives

The products of the reaction of **1** and ethylene glycol in benzene were identified by using mass spectroscopy [130,131]. The least polar compound was 164 whereas the products of the highest and intermediate polarity were 165 and 166,respectively. Reaction of simple mercaptans with 1 gave monomercaptols (Scheme 44) [132].

The antiphlogistic compound acenphth[1,2-b]oxazole-8-propionic acid (167) was prepared from 1 *via* its monoethylene ketal and 1-hydroxy-2-acenaphthenone and subsequent esterification with succinic anhydride and reaction with ammonium acetate [133].

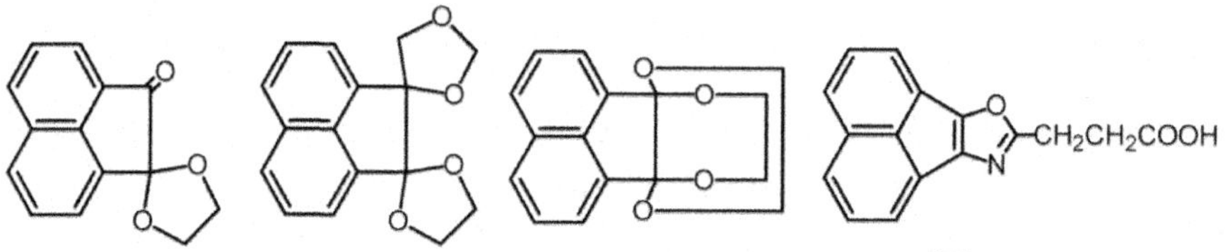

164   165   166   167

Scheme 44:

Condensation of 1 with N-(hydroxymethyl) trichloroacetamide gave 1,2-dioxo-4-trichloroacetyl-aminomethyl acenaphthene that was oxidized by dilute nitric acid in a sealed tube to give a $C_6H_2(CO_2H)_4$ (prehnitic acid) [2].

Reaction with Phosphites and Phospholanes

Reaction of 1 with dialkyl phosphites yielded the phosphonates 168 [134] whose heating gave the starting quinone. Treatment of 168 with hydrogen peroxide-sodium hydroxide gave naphthalene-1, 8-dicarboxylic acid (Scheme 45).

The reaction of P(OMe)$_3$ with 1 under air afforded P(O)(OMe)$_3$ and a 1:2 adduct 169 which was rearranged into the δ-lactone 170 by addition of water, while 169 was only obtained quantitatively under nitrogen atmosphere [135,136]. E.S.R and U.V spectra, decolorization of 1,1-diphenyl-2-picryl-hydrazyl, and initiation of styrene polymerization suggest the transient formation of radical ions. A mechanism, which involves one-electron transfer from phosphite to 1 followed by autoxidation, was proposed for the reaction under air [135]. The kinetic of the reaction to form a 1:1 adduct that cyclized was studied in anhydrous dioxan [137]. Treatment of 1 with sodium in tetrahydrofuran followed by Cl$_2$P (X) OR gave a fused di-oxophospholes 171 [138].

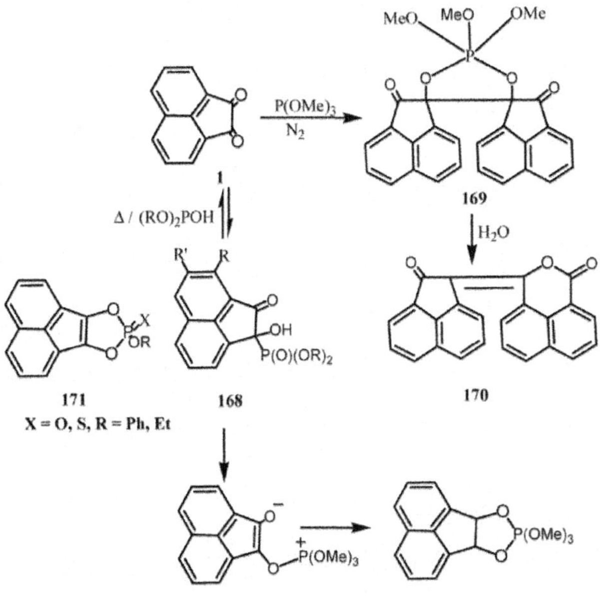

Scheme 45:

The reaction of 1 with 2-N-pyrrolidino-1,3-dimethyl-1,3,2-diazaphospholane (172) was quite vigorous in methylene chloride solution even at -70°C (Scheme 46) [139]. When the solution was allowed to reach 20°C, a deep brown mixture was produced from which the only isolable product was 2-N-pyrrolidino-2-oxo-1,3-dimethyl-1,3,2-diazaphospholane (176). A similar behavior was noted when 1 was treated with 2-dimethylamino-1,3-dimethyl-1,3,2-diazaphospholane (173). The only isolable product was 2-dimethylamino-2-oxo-1,3-dimethyl-1,3,2-diazaphospholane (177). No intermediate could be detected in these reactions, but it was assumed by analogy with previous reactions that the phosphorus of the cyclic aminophosphines attacked the oxygen of 1 to give a 1:1 dipolar adduct 174 or 175. The dipolar adduct apparently was too unstable to form the cyclic phospholene analogous to those of benzil. Instead, the 1:1 dipolar adduct lost phosphoramidate to yield a carbenoid fragment 178 which underwent further transformations. Reaction of the diazides of 1 with triphenylphosphine gave the phophazine (179) [140].

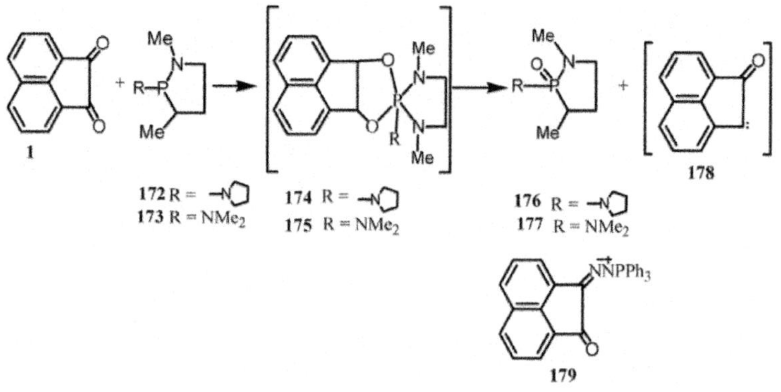

172 R = —N◯   174 R = —N◯   176 R = —N◯
173 R = NMe₂   175 R = NMe₂   177 R = NMe₂

179

Scheme 46:

Reaction with Carbenes

Addition of 1,3-diphenylimidazolidin-2-yildene to **1** gave **180** (Scheme 47) [141].

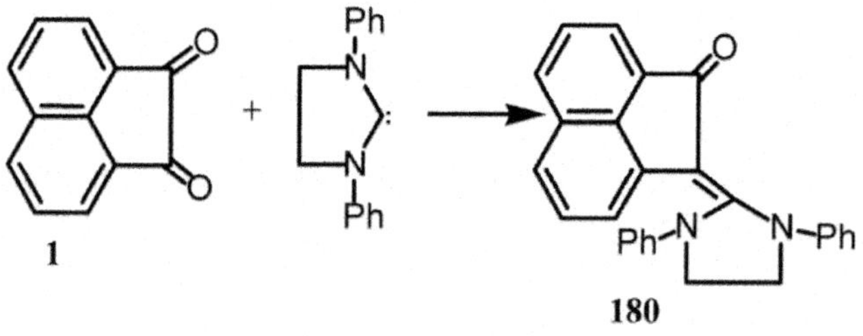

Scheme 47:

DECARBONYLATION

Stepwise elimination of carbonyl groups occurs when a vapor of **1** was passed through glow discharge plasmas to give 1,8-dehydronaphthalene, part of which dimerizes to give perylene [142].

Reaction with Nitrogen Nucleophiles

The reactions of **1** with nitrogen nucleophiles namely ammonia, amines, urea, hydroxylamines, aminoacids, o-diamines and hydrazines are included in the previous review [1]. These reactions produce heterocyclic compounds or products that are precursors for the synthesis of heterocyclic compounds. Various types of heterocyclic compounds could be prepared via the use of such nitrogen nucleophiles.

REFERENCES

1. El Ashry, E. S.; Abdel Hamid, H.; Shoukry, M. *Ind. J. Heterocycl. Chem.* 1998, *7*, 313.

2. De Diesbach, H.; Lachat, P.; Poggi, M.; Baladi , B.; Friderich, R.; Walker, H. *Helv. Chim. Acta* 1940, *23*, 1232.

3. Bader, H.; Chiang, Y. H. U.S. Pat. 1974; 3,812,115; [*Chem. Abstr.* 1974, *81*, 25456p].

4. Bader, H.; Chiang, Y. H. *Synthesis* 1976, *4*, 249.

5. Proskuryakov, V. A.; Chistyakov, A. .N.; Soboleva, T. P. U.S.S.R. Pat. 1971.

6. Randall, R. B.; Benger, M.; Groocock, C.M. *Proc. Roy. Soc.* 1938, *A165*, 432, [*Chem. Abstr.* 1938, *32*, 5374].

7. Tkacheva, G. d.; Suvorov, B. V. *Izv. Akad. Nauk Kaz. SSR, Ser. Khim.* 1987, *3*, 63, [*Chem. Abstr.*, 1988, *108*, 94341k]..

8. Tsuji, J.; Kobayashi, Y. *Jpn. Pat.* 1971, *71*, 18–969, [*Chem. Abstr.*, 1972, *76*, 34020r].

9. Kojima, H.; Takahashi, S.; Hagihara, N. *Tetrahedron Lett.* 1973, *22*, 1991.

10. Tabuchi,, T.; Nojima, M. *J. Org. Chem.* 1991, *56*, 6591.

11. Kasiwagi, I. *Bull. Chem. Soc. Jpn.* 1926, *1*, 66.

12. J-Younge, Koo; Schuster, G. B. *J. Org. Chem.* 1979, *44*, 847.

13. Schuster, G. B.; Ja-Young, K. *Gov. Rep. Announce Index (U.S.)* 1978, *78*, 116, [*Chem. Abstr.* 1979, *90*, 18601w].

14. Boujlel, K.; Simonet, J. *Tetrahedron Lett.* 1979, *12*, 1063.

15. Edwards, W. G. H; Petrow, V. *J. Chem. Soc.* 1948, 1713.

16. Eistert, B.; Schoenberg, A. *Ber.* 1962, *95*, 2416.

17. Eistert, B.; Selzer, V. *Ber.* 1963, *96*, 314.

18. Miller, A. R. *J. Org. Chem.* 1979, *44*, 1931.

19. Schwabe, I. K.; Berg, H. *Z. ElektroChem.* 1952, *56*, 952, [*Chem. Abstr.* 1954, *48*, 2005].

20. Elbel, K.; Biebrich. U.S. Pat. 1910; 965,170; [*Chem. Abstr.* 1910, *4*, 2738].

21. Kalle and co. *Ger. Pat.* 1909, *224*, 979, [*Chem. Abstr.* 1911, *5*, 213].

22. Kalle and co. *Brit.Pat.* 1909, *21*, 579, [*Chem. Abstr.* 1911, *5*, 1336].

23. Skita, A. *Ber.* 1927, *60*, 2522.

24. Braun, J.V.; Bayer, O. *Ber.* 1926, *59B*, 920.

25. Goldestien, H.; Glauser, W. *Helv. Chim. Acta* 1934, *17*, 788.

26. Jack, K. M.; Rule, H. G. *J. Chem. Soc.* 1938, 188.Ghigi, E. *Gazz. Chim. Ital.* 1938, *68*, 184, [*Chem. Abstr.*, 1938, *32*, 7910].Graebe, C.; Jequier, E. *Ann.* 1896, *290*, 202.

27. Trevoy, L.W.; Brown, W. G. *J. Am. Chem. Soc.* 1949, *71*, 1675.

28. Panaiotov, I. M.; Rashkov, I. B. *Dokl. Bolg. Akad. Nauk* 1968, *21*, 885, [*Chem. Abstr.* 1969, *70*, 28277q].

29. Bakola-Christianopoulou, M. N. *J. Organomet. Chem.* 1986, *308*, C24.

30. Bakola-Christianopoulou, M. N. *J. Mol. Catal.* 1991, *65*, 307, [*Chem. Abstr.* 1991, *115*, 28843u].

31. Yamashita, H.; Reddy, N. P.; Tanka, M. *Chem. Lett.* 1993, *2*, 315.

32. Guirado, A.; Barba, F. ; Hursthouse, M. B.; Arcas, A. *J. Org. Chem.* 1989, *54*, 3205.

33. Guirado, A.; Barba, F.; Tevar, A. *Synth. Commun.* 1984, *14*, 333, [*Chem. Abstr.* 1984, *101*, 190629m].

34. Kalinwaski, M. K.; Tenderende-Guminska, B. *J. Electroanal. Chem. Interfacial Electrochem.* 1974, *55*, 277, [*Chem. Abstr.*1975, *82*, 36634s].

35. Rozhnova, T. K.; Serazetdinova, V. A.; Sembaev, D. Kh.; Suvorov, B. V. *Zh. Anal. Khim.* 1975, *30*, 2462, [*Chem. Abstr.*1976, *85*, 71789e].

36. Ghe, A. M.; Valcher, S.; Del Monte, M. G. *Ann. Chim. (Rome)* 1970, *60*, 729, [*Chem. Abstr.* 1971, *75*, 44255q].

37. Boujlel, K.; Simonet, J. *Tetrahedron Lett.* 1979, *17*, 1497.

38. Simonet, J.; Lund, H. *Bull. Soc. Chim. Fr.* 1975, 2547, [*Chem. Abstr.* 1976, *85*, 77232q].

39. Boldt, P.; Arensmann, E.; Blenkle, M.; Kersten, H.; Tendler, H.; Trog, R. S.; Jones, P. G.; Doring, D. *Chem. Ber.* 1992, *125*, 1147.

40. Bruck, D.; Minsky, A.; Dagan, A.; Rabinovitz, M. *Tetrahedron Lett.* 1981, *22*, 3545.

41. Murayama, K.; Ono, K.; Osugi, J. *Bull. Chem. Soc. Jpn.* 1972, *45*, 847.

42. Junek, H.; Hamboeck, H.; Hornischer, B. *Monatsh. Chem.* 1967, *98*, 315.

43. Hans, J.; Albin, H.; Andre, B. M. *Monatsh. Chem.* 1975, *106*, 715.

44. Junek, H.; Hornischer, B.; Sterk, H. *Monatsh. Chem.* 1968, *99*, 2121.

45. Moureu, H.; Chovin, P.; Rivoal, G. *Bull. Soc. Chim. France* 1946, 106, [*Chem. Abstr.* 1946, *40*, 6070].

46. Moureu, H.; Chovin, P.; Rivoal, G. *Compt. Rend.* 1946, *223*, 951, [*Chem.*

Abstr. 1947, *41*, 2032].

47. Moureu, H.; Chovin, P.; Rivoal, G. *Bull. Soc. Chim. France* 1948, 99.

48. Vickery, E. H.; Eisenbraun, E. J. *Org. Prep. Proced. Int.* 1979, *11*, 259, [*Chem. Abstr.* 1980, *92*, 94121q.

49. Bedford, M. J.; Crombie, D. A. *Proc. R. Soc. Edinburgh, Sect. A* 1974, *71A*, 279, [*Chem. Abstr.* 1975, *82*, 97871p.

50. Patwardhan, B. H.; Bagavant, G. *Indian. J. Chem.* 1973, *11*, 1333.

51. Tucker, S. H. *J. Chem. Soc.* 1958, 1462.

52. Jackson, D. A.; Lacy, Ph. H.; Smith, D. C. C. *J. Chem. Soc., Perkin Trans.1* 1989, 215.

53. Koshelev, V. I.; Palkidin, V. L. *Zh. Org. Khim.* 1973, *9*, 597, [*Chem. Abstr.* 1973, *79*, 5199n].

54. Birch, A.; Crombie, D. A. *Chem. Ind.(London)* 1971, *6*, 177.

55. Rangnekar, D. W.; Mavalankar, S. V. *J. Heterocycl. Chem.* 1991, *28*, 1455.

56. Vanags, G.; Geita, L. *Zhur. Obshchei Khim.* 1956, *26*, 511, [*Chem. Abstr.* 1956, *50*, 13852a].

57. Geita, L.; Vanags, G. *Latvijas PSR Zinatnu Akad. Vestis* 1958, 127, [*Chem. Abstr.* 1959, *53*, 11371d].

58. Vanags, G.; Geita, L. *J. Gen. Chem. U.S.S.R.* 1956, *26*, 539, [*Chem. Abstr.* 1957, *51*, 2686g].

59. Ezoe, K.; Kurosawa, K. *Bull. Chem. Soc. Jpn.* 1977, *50*, 443.

60. Banerjee, P. K.; Bhattacharya, A. J. *J. Indian J. Chem., Sect. B* 1977, *15B*, 953.

61. Samanta, S. R.; Mukherjee, A. K.; Battacharyya, A. J. *Curr. Sci.* 1988, *57*, 926.

62. Guha, S. K.; Sinha, A. K. *J. Indian Chem. Soc.* 1952, *29*, 415.

63. Karishin, A. P.; Krivoshapko, N. G.; Osobik, D. I. *Khim. Geterotsikl. S oedin* 1968, *1*, 61, [*Chem. Abstr.* 1969, *70*, 12633s].

64. Karishin, A. P.; Fedorenko, T. P. *Ukrain, Khim. Zhur.* 1953, *19*, 631, [*Chem. Abstr.* 1955, *49*, 1214e].

65. Karishin, A. P. *Ukrain. Khim. Zhur.* 1952, *18*, 504, [*Chem. Abstr.* 1955, *49*, 1682i].

66. Karishin, A. P.; Kustol, D. M. *Ukrain. Khim Zhur.* 1956, *22*, 229, [*Chem. Abstr.* 1957, *51*, 365c].

67. Karishin, A. P., Timchenko; Dzhurka, G.F.; Samusenko, Yu. V.; Baklan,

T. F.; Lysenko, G. M. *Khim. Geterotskil. Soedin, Akad. Nauk Latv. SSR* 1965, *5*, 704, [*Chem. Abstr.* 1966, *64*, 9708g].

68. Karishin, A. P.; Samusenko, Yu. V. *Zh. Organ. Khim.* 1965, *1*, 1003, [*Chem. Abstr.* 1965, *63*, 11534f].

69. Karishin, A. P.; Solomakha, L. A. *Zh. Organ. Khim.* 1965, *1*, 2062, [*Chem. Abstr.* 1966, *64*, 8166b].

70. Hudson, C. B.; Robertson, A. V. *Aust. J. [Chem.* 1967, *20*, 1511, [*Chem. Abstr.* 1967, *67*, 117251g].

71. Issidorides, C. H.; Atfah, M. A.; Sabounji, J. J.; Sidani, A. R.; Haddadin, M. J. *Tetrahedron* 1978, *34*, 217.

72. Matia, M. P.; Garci-Navio, J. L.; Vaquero, J. J.; Alvarez-Builla, J. *Liebigs Ann. Chem.* 1992, *7*, 777.

73. Erwin, D.; Frank, P. *Eur. Pat.* 1989, *339*, 556, [*Chem. Abstr.* 1990, *112*, 181391c].

74. Crosby, J.; Milner, J. A. *Ger. Pat. 2* 1977, *714*, 668, [*Chem. Abstr.* 1978, *88*, 23601c].

75. Alvarez-Builla, J.; Gonzalez Trigo, G.; Ezquerra, J.; Fombella, M. E. *J. Hetrocycl. Chem.* 1985, *22*, 681.

76. Ezquerra, J.; Builla, J. A. *J. Heterocycl. Chem.* 1986, *23*, 1151.

77. Pastor, J.; Paz Matia, M.; Garcia Nario, J. L.; Vaquero, J. J.; Alvarez-Builla, J. *Heterocycles* 1989, *29*, 2369.

78. Matia, M. P.; Ezquerra, J.; Sanchez-Ferrnado, F.; Garcia Navio, J. L.; Vaquero, J. J.; Alvarez-Builla, J. *Tetrahedron* 1991, *47*, 7329.

79. Galera, C.; Vaquero, J. J.; Garcia Navio, J. L.; Alvarez-Builla, J. *J. Heterocycl. Chem.* 1986, *23*, 1889.

80. Matia, M. P.; Garcia Navio, J. L.; Vaquero, J. J.; Alvarez-Builla, J. *J. Heterocycl. Chem.* 1990, *27*, 661.

81. Nightingale, D. V.; Erickson, F. B.; Shackelford, J. M. *J. Org. Chem.* 1952, *17*, 1005.

82. De Fazi, R.; Monforte, F. *Atti accad. Lincei* 1929, *10*, 653, [*Chem. Abstr.* 1930, *24*, 24423].

83. Sircar, A. C.; Sen, S. C. *J. Indian Chem. Soc.* 1931, *8*, 605.

84. Tsuge, O.; Gunjima, T. *Asahi Garasu Kogyo Gijustsu Shorei-Kai Kenkyu Hokou* 1966, *12*, 209, [*Chem. Abstr.* 1968, *68*, 78248k].

85. Sircar, A. C.; Sen, S. C. *J. Indian Chem.. Soc.* 1936, *13*, 482.

86. White, D. N. *J. Org. Chem.* 1970, *35*, 2452.

87. Otohiko, T.; Masashi, T.; Ichiro, S. *Bull. Chem. Soc. Jpn.* 1969, *42*, 181.

88. Rice, J. E; Shih, H. C.; Hussain, N.; La Voie, E. J. *J. Org. Chem.* 1987, *52*, 849.

89. Boulos, L. S.; Abd El-Rahman, N. M. *Phosphorous, Sulfur- Silicon Relat. Elem.* 1992, *68*, 241.

90. Helena, S.; Acad, C. R. *Sci, Ser. C* 1971, *273*, 1194, [*Chem. Abstr.* 1972, *76*, 59135v].

91. Nicolaides, D. N.; Litinas, K. E.; Argyropoulos, N. G. *J. Chem. Soc., Perkin Trans. 1* 1986, 415.

92. Rice, J. E.; Czech, A.; Hussain, N.; LaVoie,, E. J. *J. Org. Chem.* 1988, *53*, 1775.

93. Lefkaditis, D. A.; Nicolaides, D. N.; Papageorgiou, G. K.; Stephanidou-Stephanatou, J. *J. Heterocycl. Chem.* 1990, *27*, 227.

94. Papageeorgiou, G.; Nicolaides, D.; Stephanidou-Stephanatou, J. *Liebigs Ann. Chem.* 1989, *4*, 397, [*Chem. Abstr.* 1989, *110*, 212325z].

95. Lefkaditis, D. A.; Argyropoulos, N. G.; Nicolaides, D.N. *Liebigs Ann. Chem.* 1986, *11*, 1863, [*Chem. Abstr.* 1987, *106*, 4947u].

96. Maxim, N. *Bull. Soc. Chim.* 1928, *43*, 769, [*Chem. Abstr.* 1928, *22*, 4121].

97. Acree, S. F. *Am. Chem. J.* 1905, *33*, 186.

98. Beschke, E.; Kitay, M. *Ann.* 1909, *369*, 200.

99. Bachmann, W. E.; Chu, E. J. *J. Am. Chem. Soc.* 1936, *58*, 1118.

100. Bartlett, P. D.; Brown, R. F. *J. Am. Chem. Soc.* 1940, *63*, 2927.

101. Brown, R. F. *J. Am. Chem. Soc.* 1954, *76*, 1279.

102. Moriconi, E. J.; O'Conner, W. F.; Kuhn, L. P.; Keneally, E. A.; Wallenberger, F. T. *J. Am. Chem. Soc.* 1959, *81*, 6472.

103. Jingshun, J. *Huaxue Xuebao* 1987, *45*, 1211, [*Chem. Abstr.* 1988, *109*, 92967r].

104. Alder, R. W.; Colclough, D.; Grams, F.; Orpen, A. G. *Tetrahedron* 1990, *46*, 7933.

105. Miller, A. R.; Curtin, D. Y. *J. Am. Chem. Soc.* 1976, *98*, 1860.

106. Levchenko, A. I.; Morzo, R. A.; Zatolokin, E. I.; Suprun, V. Z.; Beloglazova, V. V.; Gaidukova, R. G.; Stal'nova, L.K. *Kim. Atsetilena, Tr. Vses. Konf., 3rd* 1986, 311, [*Chem. Abstr.* 1973, *79*, 53803b].

107. Mitchell, R. H.; Fyles, T.; Ralph, L. M. *Can. J. Chem.* 1977, *55*, 1480, [*Chem. Abstr.* 1987, *88*, 37475b].

108. Blomberg, C.; Grootveld, H. H.; Gerner, T. H.; Bickelhaupt, F. *J.*

Organometal. Chem. 1970, *24*, 549.

109. Zsuffa, M. *Ber.* 1910, *43*, 2915, [*Chem. Abstr.* 1911, *5*, 495].

110. Tai-Shan, F.; Wang Ping, M.; Tsing Hsing, C.; Shih-Chen, S. J. *J. Chin. Soc.* 1985, *32*, 457, [*Chem. Abstr.* 1987, *106*, 196342n].

111. Matei, I. *Ber.* 1929, *62B*, 2095.

112. Salazkin, S. N.; Korshak, V. V.; Vinogradova, S. V.; Beridze, L. A.; Pankratov, V. A. *Deposited Doc., VINITI* 1976, 2833, [*Chem. Abstr.* 1978, *89*, 129169].

113. Matei, I.; Bogdan, E. *Ber.* 1938, *71B*, 2292.

114. Matei, I.; Ccea, E. *Ber.* 1944, *77B*, 714.

115. Matei, I.; Bogdan, E. *Ber.* 1934, *67B*, 1834.

116. Dewhurst, F.; Shah, P. K. J. *J. Chem. Soc., C* 1970, *12*, 1737.Graebe, C.; Guinsburg, M. *Ann.* 1903, *327*, 85.

117. Mayer, F.; Schonfelder, H. *Ber.* 1922, *55B*, 2972.

118. Rule, H. G.; Thompson, S. B. *J. Chem. Soc.* 1937, 1761.

119. Petrenko, G. P.; Terent'eva, G. N.; Usachenko, V. G. *U.S.S.R. Pat.* 1973, *406*, 827, [*Chem. Abstr.* 1974, *80*, 82499h].Dziewonski, K.; Zarkrzewska-Barnaowska, M. *Bull. Intern. Acad. Polon. Sci.* 1927, *1-2A*, 65.

120. Ghigi, E. *Gazz . Chem. Ital.* 1938, *68*, 184.

121. Graebe; Jequier. *Ann.* 1896, *290*, 202, [*Chem. Abstr.* 1938, *32*, 7910].

122. Petrenko, G. P.; Terent'eva, G. N.; Usachenko, V. G. *Zh. Org. Khim.* 1973, *9*, 2313, [*Chem. Abstr.* 1974, *80*,47690s].

123. El Kateb, A. A.; Hennaway, I.T.; R.Shabana, Osman, F. H. *Phosphorus Sulfur* 1984, *20*, 329, [*Chem. Abstr.* 1985, *103*, 53784k].

124. Singh, M. S.; Mehrotra, K. N. *Indian J. Chem., Sect.B* 1984, *23B*, 1289, [*Chem. Abstr.* 1985, *102*, 203950g].

125. Ogata, Y.; Tsuchida, M. *J. Org. Chem.* 1955, *20*, 1631.

126. Akio, T.; Yutaka, N.; Koichi, Y.; Hidetoshi, I. *Chem. Lett.* 1990, 4639.

127. Yoshinori, N. *J. Am. Chem. Soc.* 1980, *102*, 3774.

128. Brown, R. E.; Legg, K. D.; Wolf, M. W.; Singer, L. A.; Parks, J. H. *Anal. Chem.* 1974, *46*, 1690.

129. Yoon, U. C.; Kim, Y. C.; Choi, J. J.; Kim, D. U.; P. S., Mariano; Cho, I. S.; Jeon, Y. T. *J. Org. Chem.* 1992, *57*, 1422.

130. Cohen, A. I.; Harper, I. T.; Levine, S. *D. J. Chem. Soc. D* 1970, *23*, 1610.

131. Cohen, A. I.; Harper, I. T.; Puar, M. S.; Levine, S. D. *J. Org.*

Chem. 1972, *37*, 3147.

132. Schönberg, A.; Schutz, O.; Arend, G.; Peter, J. *Ber.* 1927, *60B*, 2344.

133. Seymour David, L. *Ger. Pat.2* 1971, *102*, 843.

134. Sidky, M. M.; Osman, F. H. *J. Prakt. Chem.* 1973, *315*, 881.

135. Ogata, Y.; Yamashita, M. *Bull. Chem. Soc. Jpn.* 1973, *46*, 2208.

136. Ramirez, F.; Ramanathan, N. *J. Org. Chem.* 1961, *26*, 3041.

137. Ogata, Y.; Yamashita, M. *J. Chem. Soc., Perkin Trans 2* 1972, 493.

138. Singh, M. S.; Mishra, G.; Mehrotra, K. N. *Phosphorus Sulfur Silicon Relat. Elem.* 1991, *63*, 177.

139. Ramirez, F.; Patwardhan, A. V.; Kugler, H. J.; Smith, C. P. *Tetrahedron* 1968, *24*, 2275.

140. Ried, W. *Hg. Appeal Z. Naturforschung* 1960, *15b*, 684, [*Chem. Abstr.* 1961, *55*, 21062g].

141. Burmistrov, S. I.; Kondrat'eva, S. E. *Nov. Khim. Karbenov, Mater. Vses. Soveshch. Khim. Karbenov Ikh. Analogv, 1ts 1972*(pub.1973). 240,.

142. Andras, S.; Harald, S.; Mundiyath, V. *Justus Liebigs Ann. [Chem.* 1977, *5*, 747, [*Chem.*

Chapter 7

ORGANOCATALYSIS: KEY TRENDS IN GREEN SYNTHETIC CHEMISTRY, CHALLENGES, SCOPE TOWARDS HETEROGENIZATION, AND IMPORTANCE FROM RESEARCH AND INDUSTRIAL POINT OF VIEW

Isak Rajjak Shaikh[1,2,3]

[1]Department of Chemistry, Shri Jagdishprasad Jhabarmal Tibrewala University, Vidyanagari, Jhunjhunu-Churu Road, Chudela, Jhunjhunu District, Rajasthan 333001, India

[2]Razak Institution of Skills, Education and Research (RISER), Shrinagar, Near Rafaiya Masjid and Hanuman Mandir, Nanded, Maharashtra State 431 605, India

[3]Post Graduate and Research Centre, Department of Chemistry, Poona College of Arts, Science and Commerce, Camp Area, Pune 411 001, Maharashtra State, India

ABSTRACT

This paper purports to review catalysis, particularly the organocatalysis and its origin, key trends, challenges, examples, scope, and importance. The definition of organocatalyst corresponds to a low molecular weight organic molecule which in stoichiometric amounts catalyzes a chemical reaction. In this review, the use of the term heterogenized organocatalyst will be exclusively confined to a catalytic system containing an organic molecule immobilized onto some sort of support material and is responsible for accelerating a chemical reaction. Firstly, a brief description of the field is provided putting it in a green and sustainable perspective of chemistry. Next, research findings on the use of organocatalysts on various inorganic supports including nano(porous) materials, nanoparticles, silica, and zeolite/zeolitic materials are scrutinized in brief. Then future scope, research directions, and academic and industrial applications will be outlined. A succinct account will summarize some of the research and developments in the field. This review tries to bring many outstanding researches together and shows the vitality of the organocatalysis through several aspects.

INTRODUCTION

In 1987, the United Nations Commission on Environment and Development (Brundtland Commission) [1] defined "sustainable development" as the development that meets the needs of the present without compromising the ability of future generations to meet their own needs. Two of the key aspects of sustainable development from an energy and chemical perspective are to develop more renewable forms of energy and to reduce pollution. Chemistry during the twentieth century changed the living standard of human beings. Among the greatest achievements of chemistry are petrochemical and pharmaceutical industries. But these industries are often blamed for polluting environment. The challenge for the present-day chemical industry is to continue providing applications and socioeconomic benefits in an environmentally friendly manner. Over the last few decades, green chemistry has been recognized as a culture and methodology for achieving sustainable development [2]. Green chemistry is chemistry able to promote innovative technologies that reduce or eliminate the use or generation of hazardous substances. Anastas and Warner defined the 12 principles of green chemistry (Figure 1) [3]. Catalysis (including enzyme catalysis, heterogeneous catalysis, and organocatalysis, in particular) is identified to be at the heart of greening of chemistry [4] because this branch of science is found to reduce the environmental impact of chemical processes [5].

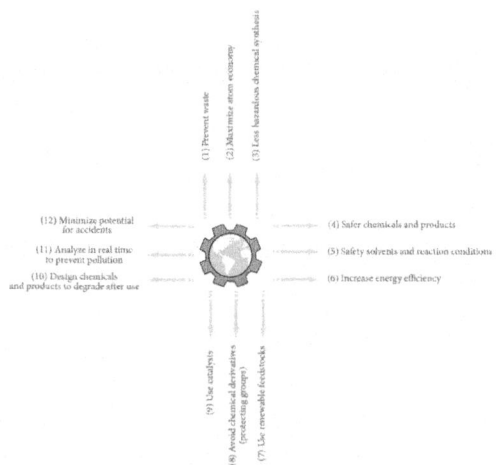

Figure 1: Inspired by the 12 principles of green chemistry, from [3].

CATALYSIS: A MULTIDISCIPLINARY AREA OF CHEMISTRY

Catalyst is one of the few conceptual words that have carried over broadly outside scientific language. Catalysis is a significant multidisciplinary area of chemistry. The term "catalysis" was first introduced in 1836 by Berzelius who tried to explain special powers of some chemical substances capable of influencing various decomposition and chemical transformations.

According to Ostwald [6], "a catalyst accelerates a chemical reaction without affecting the position of the equilibrium." A catalyst works by interacting with reactants, generating intermediates that react to give products. It affects the rate of approach to equilibrium of a reaction but not the position of the equilibrium (Figure 2). Most of the times, it also provides subtle control of chemical conversions, increasing the rate of a desired reaction pathway but not the rates of undesired side reactions (i.e., the selectivity of a chemical process).

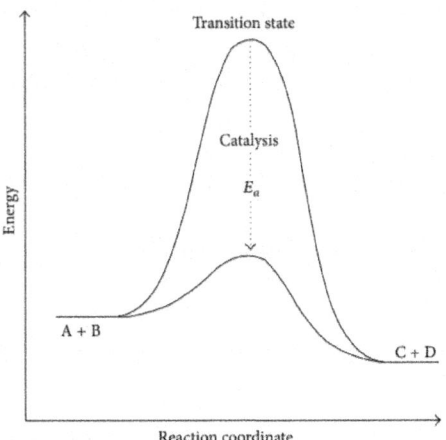

Figure 2: Schematic illustration of the effect of a catalyst on the transformation of A + B into C + D.

Catalysts are usually classified according to structure, composition, area of application, or state of aggregation. According to the state of aggregation in which the catalysts act, there are two main groups: homogeneous catalysts and heterogeneous catalysts. Homogeneous catalysts are well-defined chemical compounds or coordination complexes, which, together with the reactants, are molecularly dispersed in the reaction medium and carry out processes in a uniform gas or liquid phase. Heterogeneous catalysis takes place between different phases. Generally the catalyst is in solid state, and the reactants are gases or liquids. The immobilized catalysts obtained by attaching homogeneous catalysts to solid are in between. In supported catalysts the catalytically active substance is applied to a support material that has a large surface area and is usually porous. Arguments for and against shall always be made while discussing the merits and limitations of individual groups of catalysts for their respective applications in industries.

APPLICATION OF CATALYSIS

Catalysis has played a pivotal role in the success of the chemistry industry in the twentieth century. More than 90% of chemical manufacturing processes in the world utilize catalysts [7, 8]. Historically, the first industrially applied catalytic process was for the hydrogenation of oils and fats to produce margarine using finely divided nickel as catalyst. Thus, heterogeneous catalysis was applied first industrially. The issue of metal leaching and the avoidance of trace catalyst residues are very important from environmental health point of view. All the basic raw materials or building blocks for chemicals are manufactured by very important set of heterogeneous catalytic reactions (Table 1) [8]. Although catalysts are not consumed by the reaction itself, they may be inhibited, deactivated, or destroyed by secondary processes. The solid-state catalytic materials are the most important type of catalysts.

Table 1: List of some of the industrially applied reactions and catalysts

Reaction	Catalyst	Inventor (Year)
Sulphuric acid (lead—chamber process)	NO_x	Clement, Desormes (1806)
Chlorine production by HCl oxidation	$CuSO_4$	Deacon (1867)
Nitric acid by NH_3 oxidation	Pt, Rh nets	Ostwald (1906)
Fat hardening	Ni	Normann (1907)
Ammonia synthesis ($H_2 + N_2$)	Fe	Mittach, Haber, Bosch (1908); Production 1913 BASF
Hydrogenation of coal to hydrocarbons	Fe, Mo, Sn	Bergius (1913); Pier (1927)
Methanol synthesis from CO/H_2	Zno/Cr_2O_3	Mittach (1923)
Hydrocarbons from CO/H_2 (motor fuels)	Fe, Co, Ni	Fischer, Tropsch (1925)
Alkylation of olefins to gasoline	$AlCl_3$	Pines (1932)
Cracking of hydrocarbons	Al_2O_3/SiO_2	Houdry (1937)
Cracking in a fluidized bed	Aluminosilicates	Lewis, Gilliland (1939)
Ethylene polymerization at low pressure	Ti compounds	Ziegler, Natta (1954)
Hydrogenation, isomerization, hydroformylation	Rh-, Ru complexes	Wilkinson (1964)
Methanol carbonylation to acetic acid	Co/iodide	BASF (1960);
	Rh/iodide	Monsanto Co. (1966);
	Iridium	BP Chemicals Ltd. (2000)
Asymmetric hydrogenation	Rh/chiral phosphine	Knowles (1974)
Three-way catalyst	Pt, Rh/monolith	General Motors (1974)
Methanol conversion to hydrocarbons	Zeolites	Mobil Chemical Co. (1975)
Alpha-olefins from ethylene	Ni/chelate phosphine	Shell company (1977)
Sharpless (epi) oxidation	Ti/ROOH/Tartarate	May, Baker, Upjohn, ARCO (1981)
Selective oxidations with H_2O_2	TS-1 (titanium silicate)	Enichem (1983)
Hydroformylation	Rh/phosphine	Rhone-Poulene-Ruhrchemie (1984)
Polymerization of olefins	Zirconocene/MAO	Sinn, Kaminsky (1985)
Selective catalytic reduction (SCR)	V, W, Ti oxides/monolith	1986

Implementation of "clean" and "green" chemical technology in industries may help address the problem of environmental degradation besides producing useful chemicals [9–12]. Some catalysts also present important technology in the prevention of emissions, for example, catalytic converter for automobiles.

Keeping in mind the fragile ecosystem and the importance of preserving the environment, the heterogeneous catalysts are tested in the pollution abatement [13, 14] and to remove aqueous organic wastes, volatile organic compound, and vehicular primary emissions such as CO, unburned hydrocarbon, NOx, and soot.

Heterogeneous catalysis which takes place between different phases has now grown into an important branch of science [8–12, 15]. Exciting new opportunities are emerging in this field based on nanotechnology approaches. Hence, research efforts are being directed to develop chemistry of nanomaterials and their applications in catalysis for developing cost-effective, selective, energy sufficient, ecofriendly, and environmentally benign synthetic processes for industries. Though this involves a great scientific, economic, and technological challenge, it is essential to a sustainable and healthy way of life.

The petroleum and fine chemical and pharmaceutical industries mostly rely on catalysts to produce everything from natural gas/fuels to various commodities [15, 16]. There are a variety of new challenges in creating alternative fuels, reducing harmful by-products in manufacturing, cleaning up the environment and preventing future pollution, dealing with the causes of global warming, and creating safe chemicals and pharmaceuticals. Thus, catalysts are needed to meet these challenges of health, safety, and environmental, energy-related, and economic issues, but their complexity and diversity demand an insight on the way catalysts are studied, designed, and used. This could come into reality only through the application of an interdisciplinary approach involving new methods for synthesizing and characterizing molecular and material systems. Research in heterogeneous catalysis demands the cooperation of scientists from analytical chemistry, solid-state chemistry, (nano)materials chemistry, physical chemistry, surface science, computational and theoretical chemistry, reaction kinetics and mechanisms, reaction engineering, and so forth. Opportunities to understand and predict how catalysts work at the atomic scale are now appearing, made possible by breakthroughs in the last decade in computation, measurement techniques, characterization, spectroscopy, and imaging and especially by new developments in surface-reactivity studies, model catalyst design, and evaluation [16, 17]. Surface science and spectroscopy has its prominent role in analysing surface reaction chemistry on the catalyst [18, 19].

Fermentation of sugar to ethanol and the conversion of ethanol to acetic acid catalyzed by enzymes were an example of catalysis in antiquity. Off late, enzymatic catalysis is considered as a separate branch of (biological) catalysis. It is the most recent branch widely included in many commercial applications [20, 21]. Enzymes are also highly active under mild conditions such as temperature, pressure, and pH, which make them easier to work with on a small scale, but are also more appealing on an industrial scale on a cost and efficiency basis. Another benefit is that reactions are preferentially carried out in aqueous media therefore providing a "greener" route. Not to forget, the development of water based organic transformations is also an area of rapidly growing importance in chemistry [22, 23].

The homogeneous catalysis prior to the recognition of the effects of heterogeneous catalysts started with the action of nitrous oxides in the lead chamber process. Although the fundamental processes for refining petroleum and its conversion to basic building blocks are based on heterogeneous catalysts, many important value-added products are manufactured by homogeneous catalytic processes (Table 2) [12]. Homogeneous catalysts have been well researched, since their catalytic centers can be relatively easily defined and understood, but difficulties in separation and catalyst regeneration prevent their wider use [12, 24–28]. The most widely used homogeneous catalysts are simple acids or bases which catalyze well-known reactions such as ester and amide hydrolysis or esterification. The market share of homogeneous catalysts is estimated to be only ca. 10–15%. Many of the benefits of homogeneous catalysts, especially high selectivity, arise from tailored made catalysts involving transition metals (or lanthanides) and appropriate ligands. Such catalysts are inexpensive enough that they can be neutralized, easily separated from organic materials, and disposed of. This contributes to the huge quantity of aqueous salt waste generated by industry. Organometallic catalysts are currently widely used in industry, but the search for improved efficiencies, enantioselectivities, or recycling is still a matter of intense researches [29–33]. It is true that homogeneously catalyzed processes such as hydroformylation, carbonylation, oxidation, hydrogenation, metathesis, and hydrocyanation contribute with millions of tons to the bulk chemicals, but on the other hand the progress of homogeneous catalysis is also going on with fine chemicals [25, 33]. In spirit of that, the beginning of this century saw the renaissance of organocatalysis presenting an environmental advantage over metal based catalysts for stereoselective or asymmetric synthetic methodologies. The absence of metal in organocatalyst brings an undeniable advantage considering the principles of "green chemistry" and the economic point of view (Tables 3 and 4).

Table 2: Differences between homogeneous catalysts and heterogeneous catalysts

Homogeneous catalysts	Heterogeneous catalysts
Same phase as reaction medium	Solid phase
Difficult separation	Easily separable
No recyclability	Recyclability and regeneration
Often high rates of reaction	Low rates of reaction
No diffusion control	Diffusion control
Robust to poisoning	Poisoning, deactivation
High selectivity	Low selectivity
Short life	Long catalytic life
Mild conditions	Energy-consuming process
Well understood mechanism	Poor mechanistic understanding

Table 3: Comparison among organometallic catalysts, enzyme catalysts, and organocatalysts

Organometallic catalyst	Enzyme catalyst	Organocatalyst
Wide substrate scope; high catalytic activity; involving tedious process; potential heavy metal pollution	Limited substrate scope; high selectivity and catalytic activity; usually single enantiomer	Robust; inexpensive; readily available; nontoxic; inert toward moisture and oxygen; method especially attractive for the preparation of compound that do not tolerate metal contamination (pharmaceutical products)

Table 4: Advantages and disadvantages of organocatalysts

Advantages	Disadvantages
Easy preparation or availability	High catalyst loading
Easy handling; inert towards moisture (water) and air (oxygen)	Relatively premature field
Easy scale-up	
No metal contamination	
Easy screening	
Useful in complex (steric) reactions	

Understanding catalytic systems involve study of structure, state, and composition of catalysts and the phase in which they perform. And, for this, one needs to understand chemistry of molecules and materials useful as catalysts and or catalyst carriers.

MATERIALS FOR CATALYSIS

The diversity of research field connected to the chemistry of materials is significant for their applications in many areas including adsorption, catalysis, electrochemistry, and surface science. Nanoparticles [34], metal oxide, and mixed metal oxides [35–38] prepared by simple methods are useful catalysts. Research involving metal-organic frameworks [39, 40], mesoporous silica [41–43], microporous or open-framework inorganic [44–46], and/or mesopore modified [47] materials has steadily increased following the widespread utility of aluminosilicate zeolites [48–51], titanosilicates [52, 53], and aluminium phosphates [54, 55].

In industries, research endeavours are mainly devoted towards the study and development of zeolites or zeolitic materials with several attractive features such as catalytic active site(s), large surface area, zeolitic pore walls, and pore size control. The synthesis and characterization of the crystalline aluminosilicate materials with tunable meso-/microporosity and their strong acidity have potentially important technological implications for shape-selective catalytic reactions, ion exchange, and adsorption of organic compounds [56–58]. The pore diameters can be finely chosen, tuned, and utilized for exhibiting shape-selective catalysis depending on the molecular shape/size of organic reactant and product. The widespread interest has been developed in the zeolitic science towards the development and discovery of zeolites with desired characteristics since conventional zeolites lack fine tuning of their properties to acid/base and redox or bifunctional or multifunctional properties. Organic functionalization adds new catalytic functions to zeolites [59].

Zeolites are microporous crystalline aluminosilicates with unique properties used for many processes. Zeolites possess acid sites that are associated with the tetrahedral aluminium atoms in their framework, and therefore, the amount of acidity depends on the aluminium content of the zeolite. Zeolite acid catalysts have wide application in industrial processes. The introduction of Y-zeolites to replace amorphous silica-alumina in fluidized catalytic cracking (FCC) and in hydrocracking processes in the early 1960's was a breakthrough application in acid catalysis. With the ever increasing research efforts in this area leading to discoveries of new structures and modification schemes by academic and industrial laboratories, the potential of solid acids has evolved tremendously. However, with the exception of the ZSM-5, Mordenite, and to some extent Beta zeolite, very few of these structures have been applied in industrial processes to any significant extent. Yet one can account for relatively few applications to the huge number of new materials introduced. Notable examples are the reported uses of SAPOs in lube dewaxing and in natural gas to olefin technologies and MCM-22 in aromatics alkylation. The zeolite structures with large-pore/bimodal porosity and tunable acidity can be a remarkable benefit for catalytic processes in industries. Within industrial catalysis, there are constant innovation and changing approaches to provision of selective, economically viable, and environmentally benign processes.

Recently, enormous growth in the chemical diversity of inorganic materials with transition-metal incorporation in new compositional domains has also been explored due to their potential application in adsorption and catalysis. Inorganic polymers, transition metal phosphates, and metal-organic frameworks [39, 40] with novel catalytic, electronic, magnetic, and unique structural properties are particularly promising candidates [60–64]. The mesoporous structure with strong acidity can be a remarkable benefit for catalytic reactions involving large organic molecules, in which diffusion constraints and/or adsorption of reactant molecules onto the strong acid sites are the main concern. A lot has been reported about the pathways to develop mesostructured silica [65]. In terms of surface functionalization, the performance of mesoporous silica has been polished by several methods such as metal incorporation [66], metal deposition [67], grafting by organic functional groups [68], and immobilizing metal complex [69]. New approaches to synthesis and characterization of bimodal, that is, micro-mesoporous and or meso-macroporous materials might assist us in understanding the phenomena of diffusion, confinement effects, selectivity, and so forth [47].

ORGANOCATALYSIS: A NOVEL SYNTHETIC PHILOSO-PHY

The term "organocatalyst" is a concatenation of the words "organic" and "catalyst." The definition corresponds to a low molecular weight organic molecule which in substoichiometric amounts catalyzes a chemical reaction. The organocatalyst could be achiral or chiral and could be composed of C, H, N, S, and P. Organocatalysis has several advantages not only because of its synthetic range but also for the economic reasons. The absence of metal in organocatalyst brings an undeniable advantage considering both the principles of "green chemistry" and the economic point of view (Figure 3).

Nowadays organocatalysis is one of the hot research topics in advanced organic chemistry. It is a novel synthetic philosophy and mostly an alternative to the prevalent transition metal catalysis. Organocatalysts are often based on nontoxic organic compounds originating from biological materials. Organocatalysts can be Lewis bases, Lewis acids, Brønsted bases, and Brønsted acids. Most of the organocatalysis reported so far are explained by Lewis base mechanism. The Lewis base mechanism can be simplified as follows.

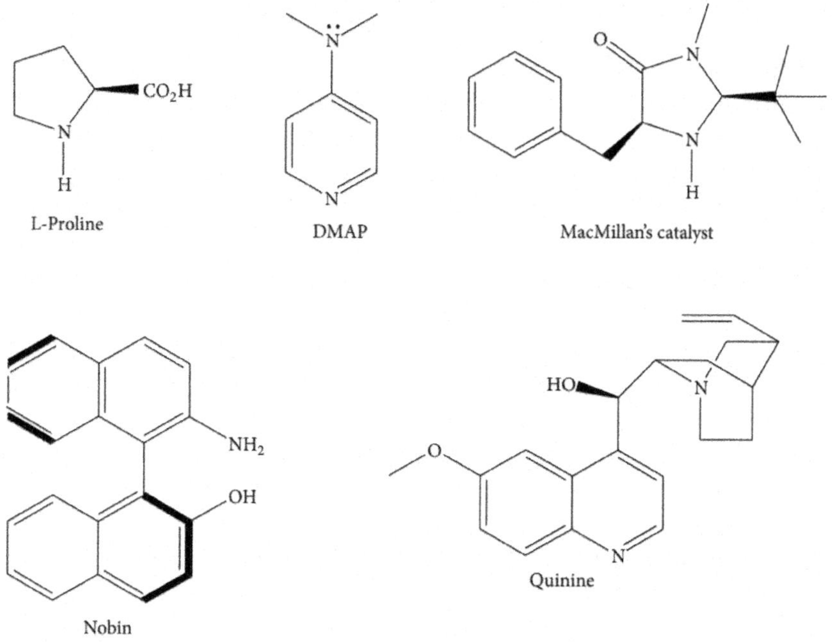

Figure 3: Few famous examples of organocatalysts: L-proline, DMAP, quinine, and MacMillan's catalyst.

Step I: Lewis base, for example, cinchona alkaloid, initiates catalysis via a nucleophilic addition to, or deprotonation of, the substrate. Step II: the chiral intermediate undergoes reaction. Step III: separation of product(s) from the catalyst occurs. Step IV: regeneration of the catalyst and its availability for a new catalytic cycle occurs. There are many organocatalytic reactions that need a second mechanism to explain the catalytic cycle involved. This is called bifunctional organocatalysis in which the organocatalyst possesses not only a Lewis base (nitrogen or phosphorus atom) but also a Brønsted acidic site. There is a lot to be done in redox catalysis and in exploitation of the synergetic effect of different catalytic active sites within one organic compound or material.

Some advantages of organocatalysts include the following.(i)Reactions performed by organocatalysts have potential application in large scale production in industries.(ii)Scope: organic reactions occur which are not in practice and not possible by other forms of catalysts, for example, asymmetric synthetic processes.(iii)Price/Availability: Low cost; alkaloids, natural amino acids, L-proline, tartaric acid, and so forth are easily available and are economically attractive.(iv)Recycling issues: immobilization of the catalyst is the simplest way to recover it.(v)Some of the examples include bioderived and biodegradable organocatalysts.(vi)Organocatalysts work at mild reaction conditions.(vii)Sometimes the final products of reactions contain high levels of metal contamination derived from catalyst degradation phenomena, which pose a serious drawback if the metal is toxic for pharmaceutical and food industries. Absence of transition-metal brings organocatalysis at the heart of greening chemistry. Organocatalysis can be categorized into our different types as follows. Type-I: activation of the reaction based on the nucleophilic/electrophilic properties of the catalyst is done (Scheme 1).

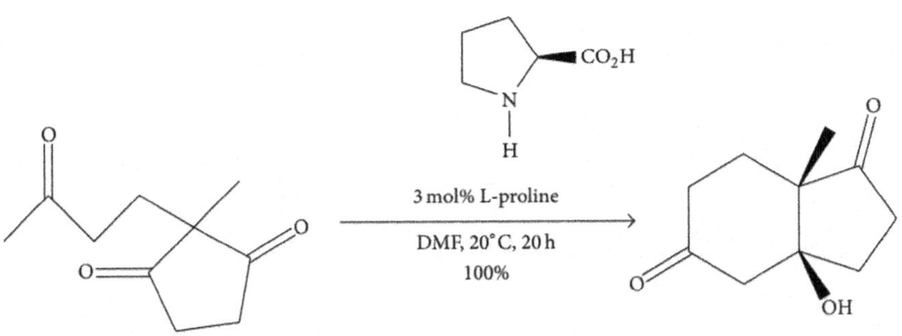

Scheme 1: Hajos and Parrish [70].

Type-II: organic molecules form reactive intermediates. The chiral catalyst is consumed in the reaction and requires regeneration in parallel catalytic cycle as it is reported in Scheme 2.

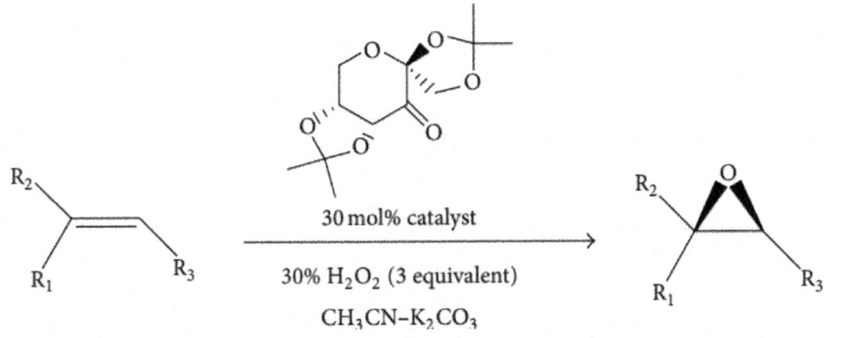

Scheme 2: Shu and Shi [71].

Type-III: phase transfer reactions occur. The chiral catalyst forms a host-guest complex with the substrate and shuttles between the standard organic solvent and a second phase. Catalytic enantioselective enolate alkylation (Scheme 3) occurs

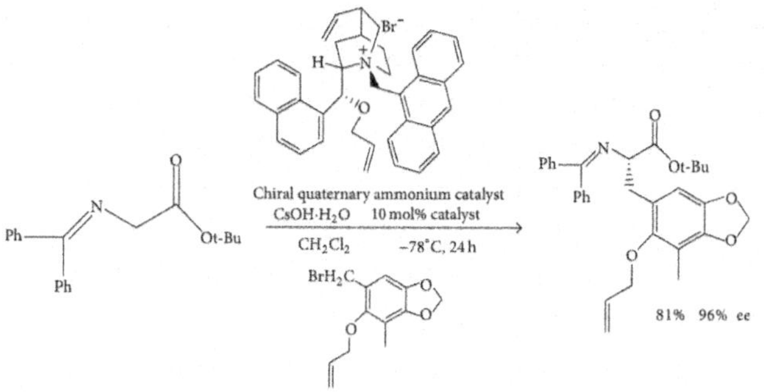

Scheme 3: Corey et al. [72].

Type-IV: molecular-cavity-accelerated asymmetric transformations occur in which the catalyst may select from competing substrates, depending on size and structure criteria (Scheme 4).

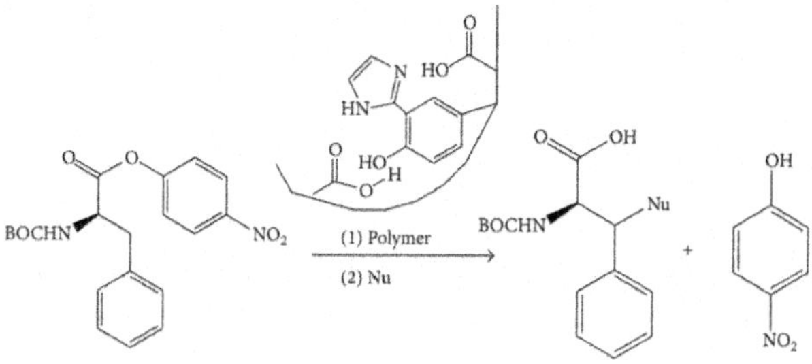

Scheme 4: Sellergren et al. [73].

DISCOVERY AND THE HISTORY OF ORGANOCATALY-SIS

Liebig's synthesis of oxamide from dicyan and water is the first organocatalytic reaction reported [74]. Acetaldehyde was further identified to behave as the then-named "ferment," now called as enzyme. Dakin in 1909 demonstrated that in a Knoevenagel type condensation between aldehydes and carboxylic acids or esters with active methylene groups, the amine catalysts could be mediated by amino acids [75]. Langenbeck is remembered for developing enamine type reactions and the application of simple amino acids and small oligopeptides as catalysts [76–78]. Natural products, in particular, strychnine, brucine, and cinchona alkaloids and amino acids (including short oligopeptides), were among the first organic catalysts tested [79, 80]. Though the field of small-molecule catalysts seems very simple, there are evidences that prove their pivotal role in synthesis of building blocks for life. L-Alanine and L-isovaline capable of catalyzing certain C–C bond formation reactions had been found on meteorites [81].

An asymmetric transformation with an organic molecule was published in 1912 by Bredig and Fiske [82]. Pracejus applied cinchona alkaloids in the asymmetric conversion of ketenes to (S)-methyl hydratropate [83,84]. An initial research by Yamada & Otani was important, too [85]. The seventies brought another milestone; Hajos and Parrish reported a L-proline catalyzed Robinson annulation in excellent enantioselectivities [70, 86,87]. Eder et al. reported organocatalytic aldol reactions in good enantioselectivities [88]. To summarize, in 1970s, Hajos and Parrish and, independently, Eder, Sauer, and Wiechert published a series of papers and patents involving such organocatalytic transformations. The asymmetric synthesis of enediones, useful building blocks in natural product total syntheses, was achieved. (S)-Proline induced the formation of (S)-enediones.

Two mechanisms [89] were proposed: (1) involving the formation of a carbinolamine intermediate, followed by the displacement of the proline moiety by nucleophilic attack of the enol from the side chain ketone; (2) involving an enaminium intermediate acting as a nucleophile in the C–C bond formation with concomitant NH···O hydrogen transfer. Wynberg firstly published various 1,2 and 1,4 additions catalyzed by cinchona bifunctional organocatalysis [90–92]. All these research endeavours were not carefully looked into till List et al. reported in 2000 proline catalyzed aldol reaction in good enantiomeric excesses [93, 94]. Asymmetric catalysis constitutes one of the most important subjects in synthetic organic chemistry. Asymmetric synthesis achieving atom economy is a challenge for organic synthesis and homogeneous catalysis using metal complexes leads the way. But the application of such methodologies in chemical industry is rather limited due to the high cost of chiral ligands and noble metals used in such transformations. Moreover, the pharmaceutical entities and food industry products do not tolerate a contamination, of even traces, of any such metals for that matter. Synthesis of chiral drugs has a major financial impact in pharmaceutical industries on a global scale. Different enantiomers or diastereomers of a molecule often have different biological activity. Most of the drugs might be soon sold as single enantiomers in the coming years. There is a huge demand for organocatalysts leading to 100% yield, 100% ee, of the desired product. The need from chemical industry, especially pharmaceutical, for reliable asymmetric transformations of molecular skeletons is higher than ever. Therefore, asymmetric organocatalysis is in a process of attaining maturity into a very powerful, practical, and broadly applicable methodological approach in the catalytic asymmetric synthesis.

SOME EXAMPLES AND DEVELOPMENTS IN CATALYST STRUCTURAL VARIANTS

Following are the famous examples of organocatalysts.

Proline and Derivatives

The proline catalyzed Robinson annulation was one of the earliest examples of an enantioselective reaction using an organic catalyst [85]. This amino acid contains both a nucleophilic secondary amino group and a carboxylic acid moiety functioning as a Brønsted acid. The availability of proline in both enantiomeric forms brings advantages over enzymatic methods. Arguably, the first example of proline catalyzed asymmetric aldol reaction is reported by Hajos and Parrish [70, 86, 87]. The Wieland-Miescher ketone is a useful synthetic building block for which a classical asymmetric procedure using (S)-proline was published forty years ago [88]. It is quite significant that the first efficient organocatalyzed asymmetric reactions were described by Hajos & Parrish [86, 87] and by Eder et al. [88], both teams from pharmaceutical companies (Hoffmann-la-Roche and Schering). The asymmetric synthesis of the Wieland-Miescher ketone is also based on proline and another early application was one of the transformations in the total synthesis of erythromycin by Woodward et al. [95]. Proline is recognized as a versatile organocatalyst of these times (Scheme 5).

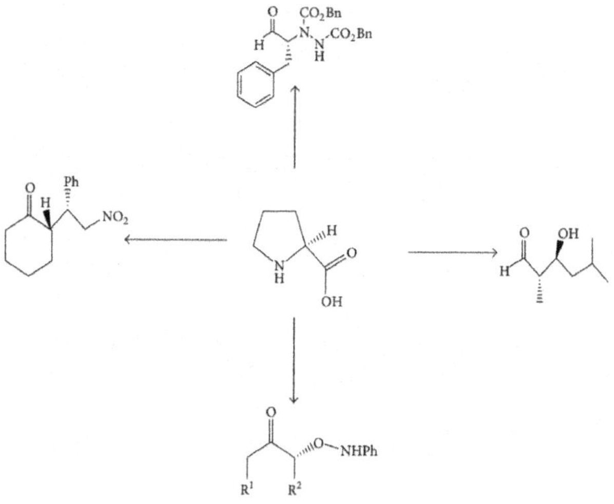

Scheme 5

In fact, the renewal of proline catalyzed transformations in early 2000 by D. W. C. MacMillan, Carlos F. Barbas III, and Benjamin List saw the renaissance of the concept and it was the starting point of the word "organocatalysis." Professor Ahrendt group reported the first enantioselective organocatalytic Diels-Alder reaction [96]. Proline was identified as an effective organocatalyst for asymmetric aldol reaction [97]. The catalytic asymmetric α-alkylation of aldehydes was reported by Vignola and List [98]. This transformation had been accomplished with the help of covalently attached auxiliaries. α-Methyl L-proline exhibits higher enantioselectivities and improved reaction rates when compared to L-proline. In recent years, proline, especially L-proline, has been used to catalyze essential transformations used in the fine chemical and pharmaceutical industries, such as the direct asymmetric aldol reactions [99–107], Diels-Alder reactions, Michael reactions [108–115], Mannich reaction [116–122], Multicomponent reactions [123, 124], and α-amination [125], α-aminoxylation [126–128], α-oxyaldehyde dimerization [100], α-functionalization of carbonyl compounds [129], and α-alkylation of aldehydes [130]. Organocatalytic cyclopropanation reactions were typically performed using catalyst-bound ylides. However, Kunz and MacMillan demonstrated that activation of olefin substrates using catalytic (S)-(−)-indoline-2-carboxylic acid is a viable route for the formation of highly enantioenriched cyclopropanes [131].

Prof. K. A. Jørgensen has made an important breakthrough in asymmetric synthesis by developing (R)- and (S)-α,α-bis[3,5-bis(trifluoromethyl)phenyl]-2-pyrrolidinemethanol trimethylsilyl ether as excellent chiral organocatalysts in the direct organocatalytic α-functionalization of aldehydes. Diarylprolinol silyl ether reagents were found to catalyze C–C, C–N, C–O, C–S, and C–Hal bond forming reactions [132–138]. Cascade or domino reactions swiftly construct complex biologically important compounds and minimize waste and also the laboratory operations [139]. Enders' group developed proline derived organocatalyst for a chemo-, diastereo-, and enantioselective three-component domino reaction to yield tetrasubstituted cyclohexene carbaldehydes [140]. The four stereogenic centers are generated in three consecutive C–C bond formations, that is, Michael/Michael/aldol condensation with high diastereo- and complete enantiocontrol. The synthesis of polyfunctional cyclohexene building blocks involving proline catalytic steps is reported in this domino reaction.

Though this segment of our review is dedicated to proline as an important class of organocatalysts, it would also be a good place to acknowledge the research contributions from professor Ahrendt et al. to the field of organocatalysis [96, 99, 100, 103, 127, 131, 141–150].

Heterogenized Proline. An explosion of research articles in the area of homogeneous organocatalysis occurred within the last decade or so. In addition to the initial proline catalyzed reactions, the word "organocatalysis" covers nowadays many other well-known organocatalysts and organic reactions. Proline immobilized onto some sort of inorganic support paves the way towards heterogenization of other organocatalysts [151–153]. A simple and efficient synthesis of polystyrene-supported proline and prolinamide is reported. Polystyrene-supported proline catalyzes the asymmetric aldol reaction between cyclohexanone and substituted benzaldehydes in water. High yields, diastereoselectivities, and ee values have been observed. The versatility of this resin was also confirmed by selenenylation of aldehydes. Both proline and prolinamide resins gave high yields. Recycling studies showed that the proline resin performs as better heterogeneous organocatalyst as compared to the prolinamide resin [153]. An aldol reaction catalyzed by proline functionalized silica gel was investigated in continuous flow microreactors by reaction-progress kinetic analysis and nonlinear chromatography [154].

Though the research community advocates asymmetric synthesis for industrial application, organocatalysts have been seldom applied in industry, due to insufficient efficiency and the difficulties in catalyst separation and recycling [155–161]. The concept of heterogenizing organocatalyst on supports combines the advantages of both homogeneous organocatalysis and heterogeneous recycling. Though this might seem simple and present opportunities in industrial chemical processes, the research in heterogeneous organocatalysis demands a multidisciplinary approach by identifying catalytic activity upon bonding organocatalyst covalently to the support; understanding noncovalent interaction as structure-reactivity and conformation deciding factor; taking into consideration the characteristic properties of the supports, confinement effects, if any, and so forth. Some recent examples will be described on the applications of supported proline as organocatalysts for their preindustrialization study. The latter half of this review, in general, irons out the detailed discussion on key trends with examples, scope, challenges, and "green" applications of heterogeneous organocatalysts.

Other Amino Acids

Barbas and coworkers found the proteinogenic amino acid tryptophan (Figure 4) to be an excellent organocatalyst for the Mannich reaction of hydroxyacetone with a variety of imines performed in DMF leading the desired anti-amino alcohols in excellent diastereoselectivities (up to >19:1) and enantioselectivities (90–98% ee). Similarly, t-butyl protected threonine catalyzed the aldol reaction of hydroxyacetone and various aldehydes in NMP to give the corresponding syn-aldol adducts in high yields and good to excellent enantio- and diastereoselectivities [162]. Zhao et al. reported [163] the development of an amino acid based small molecule capable of promoting asymmetric monosilylation of meso-1,2-diols. Furthermore, the catalyst can be recycled with equal efficiency. Three-component reaction involving condensation of aldehyde, 1,3-carbonyl compound, and (thio)urea is known as Biginelli reaction. Various organocatalysts such as tartaric acid, oxalic acid, citric acid, and lactic acid were found effective in producing Biginelli products [164–172].

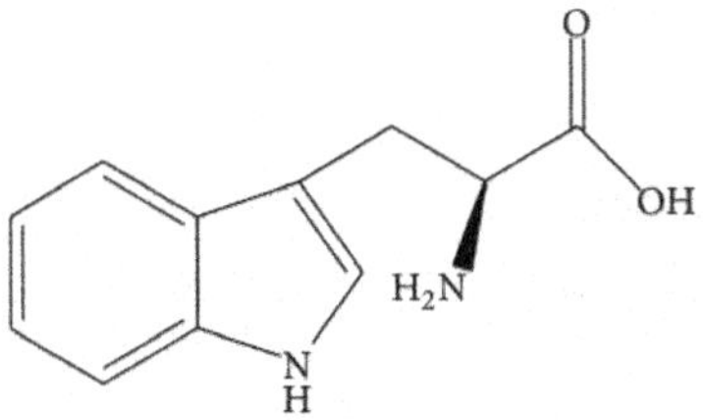

Figure 4: L-Tryptophan.

Because of the "privileged" green nature of this field, it has become an obligation for scientific community and academic institutions to be mindful and teach this branch of catalysis to students and also keep it as a driving force in the generation and dissemination of knowledge in pursuit of efficient synthetic methodologies and processes using organocatalysts. The synthesis of new chiral building blocks or complex molecular structures involving organocatalytic step(s) is being reported almost every day in scientific literature. As it is not possible here to discuss all the catalysts individually in detail, the author decides to present salient developments in organocatalysis and allied research areas.

Synthesis of prochiral cyclohexanones by amine catalyzed self-Diels-Alder reactions of α,β-unsaturated ketones in water brought coupled the concept of green chemistry with catalysis. Useful asymmetric heterodomino reactions for the highly diastereoselective synthesis of symmetrical and nonsymmetrical synthons of benzoannelated centropolyquinanes were reported [173, 174].

In recent years, trials are being made by using amino acids, especially L-proline, to catalyze essential transformations in the fine chemical and pharmaceutical industries. The interest in this field has thus increased spectacularly in the last few years. However, organocatalytic methods often require catalyst loadings as high as 30 mol% for the achievement of high conversions in reasonable reaction times [175]. Some critics suggest that low turnover numbers might limit the potential uses of organocatalysis for industrial applications. Process development and scale-up are being tested for several organocatalytic reactions confirming organocatalysis as a valuable tool for industrial scale solutions. Scientists are reporting novel organocatalytic systems in various reputed international research journals and those catalysts exhibit various valuable functions derived from a variety of structural, ionic/ electronic, and chemical properties. Broad variety of efficient syntheses will contribute to an increasing number of organocatalytic large scale reactions in the future. The organocatalysts shall be used in various fields, especially in asymmetric organic synthesis.

Amine and Chiral Vicinal Diamines

Proline and pyrrolidine derivatives, commonly known as Hayashi-Jørgensen catalysts, in the form of ferrocenyl pyrrolidine, are applied as organocatalyst wherein the ferrocene moiety controls the conformational space and a simple alkyl group effectively covers a face of the derived enamine (Figure 5) [180].

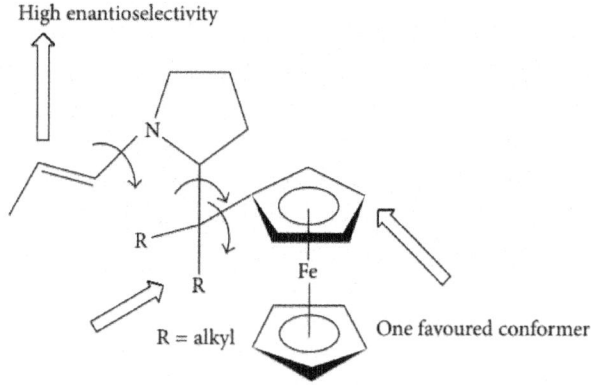

Figure 5

Sulzer-Mossé and Alexakis reported chiral amines catalyzing asymmetric conjugate addition to nitroolefins and vinyl sulfones via enamine activation [181].

A bifunctional H-bond directing aminocatalytic system is found to achieve high stereo- and regiocontrol over dienamine mediated hetero-Diels-Alder reaction [183].

Though amine catalysis is constructing a wide variety of chiral scaffolds in asymmetric synthesis, organocatalysis with amines for nonasymmetric transformations is also developing very fast [184, 185]. The [2+2] cycloaddition of enals with nitroalkenes is catalyzed by a chiral secondary amine in the presence of an achiral thiourea for the enantio- and diastereoselective synthesis of highly functionalized cyclobutanes. Mechanistically, two consecutive Michael reactions proceed through an unprecedented combination of dienamine/iminium activation mode [186]. The reaction engineering of dienamine catalysis for single-step syntheses of highly functionalized molecules such as natural products or bioactive compounds is reviewed in literature [187].

Chin and coworkers have recently reported some preliminary theoretical and experimental studies for converting a parent diamine into other chiral vicinal diamines. These diamines can be handy as ligands for chiral catalysts or they can make chiral heterocyclic rings and betalactams [188–190]. There are some C–C bonds forming reactions catalyzed by such diamine organocatalysts (Scheme 6) [191].

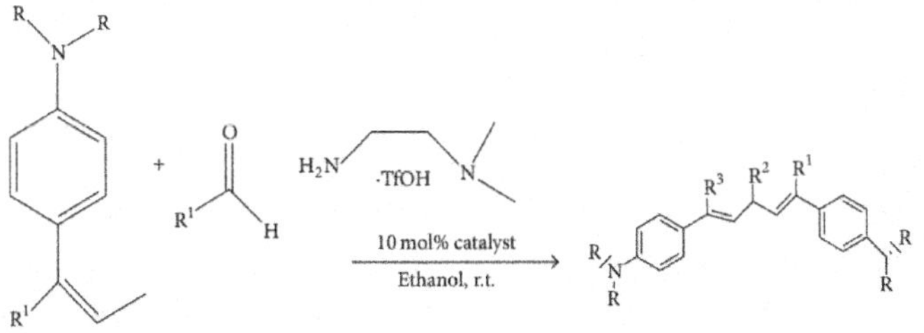

Scheme 6

Amides

Recently, the synthesis of multifunctional organocatalysts, easily obtained by the condensation of (S)-proline with 2-aminopyridine, 2,6-diaminopyridine, or 2-aminoimidazole, is reported. These chiral prolinamides promoted the aldol condensation between cyclohexanone and different aromatic aldehydes (up to 98% ee) and also catalyzed Diels-Alder and Michael reactions [192, 193].

Rationally designed pyrazole amides function as Michael donors in urea catalyzed asymmetric Michael reactions with excellent chemical and optical yields [194].

A structure-activity relationship with regard to formamides as organocatalysts is explored. This study highlights that the cis-conformation of secondary formamides is the reactive conformation in the allylation of aldehydes with allyltrichlorosilane (Scheme 7) [195].

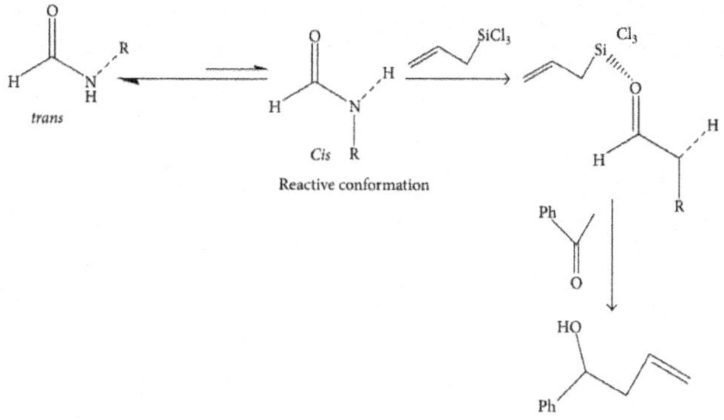

Scheme 7

Imidazolidinones

A certain class of imidazolidinone compounds (also called MacMillan imidazolidinone organocatalysts) function as catalysts for many asymmetric organic transformations such as asymmetric Diels-Alder reactions. The original such compound was derived from the chiral biomolecule phenylalanine [196, 197]. The first highly enantioselective organocatalytic Diels-Alder reaction using a chiral organocatalyst was reported in MacMillan's pioneering work where the activated iminium ion, formed through condensation of the imidazolidinone and an α,β-unsaturated aldehyde, reacted with various dienes to give [4+2] cycloadducts in excellent yields and enantioselectivities. Ahrendt et al. reported many other asymmetric reactions such as the 1,3-dipolar cycloadditions, Friedel-Crafts alkylations, α-chlorinations, α-fluorinations, and intramolecular Michael reactions using MacMillan's organocatalysts [96, 150, 198–201]. Ouellet et al. also reported the combination of imidazolidinone organocatalyst and Hantzsch ester to facilitate the first enantioselective organocatalytic hydride reduction of α,β-unsaturated aldehydes (Scheme 8) [202, 203]. This imitated nature's stereoselective enzymatic transfer hydrogenation with NADH cofactor.

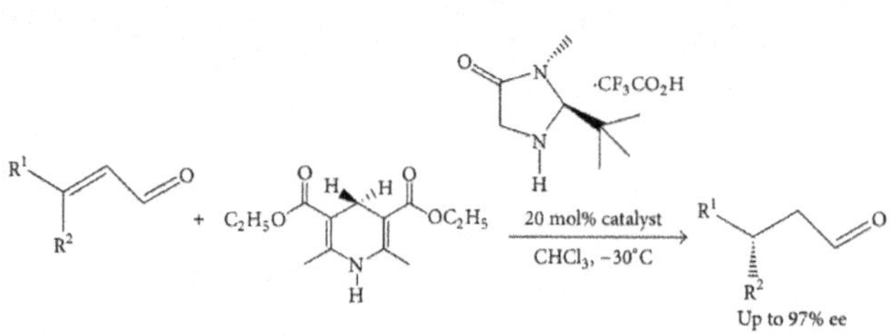

Scheme 8: Ouellet et al. [202].

Noncovalent interaction is a structure-reactivity and conformation deciding factor which requires immediate attention for understanding organocatalysis [204]. Of special interest would be the identifying and understanding of other noncovalent interactions such as charge transfer [205], p–p stacking immobilization [206], and adsorption or entrapment of organocatalysts in nanomaterials [207].

Homogeneous and Heterogenized MacMillan's Organocatalyst. An asymmetric Diels-Alder reaction was catalyzed with high efficiency and recyclability by a soluble, "self-supported" chiral organosilica polymer with embedded imidazolidinone catalytic moieties [208].

Chiral Phosphoric Acids

BINOL derived phosphoric acids catalyze nucleophilic addition reactions to imine substrates. A direct Pictet-Spengler reaction was reported using a geminally disubstituted tryptamine organocatalyst to form isoquinolines in excellent yield and enantiomeric excess [209]. Phosphoric acid catalyzing the reduction of an imine with a Hantzsch ester in good enantiomeric excess was firstly reported by Rueping et al. Hoffmann et al. reported an improvement to this methodology [210, 211].

Storer research group reported a one-pot reductive amination of a range of methyl ketones and aryl amines [212]. Reductive amination of 2-butanone over silylated phosphoric acid MacMillan TiPSY catalyst was obtained with good enantiomeric excess.

Zhou and List published on the use of chiral Brønsted acid (R)-TRIP for highly enantioselective synthesis of pharmaceutically relevant 3-substituted cyclohexylamines from 2,6-diketones wherein the achiral amine substrate accelerates the cascade reaction before getting incorporated into the final product (Scheme 9) [213].

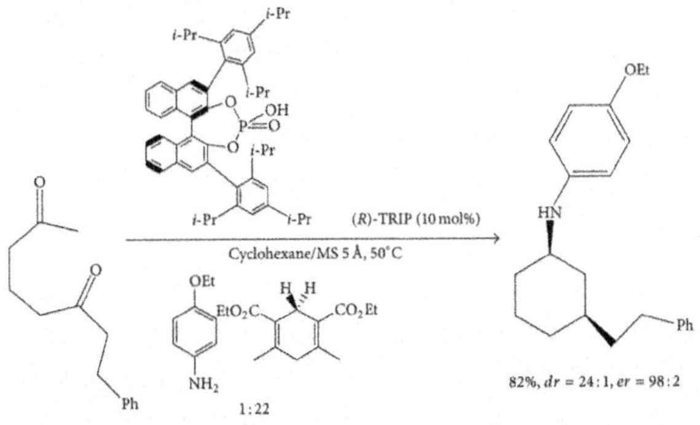

Scheme 9: Zhou and List [213].

Rowland et al. reported VAPOL derived phosphoric acid catalyzed addition of sulfonamides to BOC protected aryl imines and protected aminals in excellent enantioselectivities [214]. The (R)-TRIP catalyzed the aza-Diels-Alder reaction of aldimines with Danishefsky's diene to yield piperidinone derivatives with high enantioselectivity. Addition of acetic acid was found to improve both reactivity and enantioselectivity [215]. Six chemical bonds, five stereogenic centers, and three cycles were formed in one-pot four-component Ugi type reaction catalyzed by a chiral BINOL derived phosphoric acid [216]. BINAP is one of the earliy commercially applied ligands in industrial catalysis [217]. A microporous recyclable heterogeneous catalyst, offering excellent enantioselectivity, was made from a 1,1'-binaphthalene-2,2'-diol (binol) derived phosphoric acid chloride and was found as active as the corresponding homogeneous catalyst [218].

Chiral Diols

Huang et al. reported on use of TADDOLs (Figure 6) as Brønsted acid organocatalysts in highly stereoselective hetero-Diels-Alder reactions [219].

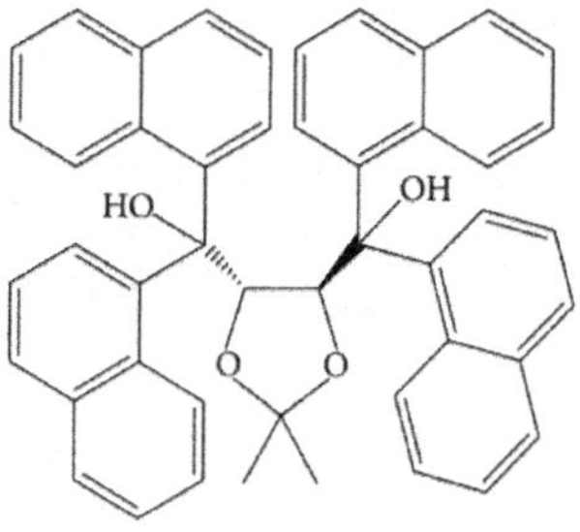

Figure 6

The α-amination of carbonyl compounds has also been accomplished by using the 1-naphthyl TADDOL derivative as a Brønsted acid organocatalyst [220]. McDougal and Schaus reported asymmetric Morita-Baylis-Hillman reaction, that is, the addition of cyclohexenone to different aldehydes catalyzed by octahydro-BINOL derived Brønsted acid [221].

Jacobsen Thioureas

Professors Vachal and Jacobsen identified chiral thioureas (Figure 7) as versatile and effective organocatalysts. 1 mol% of the thiourea was reported to have catalytic activity in the hydrocyanation (Strecker reaction) of both aldimines and ketoimines with very high enantioselectivities [222].

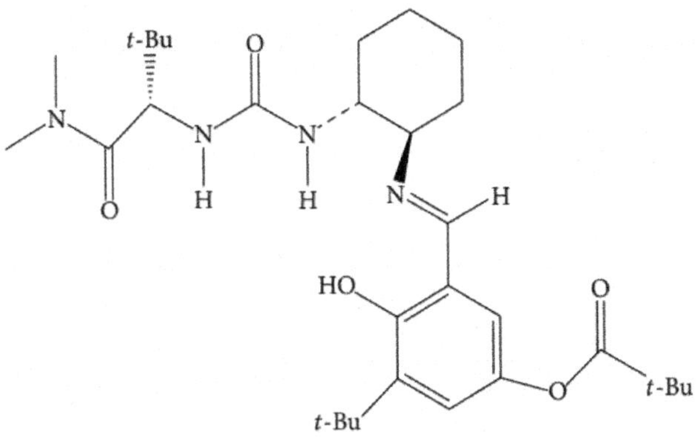

Figure 7

Imine hydrophosphonylation, particularly effective with electron-withdrawing ester substituents on the phosphate, took place in the presence of 10 mol% of the catalyst [223]. Mannich reactions of BOC protected imines have also been reported (Scheme 10) [224] with excellent yields and enantioselectivities and show heterocyclic substrate tolerance.

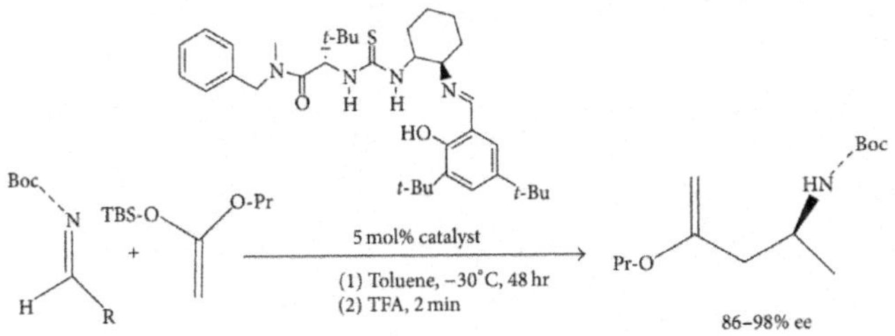

Scheme 10

A variant of thiourea organocatalyst was found to catalyze the cyanosilylation of ketones and aldehydes in high yields and enantiomeric excesses [225]. Thiourea organocatalyst was put to use in the acyl-Pictet-Spengler reaction to form tetrahydro-β-carbolines [226]. The acyl-Mannich reaction, providing a route to enantioenriched heterocycles from aromatic starting materials and trichloroethyl chloroformate (TrocCl), is also catalyzed by the thiourea organocatalyst [227]. Bifunctional thiourea-base catalyzed double-Michael addition of benzofuran-2-ones to dienones is reported for the synthesis of optically enriched spirocyclic benzofuran-2-ones [228]. In another interesting example, benzoylthiourea-pyrrolidine catalyst was synthesized and used in the asymmetric Michael addition of ketones to nitroalkenes [229]. Primary amine-thioureas were improvised to catalyze difficult Michael reactions synthesizing (S)-baclofen, (R)-baclofen, and (S)-phenibut [230].

Cinchona Alkaloids

The cinchona alkaloids catalyze many useful processes with high enantioselectivities. Asymmetric phase transfer catalysis (PTC) presents a "green" alternative to homogeneous synthetic organic chemistry methodologies. Cinchona alkaloids are synthetically modified for their usage in asymmetric PTC. O-Alkyl N-aryl methyl derivatives of cinchona alkaloid led highly enantioselective alkylation of glycine imines to generate α-amino acid derivatives. And, for the same reaction, dimeric cinchona alkaloid obtained 97–99% enantiomeric excess [231, 232]. Cinchona alkaloids deprotonate substrates with relatively acidic protons forming a contact ion pair between the resulting anion and protonated amine. This interaction generates a chiral pool around the anion and facilitates enantioselective reactions with electrophiles. The control over the formation of quaternary asymmetric centers is essential to obtain high enantiomeric excesses. The α-functionalization of ketones (Scheme 11) by the addition of TMSCN to the corresponding cyanohydrin in excellent yield and enantiomeric excess is catalyzed by the (DHQD)$_2$AQN catalyst [233].

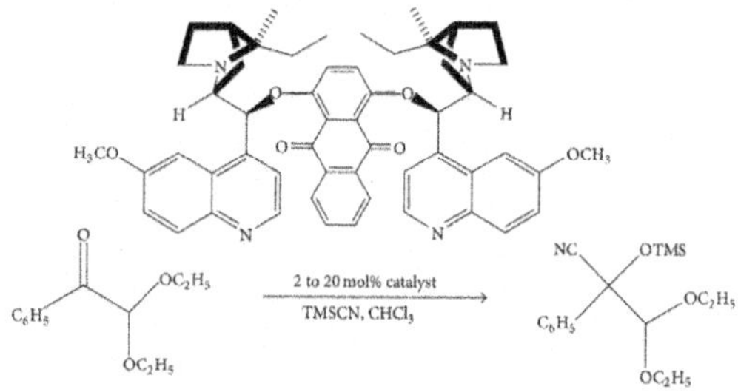

Scheme 11: Tian et al. [233].

Increasingly remote stereocenters are being targeted in asymmetric aminocatalysis. Organocatalytic allylic amination presents an alternative to the conventional palladium catalyzed methodology. Amination at the remote γ-position using (DHQ)$_2$PYR forms highly functionalized amine compounds [234].

Bella and Jørgensen's reported the first enantioselective conjugate addition of β-diketones to both aromatic and aliphatic alkynones using (DHQ)$_2$PHAL catalyst [235]. Cinchona based primary amine catalysis is found to offer high efficiency and reliability in the asymmetric functionalization of carbonyl compounds [236, 237]. Organocatalysts derived from cinchona alkaloids function as Lewis base catalyst in the cyclizations of allenoates with electron deficient olefins and imines and emerged as important synthetic tool in the preparation of biologically active and pharmaceutically interesting cyclic compounds, including natural products [238]. Recently, cinchona thioureas have been reported for the first time as catalysts in the area of asymmetric oxidations [239]. Cinchona and cinchonidine derived catalysts perform asymmetric direct aldol reaction of pyruvic aldehyde dimethyl acetal with isatin derivatives [240]. Likewise, there are various other examples reported where cinchona or its variants are used as organocatalysts for syntheses of various biologically important molecules [241–246].

Maruoka Phase Transfer Catalysts

The chiral catalyst forms a host-guest complex with the substrate and shuttles between the standard organic solvent and a second phase, example, catalytic enantioselective enolate alkylation [72]. As we know, chiral quaternary ammonium catalysts can be useful in asymmetric synthesis. Kitamura et al. reported C2-symmetric ammonium salts catalyzing monoalkylation of glycine derived Schiff bases with alkyl halides to synthesize α-alkyl-α-amino acids under remarkably low catalyst loadings [247–249].

Utilizing the solubility of aqueous quaternary ammonium salts into organic solvents, professor Maruoka developed spirotype chiral ammonium salt with two binaphthyl rings as a novel chiral phase transfer catalyst. Figure 8 depicts the organocatalyst for direct asymmetric aldol reaction of β-hydroxy-α-amino acid derivative.

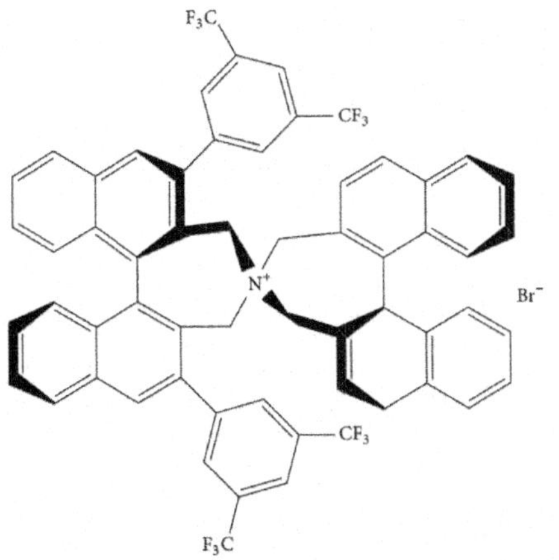

Figure 8: See [176].

Oxazaborolidines

Chiral oxazaborolidines (known as CBS oxazaborolidines after Corey, Bakshi, and Shibata) catalyze the reduction of prochiral ketones, imines, and oximes to produce chiral alcohols, amines, and amino alcohols, respectively, in high yields and excellent enantiomeric excesses. The chiral Lewis acid generated from o-tolyl-CBS-oxazaborolidine after protonation with trifluoromethanesulfonimide is very useful in the enantioselective Diels-Alder reaction (Scheme 12) [250–256].

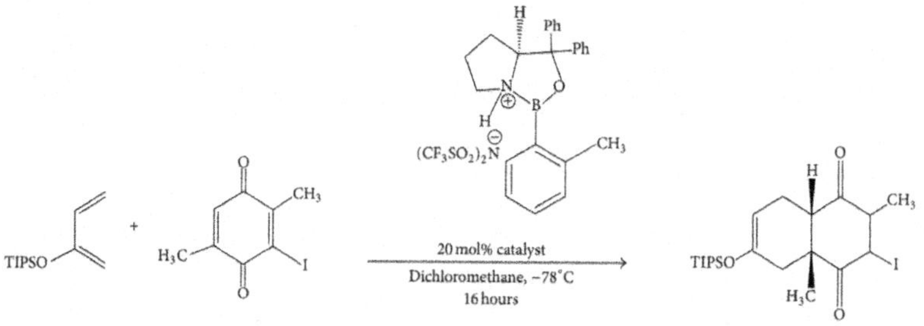

Scheme 12: See [250–256].

Shi Epoxidation Catalyst

Tu et al. reported a fructose derived ketone in efficient asymmetric epoxidation of trans-olefins (Scheme 13) [257]. Epoxidation catalyst is able to epoxidize trans-alkenes and certain cis-alkenes with good to excellent yields and selectivities [258, 259]. Marigo et al.'s group reported an asymmetric organocatalytic epoxidation of -unsaturated aldehydes with environmentally friendly oxidant hydrogen peroxide [260].

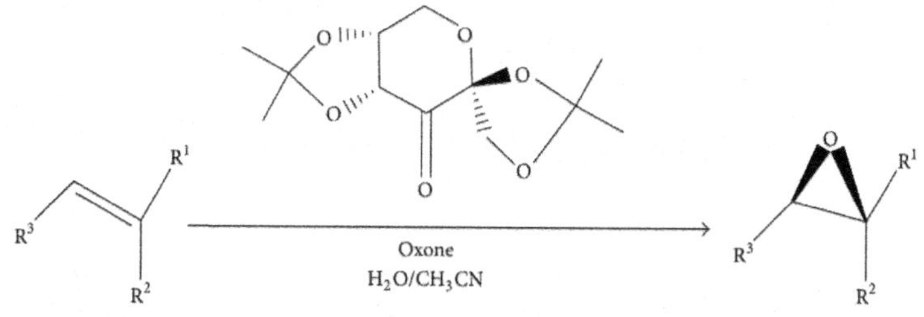

Scheme 13

N-Heterocyclic Carbene (NHC) as Organocatalysts

Rovis Catalyst

The conjugate addition of an aldehyde to an α,β-unsaturated compound (Stetter reaction) is a practical methodology for the construction of 1,4-dicarbonyl compounds bearing quaternary stereocenters. Triazolium salt in the presence of a base functions as an N-heterocyclic carbene organocatalyst (Figure 9) in highly enantioselective intramolecular Stetter reactions [261, 262].

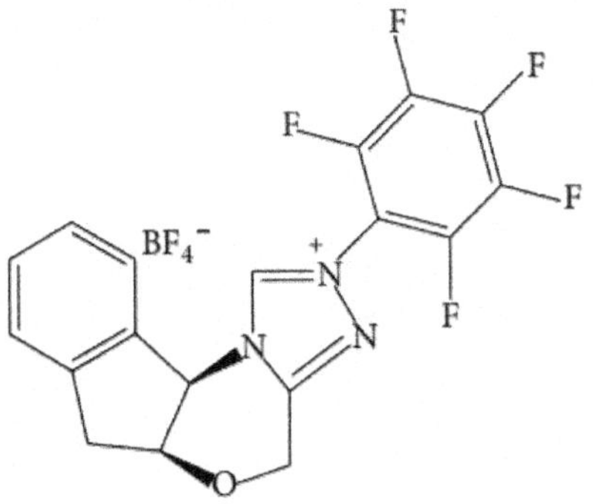

Figure 9

Bode Catalyst

Bode's studies on NHCs (Figure 10) mainly involve catalytic generation of reactive species such as enolates, homoenolates, and activated carboxylates. N-Mesityl substituent on an imidazolium or triazolium NHC precursor generates highly enantioselective annulations from simple starting materials under mild reaction conditions [263, 264]. In an interesting example, the catalysts facilitate highly enantioselective cyclopentene forming annulations of simple enals and activated enones [265]. Redox esterifications and amidations of α-functionalized aldehydes using achiral catalyst are also reported [266]. Uniting unique activation modes of N-heterocyclic carbene (NHC) catalysts with the concept of domino reactions, a new fast-growing field came into the spotlight in last couple of years or so [267–269].

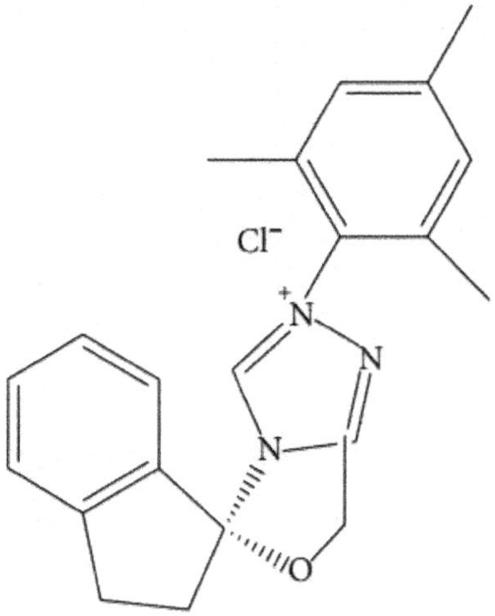

Figure 10

Domino Michael/aldol reaction between 2-mercaptoquinoline-3-carbaldehydes and enals was catalyzed by diphenylprolinol silyl ether producing 2H-thiopyrano[2,3-b]-quinolines in excellent yields and enantioselectivities [270]. Zhao et al. reported studies on "cooperative NHC and rutheniun redox catalysis" in the oxidative esterification of aldehydes [271]. Blanc et al. reported NHC mediated organocatalytic one-pot allylstannation to generate syn-diols [272]. With enals as starting aldehydes, elegant cascade processes have been developed using oxidative carbene catalysis [273–276]. Over the last few years, there are examples reported in literature where various organocatalysts rally domino reactions and attract their application in organic processes in industries [174, 175, 180, 181, 183].

Apart from the aforementioned organocatalysts, there are other organic systems which upon appropriate activation function as catalysts in useful asymmetric syntheses [155–157]. 1,2-Dicarbonyl compounds, because of their adjacent multiple reactive centers, upon activation, act as efficient pronucleophiles in asymmetric organocatalyzed sequential or domino transformations including C–C or C–N bond formation [277]. Cyclopropenone catalyzed conversion of aldoximes and primary amides into nitriles in a one-pot mild reaction widens the scope of the utilization of cyclopropenones in organic synthesis [278]. The first asymmetric organocatalytic synthesis by phosphine catalyzed [3+2] cycloaddition of allenoates onto fullerene is reported to yield enantiomerically pure carbocyclic fullerene derivatives under mild conditions [279]. A new [3+2] cycloaddition/cycloreversion strategy for organocatalytic and thermally allowed carbonyl-olefin metathesis is reported [280]. Recently, a strong evidence for halogen bond based organocatalysis is reported for the reaction of 1-chloroisochroman with ketene silyl acetals [281]. Acid catalyzed asymmetric acetalizations of aldehydes are also described [282]. beta-Isocupreidine catalyzed the enantioselective Morita-Baylis-Hillman reaction of maleimides with isatin derivatives to form 3-substituted 3-hydroxyoxindole derivatives [283, 284].

Recent advances in asymmetric catalysis resulted in utilizing the less discussed electrophilic properties of iminium type intermediates in complex annulations of indoles [285, 286] and also pyridines [287, 288]. Shibatomi and Narayama discussed the catalytic enantioselective α-chlorination of carbonyl compounds and stereospecific substitution reactions of the resulting optically active alkyl chlorides [289].

Though there is no relation between the development of two different fields of organocatalysis and gold catalysis, an interesting synthetic utility and merger of both catalytic systems in the same reaction flask is explored [290].

Following Are the Synthetic Methodologies Reported on Mannich Reactions Using Various Organocatalytic Systems. Asymmetric three-component Mannich reaction and anti-Mannich reactions were reported by organocatalysts [122, 175]. The title reaction provides facile access to enantioenriched 3,4-dihydroquinazolin-2(1H)-ones containing a quaternary stereogenic center in high yields with excellent enantioselectivities. Subsequent transformations lead to the convenient preparation of the anti-HIV drug DPC 083 and N-fused polycyclic compounds without loss of enantiomeric excess [291].

An asymmetric catalytic, desulfonylative Mannich reaction of keto imines with aldehydes, as catalyzed by diarylprolinol silyl ether, was developed. It gave the Mannich product in good yield with excellent anti- and enantioselectivity [292]. New chiral bis(betaine)s, containing two catalytically active centers, have been designed and have proven to be promising organocatalysts for the direct Mannich type reaction of azlactones with a broad spectrum of aliphatic imines [293]. Hong and Wang reviewed advances in asymmetric organocatalytic construction of 3,3'-spirocyclic oxindoles [294]. And the applications of organocatalysts in total synthesis and natural product synthesis include enantioselective synthesis of Amaryllidaceaealkaloids (+)-vittatine, (+)-epi-vittatine, and (+)-buphanisine [295, 296].

HETEROGENEOUS ORGANOCATALYSIS: NEW DIREC-TION IN RESEARCH AND DEVELOPMENT

Making of Heterogenous Catalyst from Homogeneous Catalyst: The Concept and Methodologies

With the knowledge of heterogeneous catalysis and chemistry of materials at hand, several modified and different systems are made these days and are tested for their catalytic applications in various organic transformations. Following are the most common and well-known methods for the heterogenization of homogeneous catalysts:(i)impregnation,(ii)occlusion in porous materials (ship-in-a-bottle),(iii)grafting or tethering (through covalent bond).

Impregnation is the immobilisation of catalytic element via electrostatic interactions with a solid support. One of the famous examples includes Rh-diphosphine cationic complexes impregnated on anionic resins via ion-pair formation [297]. Similarly, organocatalysts supported on solid could be envisioned to yield a recyclable catalyst.

Heterogenisation via catalyst entrapment is a typical method that is applied to zeolite supports and termed as "ship-in-bottle" catalysis. A famous example includes in situ synthesis of the transition metal complex in support cages [298, 299]. The great practical advantages of "ship-in-bottle" single-site heterogeneous catalysts were well illustrated by Zsigmond et al.'s research [300]. Similar approach, retaining steric factors, could be applicable as another methodology for heterogenization of organocatalysts.

While in grafting, the catalytic active site is directly anchored on the support. This procedure is previously used for supporting organometallic complexes (Figure 11) [177, 301–304]. The tethering technique involves a spacer that is introduced between the catalyst and the support. The characteristic chemical nature and the structure of the spacer should be chosen to avoid the steric hindrance, if any. Though structurally different, catalytic activity can be expected to be similar to catalyst in homogeneous state.

Catalyst immobilization has been widely explored with asymmetric transition metal catalysts and enzymes aiming to improve their applicability and practicability. Similarly, the immobilization strategy has also been frequently attempted for organocatalysis even before its renaissance. The preparation of supported or immobilized catalysts involve either organic polymers such as linear, non-cross-linked, or cross-linked polymers as supports or porous inorganic solids such as alumina, silica, zirconia, clays, zeolites, MCM, and SBA type mesoporous materials.

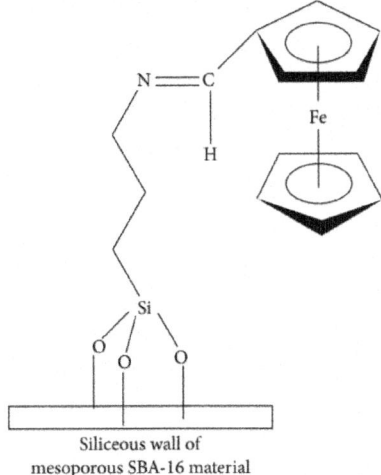

Siliceous wall of
mesoporous SBA-16 material

Figure 11: See [177].

Many inorganic supports have robust composition and offer large surface areas due to the presence of internal pores. Different routes to immobilisation of (homogeneous) catalyst can alter the nature of the catalytic active site. When the anchoring of catalytic moieties on silica is concerned, the key aspect is the functionalization of silanol groups on the surface. The pretreatment temperature of the parent silica determines the nature of the hydroxy groups (isolated, geminal, or hydrogen bound), their location, and reactivity. In nanostructured mesoporous silica of the MCM-41 type, the reactivity of the siloxanes is considered to be sufficient for covalent linking with a tether [305]. Subsequent silylation with trimethylchlorosilane neutralizes residual hydroxyl groups and increases the hydrophobicity. Various aluminosilicates and mesoporous silicas could be envisioned as suitable hosts for encapsulating and anchoring organocatalysts.

Organocatalysis has drawn a lot of interest in both academia and industry. The fact is that the asymmetric version of organocatalysis is the fastest growing field of synthetic organic chemistry. Few applications in industries include (i) epoxidation of chalcones and (ii) alkylation of cyclic ketones. Along with a commentary on very recent research endeavors in organocatalysis, the scope of heterogeneous organocatalysis in asymmetric synthesis, other important organic reactions, green chemistry, and scope for industrial applications are presented in this review.

Organic functionalization of nanostructured, nanoporous, and other nanomaterials widens the range of their applications and allows for the manipulation of the surface properties that control the interaction with various organic and inorganic guest species and thereby adds new catalytic functions to the materials. One of the basic premises behind the preparation of these kinds of materials is aiming at the development of heterogeneous and supported catalysis for their practical purposes that would have distinct advantages over widely investigated homogeneous catalysts and organometallic compounds that are vulnerable to decomposition due to oxidation of the ligand bound to the metal center.

Using a rationale, similar to that reported earlier for the immobilisation of transition-metal amino acid complexes [306–310], amino acids with functional side chains that can react with tethering silanes such as lysine, arginine, glutamic acid, cysteine, or amino acid derived organocatalysts could be covalently anchored to the inner walls of porous silica and other relevant inorganic structures. These materials have therefore been identified as promising supports for making organocatalysts heterogeneous for industrial significant reactions. This field is therefore well poised for their potential large scale applications in industry.

An organocatalyst attached to some sort of support/material is called heterogenized organocatalyst and this, as the term implies, makes the organocatalyst heterogeneous and thereby brings in all the advantages of heterogeneous catalysis to the system without losing the intrinsic catalytic activity of the organocatalyst. When used in heterogeneous system, one ensures no leaching of catalyst entity into the reaction medium and simple workup or easy separation of catalyst from the products. Such catalyst can be reused. In order to obtain heterogeneous organocatalyst, many approaches have been explored, including impregnation to layered material [311], covalent grafting to solid support [312, 313], electrostatic interaction between organocatalyst and the support [314], encapsulation to porous material [309, 310], and adsorptive method [315]. Several solid supports have been introduced for heterogenizing of organocatalysts in terms of covalent grafting such as polymer matrix, zeolites, amorphous silica, and mesoporous silica [316–319]. Due to the strong covalent bond with silica surface, the covalently anchored organic moieties are supposed to be more resistant toward leaching during the catalytic reactions. Typically, the immobilization of organocatalysts has been achieved via covalent attachment onto solid supports such as polystyrene [319–323], poly(ethylene glycol) [324, 325], and dendrimers [326]. However, most of these supported organocatalysts demonstrated reduced activity and selectivity comparing with their small molecular parent catalysts. There has also been increasing examples of supported organocatalysts via noncovalent strategies starting to show their potentials in this field. There are major progresses regarding the development of noncovalently supported asymmetric organocatalysts, with a focus on the immobilization methods, their advantages, applications, and limitations. Noncovalent immobilization includes minimal modification of the parent catalyst, facile catalyst linkage, and combinatorial flexibility for the identification of an optimal-supported catalyst. Different modes of immobilization, such as acid-base, ion-pair, hydrophobic, biphasic, and self-assembled gel type, give different types of heterogeneous organocatalysts. The catalysts obtained by different methods have their advantages and disadvantages at the same time. There is a scope for developing new methodology for heterogenization of organocatalysts.

Immobilization is an effective methodology for heterogenizing an organic molecule onto inorganic material. The two principle mechanisms of organocatalysis are of both covalent and noncovalent nature which lead to (i) the formation of covalent adduct between catalyst and substrate, and (ii) the processes that rely on noncovalent interactions such as hydrogen bonding or within the catalytic cycle is the formation of ions pairs [327]. Owing to the application of organocatalysis in industrial setup, strategies for heterogenizing organocatalysts are being explored in fine chemical industries as well as pharmaceutical industries. Though the stability of an immobilised catalyst may vary in batch operation and continuous flow operation at higher reaction temperatures, such catalysts are expected to avail the advantages of heterogeneous catalysis. Heterogeneous organocatalysis is well poised for its potential industrial applications in industry with sustainable and energy efficient approaches. In order to implement the concepts of green chemistry and engineering in the pharmaceutical industry, one must develop large scale organocatalytic reactions processes that involve(i)immobilization of organocatalysts(for recycling purposes);(ii)no leaching problems due to covalent bonding with the support unlike metal complexes;(iii)results superior to those obtained with free analogous;(iv)solution phase catalysis.The immobilisation of amino acid derived organocatalysts could be covalently anchored to the inner walls of porous silica and other relevant inorganic structures. These materials have shown promise as heterogeneous organocatalysts for industrial significant reactions, such as the Hajos-Parrish-Eder-Sauer-Wiechert reaction [88] and intramolecular [4+2] cycloadditions [328]. This approach has the potential to improve recoverability and recyclability, use far less mol% of catalyst, and potentially give rise to significant improvements in activity and selectivity. Organocatalysts can be covalently anchored to mesoporous silica by building on current methodologies that are reported in the literature. Tethering is carried out through the side chain or a functionalized side chain of a derivatised amino acid, because the mechanism of catalysis relies on the amino and carboxylate groups to be free for substrate binding. Through spatial constraints, nature of the active site, and a precise balance of substrate-catalyst-pore wall interactions, the stereoselectivity of the catalytic reaction could also be suitably enhanced. Judiciously controlling the orientation of the channels and the pore diameter of the mesoporous support could facilitate enhancements in the desired stereochemical outcome, as observed earlier with anchored transition-metal amino acid complexes. The range of substrates and scope of reactions that can be catalyzed by heterogenised amino acids and their derivatives are ever expanding and immobilisation of these organic moieties could result in the design and development of more stable, highly active heterogeneous catalysts [329–331]. The potential for expanding the scope and industrial applicability of heterogenised organocatalysts demands a greater understanding of the synergistic effects, between the support and the organocatalyst; say, for example, amino acid, at a molecular level, which could provide valuable insights on structure-property relationships and its mechanistic significance. The current organocatalysts development is more focused on solving practical asymmetric synthesis and processes [332]. Following are some interesting examples reported in the field.

Copolymer-supported heterogeneous organocatalyst is prepared and applied for asymmetric aldol addition in aqueous medium [333]. Aldol reaction between acetone and 4-nitrobenzaldehyde was catalyzed by covalently grafted lysine onto silica. The immobilisation technique was found to create well-defined and isolated catalytic sites offering improved activity and chemoselectivity when compared with the homogeneous system [331].

Enantiomerically pure iminium cations supported on zeolite Y, a microporous material, and on Al-MCM-41, a mesoporous material, were found effective for the epoxidation of aryl alkenes by using peroxymonosulfate as the stoichiometric oxidant. The catalytic reaction gave high conversions and enantioselectivities. The catalysts can be simply recycled by filtration and their activity and selectivity are found much higher than the homogeneous counterpart [334]. Chiral organocatalyst team up with heterogeneous inorganic semiconductors form C–C bond through stereoselective photocatalysis [330]. Direct stereoselective alpha-functionalization of aldehydes forming C–C, C–N, C–F, C–Br, and C–S bonds is reported over organocatalysts [335]. Solvent-free noncovalent organocatalysis is reported on enantioselective addition of nitroalkanes to alkylideneindolenines [336]. DMAP-NCP, a new network nanoporous conjugated polymer containing 4-(N,N-dimethylamino) pyridine catalytic moieties, exhibits catalytic activity in the acylation of alcohols and phenols, even under neat and continuous flow conditions for practical applications on a large scale [337]. An asymmetric Diels-Alder reaction was catalyzed with high efficiency and recyclability by a soluble, "self-supported" chiral organosilica polymer with embedded imidazolidinone catalytic moieties [338]. The role of some ionic liquids as organocatalysts or cocatalysts requires further understanding and there is a scope for future research in conceptualization of this field, its applications, and the context for heterogeneous organocatalysis (Figure 12) [178, 179, 339].

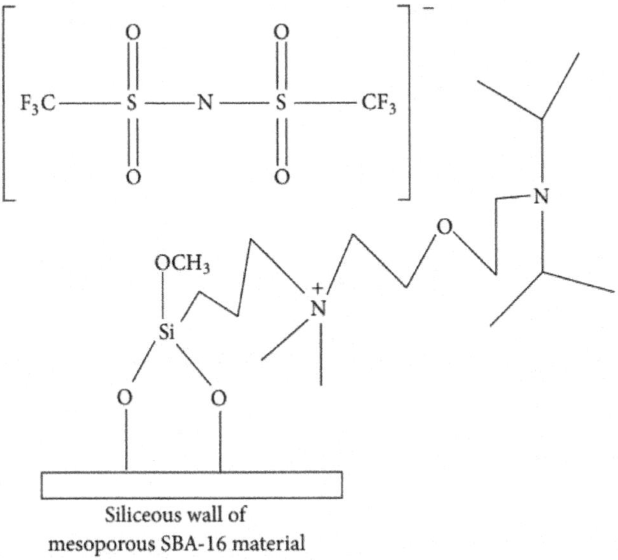

Figure 12: Heterogenization of a basic ionic liquid bearing anion on mesoporous SBA-16 silica [178, 179].

Heterogeneous Organocatalysts from Renewable Resources

Kühbeck et al. reported a critical assessment of the efficiency of neutral pH chitosan biohydrogel beads as recyclable heterogeneous organocatalyst for a variety of C−C bond formation reactions (i.e., aldol reaction, Knoevenagel condensation, nitroaldol (Henry) reaction [340]. Researchers Françoise Quignard (France) and Luca Bernardi (Italy) undertake development of new heterogeneous organocatalysts based on polysaccharides (chitosan, alginate, or carrageenan). The polysaccharides are made available from renewable resources and exploited for their intrinsic catalytic activities. For example, chitosan is known to bring basic functions while alginate and carrageenan are acidic polymers. The variety of functional groups provides polysaccharides with a surface reactivity especially appealing for specific catalysis processes.

Verma et al. reported a cost effective, environmentally benign, easily accessible, biodegradable, recyclable, and highly efficient pyridinium based bifunctional organocatalyst (ES-SO$_3$-C$_5$H$_5$NH$^+$) grafted to the chemically modified expanded starch, a biomaterial, for the synthesis of b-amino carbonyls by aza-Michael reaction of amines to electron deficient alkenes under mild reaction conditions (Scheme 14) [341].

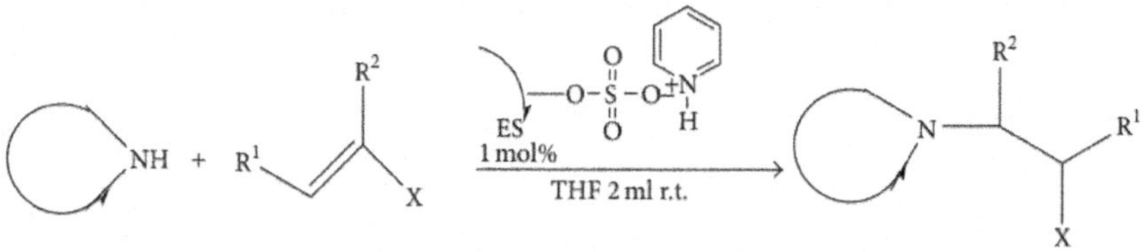

Scheme 14

Heterogenized or supported organocatalysts, for example, grafted polyHDMS, offer engineering advantages and the final macromolecular properties strongly reflect the nature of the grafts. Off late, the use of polymethylhydrosiloxane PMHS anchored cinchona alkaloid derivatives has been reported [342].

Heterogeneous Organocatalysts for Their Use under Continuous Flow Conditions

In the spirit of developing ecofriendly syntheses, researchers are interested in immobilization of various organocatalysts known to be efficient in aqueous media and in their use under continuous flow [320].

An interesting tether-free but immobilized bifunctional squaramide organocatalyst [343] has found potential application in batch and flow reactions (Scheme 15). A polystyrene-supported highly recyclable squaramide organocatalyst for the enantioselective Michael addition of 1,3 dicarbonyl compounds to β-nitrostyrenes has also found potential industrially utility [344].

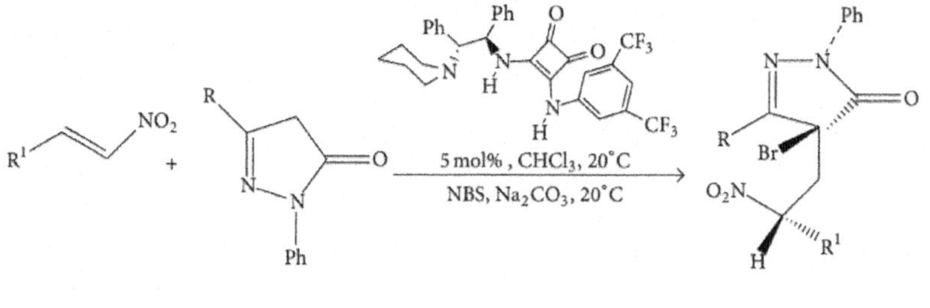

Scheme 15

Because of their stability and recyclability, solid-supported organocatalysts (Scheme 16) are therefore particularly suitable for continuous flow systems,

[154, 345, 346] which allow large scale synthesis usually accompanied by higher turnover numbers. The miniaturised flow reactors allow improved control of mass and heat transfer and the inherently low reaction volume increases safety making them valuable tools for green chemistry. Reaction-progress kinetic analysis and nonlinear chromatography were applied to investigate a model aldol reaction performed in continuous flow microreactors packed with proline functionalized silica gel. The study facilitated by an online instrumental monitoring assessed optimal operation and feed variables [154].

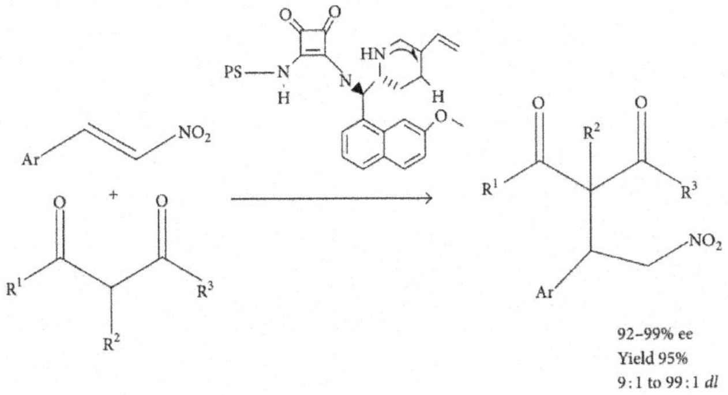

92–99% ee
Yield 95%
9:1 to 99:1 dl

Scheme 16

Enantiomerically pure iminium cations, supported on microporous zeolite Y and on mesoporous Al-MCM-41, are effective asymmetric catalysts for the epoxidation of a range of aryl alkenes, giving high conversions quickly and with enantioselectivities (Schemes 17 and 18) [334].

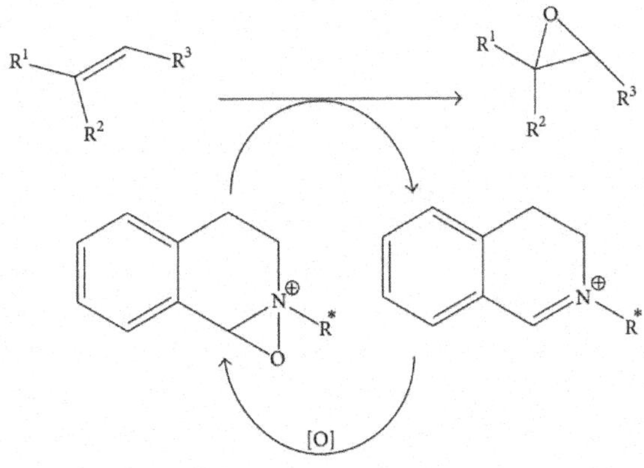

Scheme 17: Mediation of alkene epoxidation by oxaziridinium cations [334].

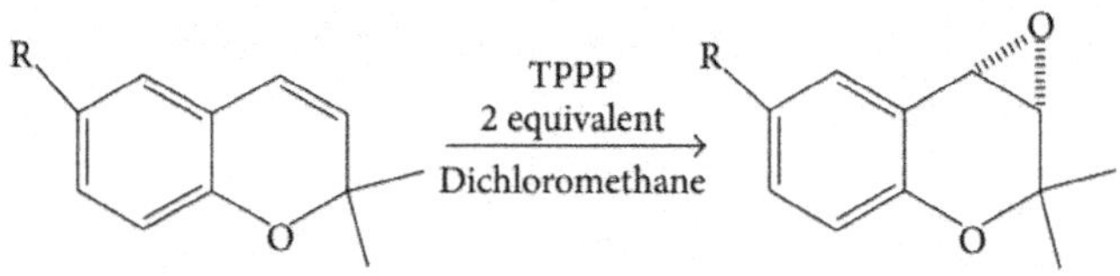

Scheme 18: Epoxidation of chromenes by MCM catalyst [334].

Epoxides are versatile synthetic intermediate. A lot has been reported on ring opening methodologies of epoxides over metal alkoxids, amides, metal halides, metal triflates, and so forth. Organocatalysts are advantageous over such metal catalysts. Schreiner published the N,N'-bis[3,5-bis(trifluoromethyl) phenyl]thiourea catalyzed ring opening of epoxides with different nucleophiles such as amines, phenols, and thiols in water with poor regioselectivity (1 : 1 to 1 : 4) [347, 348]. Bifunctional primary amine-thioureas catalyze usually difficult Michael reactions [349, 350]. Bidentate hydrogen bond donors, urea and thiourea, are found to be useful organocatalysts for the activation of carbonyl group. Hine and coworkers reported meta- andpara-substituted phenols and biphenylenediols as double hydrogen bonding organocatalysts for addition of diethylamine to phenyl glycidyl ether [351, 352]. And, there are various examples of catalysts bearing multiple hydrogen bond donors [353, 354].

PREINDUSTRIALIZATION ORGANOCATALYSIS

In 2004, a special issue of the Advanced Synthesis and Catalysis research journal was dedicated to organocatalysis [355]. In year 2006, a special issue of the Advanced Synthesis and Catalysis research journal onMultiphase Catalysis, Green Solvents, and Immobilization was published by professor Cozzi describing in detail the immobilization of organic catalysts [356]. Several synthetic strategies are developed for the polymeric immobilization of chiral organocatalysts [357–359]. Lu and Toy reviewed organic polymer supports for synthesis and for reagent and catalyst immobilization [360]. A lot has been done on supported or immobilized enamine and iminium derivatives for their applications as heterogeneous organocatalysts [361]. A heterogeneous catalyst (L-Pro LDHs) was developed using intercalation of L-proline in Mg-Al layered double hydroxides (LDH). An investigation of the thermal stability and optical stability showed that the immobilization of the chiral catalytic centers in restricted galleries enhanced the enantiomeric stability against thermal treatment and light irradiation. Asymmetric aldol reaction of benzaldehyde and acetone was carried out using L-Pro LDHs as catalyst, resulting in a good yield (90%) and a high enantiomeric excess (94%) (Figure 13). Heterogenized L-proline catalyzed the asymmetric aldol reaction using anionic clays as intercalated support [182].

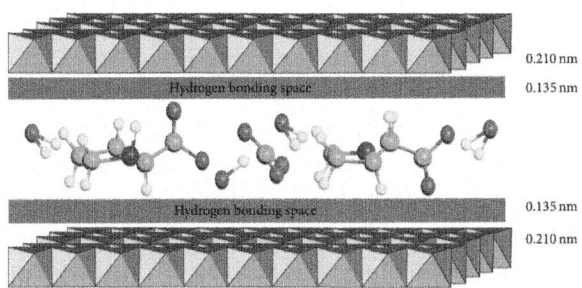

Figure 13: The schematic model of L-Pro LDHs (molar ratio of Mg/Al = 3 : 1, inter-calated yield = 59%) [182].

Calderón et al. researched [362] the aldol reaction of hydroxyacetone with different aldehydes using immobilized proline on a mesoporous support, assisted by heat and microwaves. It was found that heterogenized L-proline on MCM-41 catalyzed aldol reactions in both hydrophilic and hydrophobic solvents and provided stereoselectivities in some cases complementary to the homogeneous catalyst. The heterogeneous catalysts could be reused without significant loss of stereoselectivity. In a recent example, supramolecular assemblies of amphiphilic L-proline were regulated by compressed CO_2 as a recyclable organocatalyst for the asymmetric aldol reaction [363]. Noncovalently supported heterogeneous chiral amine is used as catalyst for asymmetric direct aldol and Michael addition reactions [364]. An amine grafted on amorphous silica performs as an effective organocatalyst for microwave-promoted Michael reaction of 1,3-dicarbonyl compounds in water [365]. Silica immobilized N-hydroxyphthalimide was found to catalyze autoxidation [366]. Silica, identified as a useful support material for homogeneous catalysts, facilitates the catalyst recovery and reusability [367]. A heterogeneous organocatalyst for cyanosilylation is also reported lately [368].

There are some interesting patents in the literature on usage of some organocatalysts in innovative chemical technologies with some industrial application. There is no need for metal based catalysis thus making a contribution to green chemistry. In this context, simple organic acids have been used as catalyst for the modification of cellulose in water on multi-ton scale [369]. In recent years, the emphasis has been on the variety of nano(porous)materials for that has found use in such catalytic systems: these include nanoparticles, dendrimers, metal oxides, inorganic polymers, and mesoporous materials such as MCM-n materials and SBA-n materials. In 2009, a patent related to the use of magnetic nanomaterial-supported glutathione and cysteine organocatalysts as "green" nanocatalysts for various reactions such as Paal-Knorr reactions, aza-Michael addition, and pyrazole synthesis was filed and later granted in year 2012 [370]. Sulfonamide based organocatalysts and method for their use are patented by inventors Rich Garrett Carter and Hua Yang. Proline sulfonamide organocatalysts performing various enantioselective or diastereoselective reactions such as the aldol reactions, conjugate additions, Michael additions, Robinson annulations, Mannich reactions, and alpha-aminooxylations, alpha-hydroxyaminations, alpha-aminations, and alkylation reactions. Among the various methodologies claimed, a sulfonamide derived catalyst was used for the construction of all carbon bicycles, such as [2.2.2]-bicycles [371].

Organocatalysts are blamed for not being recoverable after a reaction. There are interesting reports that study homogeneous or heterogeneous nanosupports for easy recovery of the organocatalysts. Caminade et al. identified perfectly defined hyperbranched polyphosphorhydrazone dendrimers, either neat or attached to siliceous solid material, as nanosupports for easy recycling of organocatalysts [326, 372–374].

CHARACTERIZATION TOOLS AND TECHNIQUES

In order for us to research catalysis and relevant aspects, we may require spectroscopic techniques, magnetic measurements, transport and thermal measurements, scattering techniques, microscopic techniques, thin-film deposition techniques, and so forth. The catalytic systems, especially the heterogenized organocatalysts, prepared can be characterized by X-ray diffraction (XRD), nitrogen sorption, scanning electron microscopy (SEM), transmission electron microscopy (TEM), FT-IR, UV-Visible, nuclear magnetic resonance (NMR), Raman, and so forth. In case of nanoporous materials, the pore diameters could be uniformly tailored. The NMR studies of organocatalysts in porous materials could also be of importance for understanding the structure and bonding within the organic and inorganic part of a catalytic system. There is always a scope for identifying and utilizing new characterization techniques for studying catalytic systems and properties, especially the surface behaviour, hydrophilic/hydrophobicity, interactions, molecular sorption, recognitions, and so forth.

FUTURE SCOPE OF HETEROGENEOUS ORGANOCATA-LYSTS

It is always a risk to predict what the future holds for any research area, but several aspects of modifications of existing catalysts and also heterogenisation of organocatalysts onto some sort of robust high surface area support including nanomaterials or nanoporous materials undoubtedly attract researchers' attention. Tremendous efforts should be directed towards the discovery and design of catalytic systems with better efficiency, new reactivity, and greater turnover numbers. Perhaps the most crucial area of research in the future will be the identification of important organic transformations at industries that are not available using other branches of catalysis. Given the huge growth and impact of organocatalysis in its present form, it will certainly be exciting to observe the development of the field upon heterogenization.

A variety of reactions such as reduction, oxidation, multicomponent, Mannich, alkylation, condensation, deprotection, cycloaddition, hydroxylation, dehydration, dehydrogenation, transesterification, reactions involving biomimetic oxygen-evolving catalysts, and other important C–C bond forming reactions are well presented on the heterogenized organocatalysts under a variety of reaction conditions.

For any large scale catalytic process, the most salient considerations are cost and safety and so critic opinion on low turnover numbers limiting the potential usage might not discourage the industrial applications. Because organocatalysts are often cheaper than metal based catalysts, they can be used in larger quantities than metal based ones for the same price. Moreover, it is widely recognized in manufacturing that the removal of toxic catalyst related impurities from the waste stream can often have a larger financial impact than the turn over number of the catalyst. In the coming years, various approaches for immobilisation will be considered, with a focus on whether a novel methodology for the immobilisation of organocatalysts could be discovered. The benefits of heterogenisation will become apparent on evaluation of the catalytic potential. Apart from the nature of the support, the degree of catalyst loading and selection of reaction media or solvent tune the catalytic behaviour. Alternatively, immobilized organocatalysts can be prepared that are linked to the solvents like ionic liquids that allow the catalytic entity to be separate from the reaction medium by simple extraction or filtration. In another approach, the immobilized organocatalyst might permit us to perform the reactions in biphasic or different conditions. This catalytic system and reaction conditions could find industrial application. Scientific and technical tasks that researchers can undertake are the immobilization of the organocatalyst on solvent and or onto the walls of a reactor. Future catalysts development should be more oriented toward real problems in synthetic processes. It would be also interesting to see (i) the explorations of noncovalent interactions and entrapment in nanosized materials in developing supported organocatalysts; (ii) fine tuning of the screened catalysts and their thermal, mechanical, and catalytic stability; (iii) evaluation of catalyst deactivation and its rejuvenation mechanisms.

CONCLUSIONS

The literature survey and ideas jotted above do not fully claim to conceptualize all the aspects of this new field of heterogeneous organocatalysis. The aim of this review is to illustrate the significance of a variety of organic molecules as catalysts employed for a wide variety of organic transformations. The most highlighted fact is that these organocatalysts represent an interesting scope for heterogenization. In the review of literature, the basic concept of organocatalyst, strategies for heterogenization, preparation methods, and various applications of heterogeneous organocatalysts have been discussed. Like in other heterogeneous catalytic systems, the materials or supports could also play very important role in the total catalyst system used as both catalyst and catalyst carriers. The emphasis has been on the variety of nano(porous) materials that have found or could find use in such catalytic systems: these include nanoparticles, dendrimers, metal oxides, inorganic polymers, and mesoporous materials such as MCM-n materials and SBA-n materials (this is an Open Access review under the terms of the Creative Commons Attribution License, which permits readers to use and distribute this e-review, provided that the original research work is properly cited. All rights are therefore reserved to the author and/or respective authors and researchers cited herein (2013)).

In this review, the author focused on applications of organocatalysts in asymmetric synthesis, synthesis of fine chemicals, and green chemistry. The preceding pages are an account of recent research and developments made in the use of organic molecules as catalysts. It is highlighted that organocatalysts are versatile which not only catalyze the most fundamental reactions in organic chemistry but also finds applications in processes of commercial importance in chemical industries. Several very useful applications of organocatalysts are in production of agrochemicals, clean energy, drugs, polymers, petrochemicals, and so forth. In view of the globally increasing interest towards energy efficient and environmentally friendly chemical technology for industries, research on the development of robust, recyclable, selective, metal-free, easy-to-scale up catalysts and their useful preparation methods are thus the need of the hour.

Great strides are still to be made in the synthesis and characterization of supported or immobilized organocatalysts in order to apply them as heterogeneous catalysts that emphasize applications in existing industrial processes. These catalysts are going to be modified in future and make their catalytic activities accessible in industrial setup for sustainable and environmentally benign technologies.

The author has also outlined recent progress in the use of heterogeneous organocatalysts by briefing the recent publications in scientific publications. Year 2012 saw thematic series, dedicated to showcase the current state of the art of organocatalysis, published by the Beilstein Journal of Organic Chemistry. For those readers who are expecting an update on the recent trends in organocatalysis, the author lists herein [375–383] excellent reviews published in year 2013 on some very specialized topics of organocatalysis.

To summarize, today, the growing need for the atom efficiency, energy efficiency, and environmental concerns stemming from the increase in chemical products demand necessitate the development of the "sustainable or environmentally benign" production methods. Asymmetric synthesis achieving atom economy is a challenge for organic synthesis and homogeneous catalysis using metal complexes leads the way. But the pharmaceutical entities and food industry products do not tolerate a contamination, of even traces, of any such metals for that matter. That why there is a huge demand for metal-free organocatalysts leading to 100% yield, 100% ee, of the desired product. The need from chemical industry, especially pharmaceutical, for reliable asymmetric transformations of molecular skeletons is higher than ever. Therefore, (asymmetric) organocatalysis is attracting widespread attention and has become an important field of research. And, moreover, heterogeneous organocatalysis is bringing in some important breakthroughs for industry, especially through flow chemistry.

This review encloses 376 useful references for providing readers with a well described state of the art. The studies to address the scope of this catalytic science are underway. The scientific discoveries and innovations involving organocatalysts are reported almost on a daily basis. Many researchers are joining this field, especially to understand the catalytic phenomena and develop heterogeneous catalysts for asymmetric synthetic methodologies for chemical industry. Let us catalyze the green chemistry revolution. As rightly said by Professor Benjamin List, "the research in organocatalysis constitutes the tip of the iceberg of a novel catalytic principle, of which the entire scope still remains to be fully uncovered."

ACKNOWLEDGMENT

The author, Isak R. Shaikh, thanks National Chemical Laboratory (http://www. ncl-india.org/), Pune, India, for providing the library facilities.

REFERENCES

1.   "Our Common Future: Report of the World Commission on Environment

and Development, United Nations (UN) Commission on Environment and Development (Brundtland Commission), 1987," Published as Annex to General Assembly document A/42/427, Development and International Co-operation: Environment, August 1987.

2.   M. Lancaster, Green Chemistry: An Introductory Text, The Royal Society of Chemistry, Cambridge, UK, 2002.

3.   P. T. Anastas and J. C. Warner, Green Chemistry: Theory and Practice, Oxford University Press, New York, NY, USA, 1998.

4.   J. H. Clark, "Catalysis for green chemistry," Pure and Applied Chemistry, vol. 73, no. 1, pp. 103–111, 2001.·

5.   J. H. Clark and C. N. Rhodes, Clean Synthesis Using Porous Inorganic Solid Catalysts, RSC Clean Technology Monographs, Cambridge, UK, 2000.

6.   W. Ostwald, "Die Überwindung des wissenschaftlichen Materialismus," in Verhandlungen der Gesellschaft Deutscher Naturforscher und Ärzte, pp. 155–168, 1895.

7.   R. A. Sheldon, "Consider the environmental quotient," Chemtech, vol. 25, pp. 38–47, 1994.

8.   R. A. Sheldon and H. van Bekkum, Eds., Fine Chemicals Through Heterogeneous Catalysis, Wiley-VCH, Weinheim, Germany, 2001.

9.   R. A. Sheldon and R. S. Downing, "Heterogeneous catalytic transformations for environmentally friendly production," Applied Catalysis A, vol. 189, no. 2, pp. 163–183, 1999. ·

10.   R. A. Sheldon, "The e factor: fifteen years on," Green Chemistry, vol. 9, no. 12, pp. 1273–1283, 2007.· ·

11.   R. A. Sheldon, "E factors, green chemistry and catalysis: an odyssey," Chemical Communications, no. 29, pp. 3352–3365, 2008.· ·

12.   R. A. Sheldon, "Selective catalytic synthesis of fine chemicals: opportunities and trends," Journal of Molecular Catalysis A, vol. 107, no. 1–3, pp. 75–83, 1996. ·

13.   R. A. Sheldon, "Catalysis and pollution prevention," Chemistry & Industry, vol. 1, pp. 12–15, 1997.

14.   R. A. Sheldon, "Catalysis: the key to waste minimization," Journal of Chemical Technology and Biotechnology, vol. 68, pp. 381–388, 1997.

15.   G. Ertl, H. Knozinger, and J. Weitkamp, Eds., Handbook of Heterogeneous Catalysis, Wiley-Vch, Weinheim, Germany, 1997.

16.   R. A. van Santen, P. W. N. M. van Leeuwen, J. A. Moulijn, and B. A. Averill, Eds., Catalysis: An Intergrated Approach, Elsevier, Amsterdam,

The Netherlands, 2nd edition, 1999.

17. J. R. Anderson and K. C. Pratt, Introduction To Characterisation and Testing of Catalysts, Academic Press, Sydney, Australia, 1985.

18. J. M. Walls and R. Smith, Eds., Surface Science Techniques, Pergamon, 1994.

19. J. W. Niemantsverdriet, Spectroscopy in Catalysis, Wiley-Vch, Weinheim, Germany, 2007.

20. J. J. Bozell and G. R. Petersen, "Technology development for the production of biobased products from biorefinery carbohydrates—the US Department of Energy›s "top 10" revisited," Green Chemistry, vol. 12, no. 4, pp. 539–554, 2010. · ·

21. R. A. Gross, M. Ganesh, and W. Lu, "Enzyme-catalysis breathes new life into polyester condensation polymerizations," Trends in Biotechnology, vol. 28, no. 8, pp. 435–443, 2010. · ·

22. C. J. Li and T. H. Chan, Organic Reactions in Aqueous Media, Wiley-VCH, New York, NY, USA, 1997.

23. P. A. Grieco, Organic Synthesis in Water, Kluwer Academic, Dordrecht, The Netherlands, 1997.

24. B. M. Trost, "Atom economy—a challenge for organic synthesis: homogeneous catalysis leads the way,"Angewandte Chemie International Edition, vol. 34, no. 3, pp. 259–281, 1995. ·

25. B. Cornils and W. A. Herrmann, in Applied Homogeneous Catalysis with Organometallic Compounds, B. Cornils and W. A. Herrmann, Eds., Wiley-VCH, Weinheim, Germany, 2nd edition, 2002.

26. B. Cornils and W. A. Herrmann, "Concepts in homogeneous catalysis: the industrial view," Journal of Catalysis, vol. 216, no. 1-2, pp. 23–31, 2003. · ·

27. S. Bhaduri and D. Mukesh, Homogeneous Catalysis, John Wiley & Sons, 2002.

28. G. W. Parshall and S. D. Ittel, Homogeneous Catalysis, John Wiley & Sons, New York, NY, USA, 1992, Emphasis on Industrial Applications.

29. B. Cornils and W. A. Herrmann, Eds., Applied Homogeneous Catalysis by Organometallic Complexes, Verlag Chemie, Weinheim, Germany, 1996.

30. I. Ojima, Ed., Catalytic Asymmetric Synthesis, Verlag VCH, Weinheim, Germany, 1993.

31. R. Noyori, Asymmetric Catalysis in Organic Synthesis, John Wiley & Sons, New York, NY, USA, 1994.

32. H.-U. Blaser, A. Indolese, and A. Schnyder, "Applied homogeneous catalysis by organometallic complexes," Current Science, vol. 78, no. 11, pp. 1336–1344, 2000. ·

33. M. Beller and H.-U. Blaser, Eds., Organometallics as Catalysts in the Fine Chemical Industry, vol. 42 of Topics in Organometallic Chemistry, Springer, 2012.

34. B. R. Cuenya, "Synthesis and catalytic properties of metal nanoparticles: size, shape, support, composition, and oxidation state effects," Thin Solid Films, vol. 518, no. 12, pp. 3127–3150, 2010. · ·

35. E. Victor Henrich and P. A. Cox, The Surface Science of Metal Oxides, Cambridge University Press, Cambridge, UK, 1994.

36. H. H. Kung, "Transition metal oxides," in Surface Chemistry and Catalysis, Elsevier, Amsterdam, The Netherlands, 1989.

37. K. J. Klabunde, M. Fazlul Hoq, F. Mousah, and H. Matsuhashi, "Metal Oxides and their physico-chemical properties in Catalysis and Synthesis," in Preparative Chemistry Using Supported Reagents, Academic Press, London, UK, 1987.

38. H. Schäfer, Chemiker-Zeitung, vol. 101, no. 7/8, p. 325, 1977.

39. H.-C. Zhou, J. R. Long, and O. M. Yaghi, "Introduction to metal-organic frameworks," Chemical Reviews, vol. 112, no. 2, pp. 673–674, 2012. · ·

40. D. Farrusseng, Ed., Metal-Organic Frameworks. Applications From Catalysis to Gas Storage, Wiley-VCH, Weinheim, Germany, 2011.

41. N. Rahmat, A. Z. Abdullah, and A. R. Mohamed, "A review: mesoporous Santa Barbara amorphous-15, types, synthesis and its applications towards biorefinery production," American Journal of Applied Sciences, vol. 7, no. 12, pp. 1579–1586, 2010. ·

42. S.-H. Wu, C.-Y. Mou, and H.-P. Lin, "Synthesis of mesoporous silica nanoparticles," Chemical Society Reviews, vol. 42, pp. 3862–3875, 2013. ·

43. A. Bernardos and L. Kouřimská, "Applications of mesoporous silica materials in food: a review," Czech Journal of Food Sciences, vol. 31, no. 2, pp. 99–107, 2013.

44. M. N. Timofeeva, V. N. Panchenko, Z. Hasan, and S.-H. Jhung, "Catalytic potential of the wonderful chameleons: nickel phosphate molecular sieves," Applied Catalysis A, vol. 455, pp. 71–85, 2013. ·

45. I. R. Shaikh and S.-E. Park, "Microwave synthesis and catalytic applications of novel cobalt incorporated nickel phosphate," Diffusion and Defect Data Pt.B: Solid State Phenomena, vol. 119, pp. 279–282,

2007.· ·

46. I. R. Shaikh, Nanoporous nickel phosphate and silica materials for heterogeneous oxidation catalysis [Ph.D. thesis], Inha University, Inchon, Republic of Korea, 2007, Ph.D. thesis incomplete, unpublished and in conflict between S-E Park and Isak R. Shaikh.

47. Y. Tao, H. Kanoh, L. Abrams, and K. Kaneko, "Mesopore-modified zeolites: preparation, characterization, and applications," Chemical Reviews, vol. 106, no. 3, pp. 896–910, 2006. · ·

48. M. E. Davis, "Ordered porous materials for emerging applications,"Nature, vol. 417, no. 6891, pp. 813–821, 2002. · ·

49. K. Tsuji, C. W. Jones, and M. E. Davis, "Organic-functionalized molecular sieves (OFMSs): I. Synthesis and characterization of OFMSs with polar functional groups," Microporous and Mesoporous Materials, vol. 29, no. 3, pp. 339–349, 1999. ·

50. C. W. Jones, K. Tsuji, and M. E. Davis, "Organic-functionalized molecular sieves as shape-selective catalysts," Nature, vol. 393, no. 6680, pp. 52–54, 1998. · ·

51. M. E. Davis, "The quest for extra-large pore, crystalline molecular sieves," Chemistry—A European Journal, vol. 3, pp. 1745–1750, 1997. ·

52. B. Notori, "Microporous crystalline titanium silicates," Advances in Catalysis, vol. 41, pp. 253–334, 1996.·

53. R. J. Saxon, "Crystalline microporous titanium silicates," Topics in Catalysis, vol. 9, no. 1-2, pp. 43–57, 1999.

54. J. M. Thomas and R. Raja, "Design of a "green" one-step catalytic production of ε-caprolactam (precursor of nylon-6)," Proceedings of the National Academy of Sciences of the United States of America, vol. 102, no. 39, pp. 13732–13736, 2005. · ·

55. M. Hartmann and L. Kevan, "Transition-metal ions in aluminophosphate and silicoaluminophosphate molecular sieves: location, interaction with adsorbates and catalytic properties," Chemical Reviews, vol. 99, no. 3, pp. 635–663, 1999. ·

56. C. S. Cundy and P. A. Cox, "The hydrothermal synthesis of zeolites: history and development from the earliest days to the present time," Chemical Reviews, vol. 103, no. 3, pp. 663–701, 2003. View at Publisher· ·

57. A. Corma, "From microporous to mesoporous molecular sieve materials and their use in catalysis,"Chemical Reviews, vol. 97, no. 6, pp. 2373–2419, 1997. ·

58. G. De, M. Gusso, L. Tapfer et al., "Annealing behavior of silver, copper,

and silver-copper nanoclusters in a silica matrix synthesized by the sol-gel technique," Journal of Applied Physics, vol. 80, no. 12, pp. 6734–6739, 1996. ·

59.   Y. Wan and D. Zhao, "On the controllable soft-templating approach to mesoporous silicates," Chemical Reviews, vol. 107, no. 7, pp. 2821–2860, 2007. · ·

60.   A. K. Cheetham, G. Ferey, and T. Loiseau, Angewandte Chemie International Edition, vol. 38, p. 3298, 1999.

61.   G. Ferey, "Microporous solids: from organically templated inorganic skeletons to hybrid frameworks . . . ecumenism in chemistry," Chemistry of Materials, vol. 13, pp. 3084–3098, 2001. ·

62.   C. N. R. Rao, S. Natarajan, and R. Vaidhyanathan, "Metal carboxylates with open architectures,"Angewandte Chemie International Edition, vol. 43, no. 12, pp. 1466–1496, 2004. · ·

63.   F. Schüth and W. Schmidt, "Microporous and mesoporous materials," Advanced Materials, vol. 14, pp. 629–638, 2002.

64.   F. Schuth, K. Sing, and J. Weitkamp, Eds., Handbook of Porous Solids, vol. I-V, Wiley-VCH, Weinheim, Germany, 2002.

65.   Y. Wan and D. Y. Zhao, "On the controllable soft-templating approach to mesoporous silicates," Chemical Reviews, vol. 107, no. 7, pp. 2821–2860, 2007. · ·

66.   V. Parvulescu, C. Anastasescu, and B. L. Su, "Bimetallic Ru-(Cr, Ni, or Cu) and La-(Co or Mn) incorporated MCM-41 molecular sieves as catalysts for oxidation of aromatic hydrocarbons," Journal of Molecular Catalysis A, vol. 211, pp. 143–148, 2004. ·

67.   S. Vetrivel and A. Pandurangan, "Co and Mn impregnated MCM-41: their applications to vapour phase oxidation of isopropylbenzene," Journal of Molecular Catalysis A, vol. 227, pp. 269–278, 2005. ·

68.   Y. Wan, D. Zhang, N. Hao, and D. Zhao, "Organic groups functionalised mesoporous silicates,"International Journal of Nanotechnology, vol. 4, p. 66, 2007. ·

69.   V. Ayala, A. Corma, M. Iglesias, and F. Sanchez, "Mesoporous MCM41-heterogenised (salen)Mn and Cu complexes as effective catalysts for oxidation of sulfides to sulfoxides: isolation of a stable supported Mn(V)=O complex, responsible of the catalytic activity," Journal of Molecular Catalysis A, vol. 221, pp. 201–208, 2004.

70.   Z. G. Hajos and D. R. Parrish, "Asymmetric synthesis of bicyclic intermediates of natural product chemistry," The Journal of Organic

Chemistry, vol. 39, no. 12, pp. 1615–1621, 1974. ·

71. L. Shu and Y. Shi, "An efficient ketone-catalyzed epoxidation using hydrogen peroxide as oxidant," The Journal of Organic Chemistry, vol. 65, pp. 8807–8810, 2000. ·

72. E. J. Corey, F. Xu, and M. C. Noe, "A rational approach to catalytic enantioselective enolate alkylation using a structurally rigidified and defined chiral quaternary ammonium salt under phase transfer conditions," Journal of the American Chemical Society, vol. 119, pp. 12414–12415, 1997. ·

73. B. Sellergren, R. N. Karmalkar, and K. J. Shea, "Enantioselective ester hydrolysis catalyzed by imprinted polymers. 2," Journal of Organic Chemistry, vol. 65, no. 13, pp. 4009–4027, 2000. · ·

74. J. von Liebig, "Ueber die Bildung des Oxamids aus Cyan," Annalen der Chemie und Pharmacie, vol. 113, no. 2, pp. 246–247, 1860. ·

75. H. D. Dakin, "The catalytic action of amino-acids, peptones and proteins in effecting certain syntheses,"Journal of Biological Chemistry, vol. 7, p. 49, 1909.

76. W. Langenbeck, "Über organische Katalysatoren. III. Die Bildung von Oxamid aus Dicyan bei Gegenwart von Aldehyden," Justus Liebigs Annalen der Chemie, vol. 469, p. 16, 1929. ·

77. W. Langenbeck, Die Organische Katalysatoren und ihre Beziehungen zu den Fermenten, vol. 2, Springer, 1949.

78. W. Langenbeck, Fortschritte Der Chemischen Forschung, vol. 6, Springer, Berlin, Germany, 1966.

79. M. M. Vavon and P. Peignier, "L›application des alcaloïdes dans la synthèse organique," Bulletin de la Société Chimique de France, vol. 45, p. 293, 1929.

80. R. Wegler, "Über die mit verschiedener Reaktionsgeschwindigkeit erfolgende Veresterung der optischen Antipoden eines Racemates durch opt. akt. Katalysatoren," Justus Liebigs Annalen der Chemie, vol. 498, pp. 62–73, 1932.

81. S. Pizzarello and A. L. Weber, "Prebiotic amino acids as asymmetric catalysts," Science, vol. 303, no. 5661, p. 1151, 2004. · ·

82. G. Bredig and P. S. Fiske, Biochemische Zeitschrift, vol. 46, p. 7, 1912.

83. H. Pracejus, "Organische Katalysatoren, LXI. Asymmetrische Synthesen mit Ketenen, I. Alkaloid-katalysierte asymmetrische Synthesen von α-Phenyl-propionsäureestern," Justus Liebigs Annalen der Chemie, vol. 634, pp. 9–22, 1960. ·

84. H. Pracejus, "Asymmetrische Synthesen mit Ketenen, II. Stereospezifische Addition von α-Phenyl-äthylamin an Phenyl-methyl-keten," Justus Liebigs Annalen der Chemie, vol. 634, pp. 23–29, 1960. ·

85. S.-I. Yamada and G. Otani, "Asymmetric synthesis with amino acid II asymmetric synthesis of optically active 4,4-disubstituted-cyclohexenone," Tetrahedron Letters, vol. 10, no. 48, pp. 4237–4240, 1969. ·

86. Z. G. Hajos and D. R. Parrish, German Patent DE, 2102623 1971.

87. Z. G. Hajos and D. R. Parrish, United States Patent US, 3975442, 1971.

88. U. Eder, G. Sauer, and R. Wiechert, "New type of asymmetric cyclization to optically active steroid CD partial structures," Angewandte Chemie International Edition, vol. 10, pp. 496–497, 1971. ·

89. C. Allemann, R. Gordillo, F. R. Clemente, P. H.-Y. Cheong, and K. N. Houk, "Theory of asymmetric organocatalysis of aldol and related reactions: rationalizations and predictions," Accounts of Chemical Research, vol. 37, no. 8, pp. 568–569, 2004. · ·

90. H. Wynberg, "Asymmetric catalysis by alkaloids," Topics in Stereochemistry, vol. 16, pp. 87–129, 1986.

91. H. Wynberg, "Catalytic asymmetric synthesis of chiral 4-substituted 2-oxetanones," The Journal of Organic Chemistry, vol. 50, pp. 1977–1979, 1985. ·

92. H. Wynberg, "Asymmetric synthesis of (S)- and (R)-malic acid from ketene and chloral," Journal of the American Chemical Society, vol. 104, pp. 166–168, 1982. ·

93. B. List, R. A. Lerner, and C. F. Barbas III, "Proline-catalyzed direct asymmetric aldol reactions," Journal of the American Chemical Society, vol. 122, pp. 2395–2396, 2000. ·

94. T. Bui and C. F. Barbas III, "A proline-catalyzed asymmetric Robinson annulation reaction," Tetrahedron Letters, vol. 41, no. 36, pp. 6951–6954, 2000. ·

95. R. B. Woodward, E. Logusch, K. P. Nambiar et al., "Asymmetric total synthesis of erythromcin. 1. Synthesis of an erythronolide A secoacid derivative via asymmetric induction," Journal of the American Chemical Society, vol. 103, no. 11, pp. 3210–3213, 1981. ·

96. K. A. Ahrendt, C. J. Borths, and D. W. C. MacMillan, "New strategies for organic catalysis: the first highly enantioselective organocatalytic diels—Alder reaction," Journal of the American Chemical Society, vol. 122, no. 17, pp. 4243–4244, 2000. · ·

97. B. List, R. A. Lerner, and C. F. Barbas III, "Proline-catalyzed direct asymmetric aldol reactions," Journal of the American Chemical Society, vol. 122, no. 10, pp. 2395–2396, 2000. · ·

98. N. Vignola and B. List, "Catalytic asymmetric intramolecular α-alkylation of aldehydes," Journal of the American Chemical Society, vol. 126, no. 2, pp. 450–451, 2004. ·

99. A. B. Northrup and D. W. C. MacMillan, "Two-step synthesis of carbohydrates by selective aldol reactions," Science, vol. 305, no. 5691, pp. 1752–1755, 2004. · ·

100. A. B. Northrup, I. K. Mangion, F. Hettche, and D. W. C. MacMillan, "Enantioselective organocatalytic direct aldol reactions of α-oxyaldehydes: step one in a two-step synthesis of carbohydrates," Angewandte Chemie International Edition, vol. 43, no. 16, pp. 2152–2154, 2004. · ·

101. L. C. Dias, L. J. Steil, and V. D. A. Vasconcelos, "A short and efficient synthesis of (+)-prelactone B," Tetrahedron Asymmetry, vol. 15, no. 1, pp. 147–150, 2004. · ·

102. P. M. Pihko and A. Erkkilä, "Enantioselective synthesis of prelactone B using a proline-catalyzed crossed-aldol reaction," Tetrahedron Letters, vol. 44, no. 41, pp. 7607–7609, 2003. · ·

103. A. B. Northrup and D. W. C. MacMillan, "The first direct and enantioselective cross-aldol reaction of aldehydes," Journal of the American Chemical Society, vol. 124, no. 24, pp. 6798–6799, 2002. · ·

104. K. Sakthivel, W. Notz, T. Bui, and C. F. Barbas III, "Amino acid catalyzed direct asymmetric aldol reactions: a bioorganic approach to catalytic asymmetric carbon-carbon bond-forming reactions," Journal of the American Chemical Society, vol. 123, no. 22, pp. 5260–5267, 2001. · ·

105. B. List, P. Pojarliev, and C. Castello, "Proline-catalyzed asymmetric aldol reactions between ketones and α-unsubstituted aldehydes," Organic Letters, vol. 3, no. 4, pp. 573–575, 2001. · ·

106. B. List, R. A. Lerner, and C. F. Barbas III, "Proline-catalyzed direct asymmetric aldol reactions," Journal of the American Chemical Society, vol. 122, no. 10, pp. 2395–2396, 2000. · ·

107. B. List, R. A. Lerner, and C. F. Barbas III, "Enantioselective aldol cyclodehydrations catalyzed by antibody 38C2," Organic Letters, vol. 1, no. 1, pp. 59–61, 1999. ·

108. J. M. Betancort and C. F. Barbas III, "Direct asymmetric organocatalytic michael reactions of α,α-disubstituted aldehydes with β-nitrostyrenes for the synthesis of quaternary carbon-containing products," Organic Letters,

vol. 6, pp. 2527–2530, 2004. ·

109. A. Alexakis and O. Andrey, "Diamine-catalyzed asymmetric Michael additions of aldehydes and ketones to nitrostyrene," Organic Letters, vol. 4, no. 21, pp. 3611–3614, 2002. · ·

110. B. List, P. Pojarliev, and H. J. Martin, "Efficient proline-catalyzed Michael additions of unmodified ketones to nitro olefins," Organic Letters, vol. 3, no. 16, pp. 2423–2425, 2001. · ·

111. D. Enders and A. Seki, "Proline-catalyzed enantioselective Michael additions of ketones to nitrostyrene,"Synlett, no. 1, pp. 26–28, 2002. ·

112. J. M. Betancort and C. F. Barbas III, "Catalytic direct asymmetric Michael reactions: taming naked aldehyde donors," Organic Letters, vol. 3, no. 23, pp. 3737–3740, 2001. · ·

113. J. M. Betancort, K. Sakthivel, R. Thayumanavan, and C. F. Barbas III, "Catalytic enantioselective direct Michael additions of ketones to alkylidene malonates," Tetrahedron Letters, vol. 42, no. 27, pp. 4441–4444, 2001. · ·

114. S. Hanessian and V. Pham, "Catalytic asymmetric conjugate addition of nitroalkanes to cycloalkenones,"Organic Letters, vol. 2, no. 19, pp. 2975–2978, 2000. ·

115. M. Yamaguchi, Y. Igarashi, R. S. Reddy, T. Shiraishi, and M. Hirama, "Asymmetric Michael addition of nitroalkanes to prochiral acceptors catalyzed by proline rubidium salts," Tetrahedron, vol. 53, no. 32, pp. 11223–11236, 1997. · ·

116. N. S. Chowdari, J. T. Suri, and C. F. Barbas III, "Asymmetric synthesis of quaternary α-and β-amino acids and β-lactams via proline-catalyzed Mannich reactions with branched aldehyde donors," Organic Letters, vol. 6, no. 15, pp. 2507–2510, 2004. · ·

117. N. S. Chowdari, D. B. Ramachary, and C. F. Barbas III, "Organocatalysis in ionic liquids: highly efficient L-proline-catalyzed direct asymmetric Mannich reactions involving ketone and aldehyde nucleophiles,"Synlett, no. 12, pp. 1906–1909, 2003. ·

118. B. List, P. Pojarliev, W. T. Biller, and H. J. Martin, "The proline-catalyzed direct asymmetric three-component Mannich reaction: scope, optimization, and application to the highly enantioselective synthesis of 1,2-amino alcohols," Journal of the American Chemical Society, vol. 124, no. 5, pp. 827–833, 2002. · ·

119. A. Córdova, W. Notz, G. Zhong, J. M. Betancort, and C. F. Barbas III, "A highly enantioselective amino acid-catalyzed route to

functionalized α-amino acids," Journal of the American Chemical Society, vol. 124, no. 9, pp. 1842–1843, 2002. · ·

120. A. Córdova, S.-I. Watanabe, F. Tanaka, W. Notz, and C. F. Barbas III, "A highly enantioselective route to either enantiomer of both α- and β-amino acid derivatives," Journal of the American Chemical Society, vol. 124, no. 9, pp. 1866–1867, 2002. · ·

121. A. Corodova and C. F. Barbas III, "anti-Selective SMP-catalyzed direct asymmetric Mannich-type reactions: synthesis of functionalized amino acid derivatives," Tetrahedron Letters, vol. 43, pp. 7749–7752, 2002. ·

122. B. List, "The direct catalytic asymmetric three-component Mannich reaction," Journal of the American Chemical Society, vol. 122, no. 38, pp. 9336–9337, 2000. · ·

123. F. Shi, W. Tan, R.-Y. Zhu, G.-J. Xing, and S.-J. Tu, "catalytic asymmetric five-component tandem reaction: diastereo- and enantioselective synthesis of densely functionalized tetrahydropyridines with biological importance," Advanced Synthesis & Catalysis, vol. 355, no. 8, pp. 1605–1622, 2013. ·

124. J. W. Yang, M. T. Hechavarria Fonseca, and B. List, "Catalytic asymmetric reductive Michael cyclization,"Journal of the American Chemical Society, vol. 127, no. 43, pp. 15036–15037, 2005. · ·

125. B. List, "Direct catalytic asymmetric α-amination of aldehydes," Journal of the American Chemical Society, vol. 124, pp. 5656–5657, 2002. ·

126. G. Zhong, "A facile and rapid route to highly enantiopure 1,2-diols by novel catalytic asymmetric α-aminoxylation of aldehydes," Angewandte Chemie International Edition, vol. 42, no. 35, pp. 4247–4250, 2003. · ·

127. S. P. Brown, M. P. Brochu, C. J. Sinz, and D. W. C. MacMillan, "The direct and enantioselective organocatalytic α-oxidation of aldehydes," Journal of the American Chemical Society, vol. 125, no. 36, pp. 10808–10809, 2003. · ·

128. A. Bøgevig, H. Sunden, and A. Córdova, "Direct catalytic enantioselective alpha-aminoxylation of ketones: a stereoselective synthesis of alpha-hydroxy and alpha,alpha›-dihydroxy ketones," Angewandte Chemie International Edition, vol. 43, pp. 1109–1112, 2004. ·

129. J. Seayad and B. List, "Asymmetric organocatalysis," Organic & Biomolecular Chemistry, vol. 3, pp. 719–724, 2005. ·

130. R. R. Shaikh, A. Mazzanti, M. Petrini, G. Bartoli, and P. Melchiorre, "Proline-catalyzed asymmetric formal α-alkylation of aldehydes via vinylogous iminium ion intermediates generated from arylsulfonyl

indoles," Angewandte Chemie International Edition, vol. 47, no. 45, pp. 8707–8710, 2008. · ·

131. R. K. Kunz and D. W. C. MacMillan, "Enantioselective organocatalytic cyclopropanations. The identification of a new class of iminium catalyst based upon directed electrostatic activation," Journal of the American Chemical Society, vol. 127, no. 10, pp. 3240–3241, 2005. · ·

132. J. Franzén, M. Marigo, D. Fielenbach, T. C. Wabnitz, A. Kjærsgaard, and K. A. Jørgensen, "A general organocatalyst for direct α-functionalization of aldehydes: stereoselective C–C, C–N, C–F, C–Br, and C–S bond-forming reactions. Scope and mechanistic insights," Journal of the American Chemical Society, vol. 127, p. 18296, 2005. ·

133. A. Bøgevig, K. Juhl, N. Kumaragurubaran, W. Zhuang, and K. A. Jørgensen, "Direct organo-catalytic asymmetric α-amination of aldehydes—a simple approach to optically active α-amino aldehydes, α-amino alcohols, and α-amino acids," Angewandte Chemie International Edition, vol. 41, pp. 1790–1793, 2002.

134. M. Marigo and K. A. Jørgensen, "α-Heteroatom functionalization," in Enantioselective Organocatalysis, P. I. Dalko, Ed., Chapter 2.2, Wiley-VCH, Weinheim, Germany, 2007.

135. M. Marigo, T. Schulte, J. Franzén, and K. A. Jørgensen, "Asymmetric multicomponent domino reactions and highly enantioselective conjugated addition of thiols to α,β-unsaturated aldehydes," Journal of the American Chemical Society, vol. 127, pp. 15710–15711, 2005.

136. M. Marigo, J. Franzén, T. B. Poulsen, W. Zhuang, and K. A. Jørgensen, "Asymmetric organocatalytic epoxidation of α,β-unsaturated aldehydes with hydrogen peroxide," Journal of the American Chemical Society, vol. 127, pp. 6964–6965, 2005. ·

137. A. Carlone, G. Bartoli, M. Bosco, L. Sambri, and P. Melchiorre, "Organocatalytic asymmetric hydrophosphination of α,β-unsaturated aldehydes," Angewandte Chemie International Edition, vol. 46, no. 24, pp. 4504–4506, 2007. · ·

138. I. Ibrahem, R. Rios, J. Vesely et al., "Enantioselective organocatalytic hydrophosphination of α,β-unsaturated aldehydes," Angewandte Chemie International Edition, vol. 46, no. 24, pp. 4507–4510, 2007. · ·

139. X. Yu and W. Wang, "Organocatalysis: asymmetric cascade reactions catalysed by chiral secondary amines," Organic & Biomolecular Chemistry, vol. 6, pp. 2037–2046, 2008. ·

140. D. Enders, M. R. M. Hüttl, C. Grondal, and G. Raabe, "Control of four stereocentres in a triple cascade organocatalytic reaction," Nature, vol.

441, pp. 861–863, 2006.

141. A. E. Allen and D. W. C. MacMillan, "Enantioselective α-arylation of aldehydes via the productive merger of iodonium salts and organocatalysis," Journal of the American Chemical Society, vol. 133, no. 12, pp. 4260–4263, 2011. · ·

142. A. E. Allen and D. W. C. MacMillan, "The productive merger of iodonium salts and organocatalysis: a non-photolytic approach to the enantioselective α-trifluoromethylation of aldehydes," Journal of the American Chemical Society, vol. 132, no. 14, pp. 4986–4987, 2010. · ·

143. H.-W. Shih, M. N. Vander Wal, R. L. Grange, and D. W. C. MacMillan, "Enantioselective α-benzylation of aldehydes via photoredox organocatalysis," Journal of the American Chemical Society, vol. 132, no. 39, pp. 13600–13603, 2010. · ·

144. D. W. C. MacMillan, "The advent and development of organocatalysis," Nature, vol. 455, no. 7211, pp. 304–308, 2008. · ·

145. T. D. Beeson, A. Mastracchio, J.-B. Hong, K. Ashton, and D. W. C. MacMillan, "Enantioselective organocatalysis using SOMO activation," Science, vol. 316, no. 5824, pp. 582–585, 2007. · ·

146. S. G. Ouellet, A. Walji, and D. W. C. MacMillan, "Enantioselective organocatalytic transfer hydrogenation reactions using Hantzsch esters," Accounts of Chemical Research, vol. 40, pp. 1327–1339, 2007. ·

147. Y. Huang, A. M. Walji, C. H. Larsen, and D. W. C. MacMillan, "Enantioselective organo-cascade catalysis," Journal of the American Chemical Society, vol. 127, no. 43, pp. 15051–15053, 2005. · ·

148. S. P. Brown, M. P. Brochu, C. J. Sinz, and D. W. C. MacMillan, "The direct and enantioselective organocatalytic α-oxidation of aldehydes," Journal of the American Chemical Society, vol. 125, no. 36, pp. 10808–10809, 2003. · ·

149. N. A. Para and D. W. C. MacMillan, "New strategies in organic catalysis: the first enantioselective organocatalytic Friedel-Crafts alkylation," Journal of the American Chemical Society, vol. 123, no. 18, pp. 4370–4371, 2001. · ·

150. W. S. Jen, J. J. M. Wiener, and D. W. C. MacMillan, "New strategies for organic catalysis: the first enantioselective organocatalytic 1,3-dipolar cycloaddition," Journal of the American Chemical Society, vol. 122, no. 40, pp. 9874–9875, 2000. · ·

151. F. Calderón, R. Fernández, F. Sánchez, and A. Fernández-Mayoralas, "Asymmetric aldol reaction using immobilized proline on mesoporous

support," Advanced Synthesis & Catalysis, vol. 347, no. 10, pp. 1395–1403, 2005.

152. Z. An, W. Zhang, H. Shi, and J. He, "An effective heterogeneous l-proline catalyst for the asymmetric aldol reaction using anionic clays as intercalated support," Journal of Catalysis, vol. 241, no. 2, pp. 319–327, 2006. · ·

153. M. Gruttadauria, F. Giacalone, and R. Noto, "Supported proline and proline-derivatives as recyclable organocatalysts," Chemical Society Reviews, vol. 37, no. 8, pp. 1666–1688, 2008. · ·

154. O. Bortolini, A. Cavazzini, P. P. Giovannini et al., "A combined kinetic and thermodynamic approach for the interpretation of continuous-flow heterogeneous catalytic processes," Chemistry—A European Journal, vol. 19, no. 24, pp. 7802–7808, 2013. ·

155. A. Berkessel and H. Groeger, Eds., Asymmetric Organocatalysis: From Biomimetic Concepts To Applications in Asymmetric SynthesisSpecial Issue: Asymmetric Organocatalysis, Wiley-VCH, Weinheim, Germany, 2005.

156. A. Berkessel and H. Groeger, Metal-Free Organic Catalysts in Asymmetric Synthesis, Wiley-VCH, Weinheim, Germany, 2004.

157. K. N. Houk and B. List, "Special issue: asymmetric organocatalysis," Accounts of Chemical Research, vol. 37, no. 8, p. 487, 2004. ·

158. F. Xu, M. Zacuto, N. Yoshikawa et al., "Asymmetric synthesis of telcagepant, a CGRP receptor antagonist for the treatment of migraine," Journal of Organic Chemistry, vol. 75, no. 22, pp. 7829–7841, 2010. · ·

159. H. U. Blaser and E. Schmidt, "Introduction," in Asymmetric Catalysis on Industrial Scale, H. U. Blaser and E. Schmidt, Eds., pp. 1–19, Wiley-VCH, Weinheim, Germany, 2004.

160. H. U. Blaser, B. Pugin, and F. Spindler, "Progress in enantioselective catalysis assessed from an industrial point of view," Journal of Molecular Catalysis A, vol. 231, pp. 1–20, 2005. ·

161. H. U. Blaser, "Enantioselective catalysis in fine chemicals production," Chemical Communications, pp. 293–296, 2003. ·

162. S. S. V. Ramasastry, H. Zhang, F. Tanaka, and C. F. Barbas III, "Direct catalytic asymmetric synthesis of anti-1,2-amino alcohols and syn-1,2-diols through organocatalytic anti-Mannich and syn-aldol reactions," Journal of the American Chemical Society, vol. 129, pp. 288–

289, 2007. ·

163. Y. Zhao, J. Rodrigo, A. H. Hoveyda, and M. L. Snapper, "Enantioselective silyl protection of alcohols catalysed by an amino-acid-based small molecule," Nature, vol. 443, no. 7107, pp. 67–70, 2006. · ·

164. S. Saha and J. N. Moorthy, "Enantioselective organocatalytic Biginelli reaction: dependence of the catalyst on sterics, hydrogen bonding, and reinforced chirality," The Journal of Organic Chemistry, vol. 76, pp. 396–402, 2011. ·

165. S. Gore, S. Baskaran, and B. Koenig, "Efficient synthesis of 3,4-dihydropyrimidin-2-ones in low melting tartaric acid-urea mixtures," Green Chemistry, vol. 13, no. 4, pp. 1009–1013, 2011. · ·

166. N. Li, X. H. Chen, J. Song, S. W. Luo, W. Fan, and L. Z. Gong, "Highly enantioselective organocatalytic Biginelli and Biginelli-like condensations: reversal of the stereochemistry by tuning the 3,3′-disubstituents of phosphoric acids," Journal of the American Chemical Society, vol. 132, p. 10953, 2010.·

167. N. Lu, D. Chen, G. Zhang, and Q. Liu, "Theoretical investigation on enantioselective Biginelli reaction catalyzed by natural tartaric acid," International Journal of Quantum Chemistry, vol. 111, pp. 2031–2038, 2010. ·

168. M. Lei, "An efficient and environmentally friendly procedure for synthesis of pyrimidinone derivatives by use of a Biginelli-type reaction," Monatshefte für Chemie, vol. 141, pp. 1005–1008, 2010. ·

169. Y. Y. Wu, Z. Chai, X. Y. Liu, G. Zhao, and S. W. Wang, "Synthesis of substituted 5-(Pyrrolidin-2-yl)tetrazoles and their application in the asymmetric Biginelli reaction," European Journal of Organic Chemistry, vol. 6, pp. 904–911, 2009. ·

170. M. M. Savant, A. M. Pansuriya, C. V. Bhuva, N. P. Kapuriya, and Y. T. Naliapara, "Etidronic acid: a new and efficient catalyst for the synthesis of novel 5-nitro-3,4-dihydropyrimidin-2(1H)-ones," Catalysis Letters, vol. 132, no. 1-2, pp. 281–284, 2009. · ·

171. J. N. Sangshetti, N. D. Kokare, and D. B. Shinde, "Oxalic acid as a versatile catalyst for one pot facile synthesis of 3,4-dihydropyrimidin-2-(1H)-ones and their thione analogues," Journal of Heterocyclic Chemistry, vol. 45, no. 4, pp. 1191–1194, 2008. · ·

172. E. Ramu, V. Kotra, N. Bansal, R. Varala, and S. R. Adapa, "Green approach for the efficient synthesis of Biginelli compounds promoted by citric acid under solvent-free conditions," Rasayan Journal of Chemistry, vol. 1, pp. 188–194, 2008.

173. D. B. Ramachary, N. S. Chowdari, and C. F. Barbas III, "Amine-catalyzed direct self Diels-Alder reactions of α,β-unsaturated ketones in water: synthesis of pro-chiral cyclohexanones," Tetrahedron Letters, vol. 43, no. 38, pp. 6743–6746, 2002. · ·

174. D. B. Ramachary, K. Anebouselvy, N. S. Chowdari, and C. F. Barbas III, "Direct organocatalytic asymmetric heterodomino reactions: the Knoevenagel/Diels-Alder/epimerization sequence for the highly diastereoselective synthesis of symmetrical and nonsymmetrical synthons of benzoannelated centropolyquinanes," Journal of Organic Chemistry, vol. 69, no. 18, pp. 5838–5849, 2004. · ·

175. R. Martín-Rapún, X. Fan, S. Sayalero, M. Bahramnejad, F. Cuevas, and M. A. Pericàs, "Highly active organocatalysts for asymmetric anti-mannich reactions," Chemistry—A European Journal, vol. 17, no. 32, pp. 8780–8783, 2011. · ·

176. T. Ooi, M. Taniguchi, M. Kameda, and K. Maruoka, "Direct asymmetric aldol reactions of glycine schiff base with aldehydes catalyzed by chiral quaternary ammonium salts," Angewandte Chemie International Edition, vol. 41, pp. 4542–4544, 2002.

177. D. R. Burri, I. R. Shaikh, S.-C. Han, and S.-E. Park, "Facile heterogenization of homogeneous ferrocene catalyst on SBA-16," Studies in Surface Science and Catalysis, vol. 165, pp. 647–650, 2007. · ·

178. I. R. Shaikh, "Heterogenization of a basic ionic liquid bearing NTf2- anion on mesoporous SBA-16 and its use as catalyst in knoevenagel reaction," in Proceedings of the National Conference on Drug Designing and Discovery, pp. 81–85, Devchand College, Arjun Nagar, Ta. Kagal, District Kolhapur, India, September 2013, http://devchandcollege.org/e%20proceeding%203D13.html.

179. I. R. Shaikh and A. A. Shaikh, "Heterogenization of ionic liquid containing Hünig base on mesoporous SBA-16 and its use as catalyst in knovenagel condensation," in Proceedings of the National Conference on Frontiers of Physical, Chemical and Biological Sciences (FPCBS ‹13), University of Pune, Ganeshkhind, Pune, India, October 2013, published as special issue in: Environment Observer, vol. 13, pp. 35-36, 2013, http://www.seeram.org/.

180. D. Petruzziello, M. Stenta, A. Mazzanti, and P. G. Cozzi, "A rational approach towards a new ferrocenyl pyrrolidine for stereoselective enamine catalysis," Chemistry—A European Journal, vol. 19, no. 24, pp. 7696–7700, 2013. ·

181. S. Sulzer-Mossé and A. Alexakis, "Chiral amines as organocatalysts

for asymmetric conjugate addition to nitroolefins and vinyl sulfonesviaenamine activation," Chemical Communications, no. 30, pp. 3123–3135, 2007. ·

182. Z. An, W. Zhang, H. Shi, and J. He, "An effective heterogeneous l-proline catalyst for the asymmetric aldol reaction using anionic clays as intercalated support," Journal of Catalysis, vol. 241, no. 2, pp. 319–327, 2006. · ·

183. L. Albrecht, G. Dickmeiss, C. F. Weise, C. Rodríguez-Escrich, and K. A. Jørgensen, "Dienamine-mediated inverse-electron-demand hetero-diels-alder reaction by using an enantioselective H-bond-directing strategy," Angewandte Chemie International Edition, vol. 51, no. 52, pp. 13109–13113, 2012.

184. Q. Ren and J. Wang, "Recent developments in amine-catalyzed non-asymmetric transformations," Asian Journal of Organic Chemistry, vol. 2, no. 7, pp. 542–557, 2013.

185. T. Li, J. Zhu, D. Wu et al., "A strategy enabling enantioselective direct conjugate addition of inert aryl methane nucleophiles to enals with a chiral amine catalyst under mild conditions," Chemistry—A European Journal, vol. 19, no. 28, pp. 9147–9150, 2013.

186. G. Talavera, E. Reyes, J. L. Vicario, and L. Carrillo, "Cooperative dienamine/hydrogen-bonding catalysis: enantioselective formal [2+2] cycloaddition of enals with nitroalkenes," Angewandte Chemie International Edition, vol. 51, no. 17, pp. 4104–4107, 2012.

187. D. B. Ramachary and Y. V. Reddy, "Dienamine catalysis: an emerging technology in organic synthesis,"European Journal of Organic Chemistry, no. 5, pp. 868–887, 2012.

188. J. Chin, F. Mancin, N. Thavarajah, D. Lee, A. Lough, and D. S. Chung, "Controlling diaza-Cope rearrangement reactions with resonance-assisted hydrogen bonds," Journal of the American Chemical Society, vol. 125, pp. 15276–15277, 2003.

189. H. J. Kim, H. Kim, G. Alhakimi et al., "Preorganization in highly enantioselective diaza-Cope rearrangement reaction," Journal of the American Chemical Society, vol. 127, pp. 16370–16371, 2005.·

190. H. J. Kim, W. Kim, A. J. Lough, B. M. Kim, and J. Chin, "A cobalt(III)-salen complex with an axial substituent in the diamine backbone: stereoselective recognition of amino alcohols," Journal of the American Chemical Society, vol. 127, pp. 16776–16777, 2005.

191. L. Cui, Y. Zhu, S. Luo, and J.-P. Cheng, "Primary-tertiary diamine/brønsted acid catalyzed C–C coupling between para-vinylanilines and

aldehydes," Chemistry—A European Journal, vol. 19, no. 29, pp. 9481–9484, 2013.

192. M. Orlandi, M. Benaglia, L. Raimondi, and G. Celentano, "2-Aminoimidazolyl and 2-aminopyridyl (S)-prolinamides as versatile multifunctional organic catalysts for aldol, Michael and Diels Alder reactions,"European Journal of Organic Chemistry, vol. 12, pp. 2346–2354, 2013.

193. H. Wang, Y. Wang, H. Song, Z. Zhou, and C. Tang, "Bifunctional squaramide-catalyzed one-pot sequential Michael addition/dearomative bromination: convenient access to optically active brominated pyrazol-5(4H)-ones with adjacent quaternary and tertiary stereocenters," European Journal of Organic Chemistry, vol. 2013, no. 22, pp. 4844–4851, 2013.

194. B. Tan, G. Hernández-Torres, and C. F. Barbas III, "Rationally designed amide donors for organocatalytic asymmetric Michael reactions," Angewandte Chemie International Edition, vol. 51, no. 22, pp. 5381–5385, 2012.

195. D. Nguyen, R. K. Akhani, C. I. Sheppard, and S. L. Wiskur, "Structure-activity relationship of formamides as organocatalysts: the significance of formamide structure and conformation," European Journal of Organic Chemistry, vol. 12, pp. 2279–2283, 2013.

196. G. Lelais and D. W. C. MacMillan, "Modern strategies in organic catalysis: the advent and development of iminium activation," Aldrichimica Acta, vol. 39, no. 3, pp. 79–87, 2006.

197. S. Lee and D. W. C. MacMillan, "Organocatalytic vinyl and Friedel-Crafts alkylations with trifluoroborate salts," Journal of the American Chemical Society, vol. 129, no. 50, pp. 15438–15439, 2007.

198. N. A. Paras and D. W. C. MacMillan, "New strategies in organocatalysis: The first enantioselective organocatalytic Friedel-Crafts alkylation," Journal of the American Chemical Society, vol. 123, pp. 4370–4371, 2001.

199. M. P. Brochu, S. P. Brown, and D. W. C. MacMillan, "Direct and enantioselective organocatalytic α-chlorination of aldehydes," Journal of the American Chemical Society, vol. 126, no. 13, pp. 4108–4109, 2004.

200. T. D. Beeson and D. W. C. MacMillan, "Enantioselective organocatalytic α-fluorination of aldehydes,"Journal of the American Chemical Society, vol. 127, no. 24, pp. 8826–8828, 2005.

201. M. T. H. Fonseca and B. List, "Catalytic asymmetric intramolecular Michael reaction of aldehydes,"Angewandte Chemie International Edition, vol. 43, no. 30, pp. 3958–3960, 2004. · ·

202. S. G. Ouellet, J. B. Tuttle, and D. W. MacMillan, "Enantioselective organocatalytic hydride reduction,"Journal of the American Chemical Society, vol. 127, pp. 32–33, 2005.

203. MacMillan Imidazolidinone OrganoCatalysts are a trademark of Materia, Inc.,https://www.princeton.edu/chemistry/macmillan/publications/aldrichimica.pdf.

204. M. C. Holland, S. Paul, W. B. Schweizer et al., "Noncovalent interactions in organocatalysis: modulating conformational diversity and reactivity in the MacMillan catalyst," Angewandte Chemie International Edition, vol. 52, no. 31, pp. 7967–7971, 2013.

205. G. Chollet, F. Rodriguez, and E. Schulz, "A new method for recycling asymmetric catalysts via formation of charge transfer complexes," Organic Letters, vol. 8, pp. 539–542, 2006.

206. L. Xing, J. H. Xie, Y. S. Chen, L. X. Wang, and Q. L. Zhou, "Simply modified chiral diphosphine: catalyst recycling via non-covalent absorption on carbon nanotubes," Advanced Synthesis & Catalysis, vol. 350, pp. 1013–1016, 2008.

207. R. Akiyama and S. Kobayashi, "Microencapsulated and related catalysts for organic chemistry and organic synthesis," Chemical Reviews, vol. 109, no. 2, pp. 594–642, 2009.

208. C. A. Wang, Y. Zhang, J. Y. Shi, and W. Wang, "A self-supported polymeric MacMillan catalyst for homogeneous organocatalysis and heterogeneous recycling," Chemistry—A European Journal, vol. 8, no. 6, pp. 1110–1114, 2013.

209. J. Seayad, A. M. Seayad, and B. List, "Catalytic asymmetric Pictet-Spengler reaction," Journal of the American Chemical Society, vol. 128, no. 4, pp. 1086–1087, 2006.

210. S. Hoffmann, A. M. Seayad, and B. List, "A powerful Brønsted acid catalyst for the organocatalytic asymmetric transfer hydrogenation of imines," Angewandte Chemie International Edition, vol. 44, no. 45, pp. 7424–7427, 2005.

211. M. Rueping, E. Sugiono, C. Azap, T. Theissmann, and M. Bolte, "Enantioselective Brønsted acid catalyzed transfer hydrogenation: organocatalytic reduction of imines," Organic Letters, vol. 7, no. 17, pp. 3781–3783, 2005.

212. R. I. Storer, D. E. Carrera, Y. Ni, and D. W. C. MacMillan, "Enantioselective organocatalytic reductive amination," Journal of the American Chemical Society, vol. 128, no. 1, pp. 84–86, 2006

213. J. Zhou and B. List, "Organocatalytic asymmetric reaction cascade to substituted cyclohexylamines,"Journal of the American Chemical Society, vol. 129, no. 24, pp. 7498–7499, 2007.

214. G. B. Rowland, H. Zhang, E. B. Rowland, S. Chennamadhavuni, Y. Wang, and J. C. Antilla, "Brønsted acid-catalyzed imine amidation," Journal of the American Chemical Society, vol. 127, no. 45, pp. 15696–15697, 2005.

215. T. Akiyama, Y. Tamura, J. Itoh, H. Morita, and K. Fuchibe, "Enantioselective aza-Diels-Alder reaction catalyzed by a chiral Brønsted acid: effect of the additive on the enantioselectivity," Synlett, no. 1, Article ID Y07205ST, pp. 141–143, 2006.

216. Y. Su, M. J. Bouma, L. Alcaraz et al., "Organocatalytic enantioselective one-pot four-component ugi-type multicomponent reaction for the synthesis of epoxy-tetrahydropyrrolo[3,4-b]pyridin-5-ones," Chemistry—A European Journal, vol. 18, no. 40, pp. 12624–12627, 2012.

217. D. J. Bayston, J. L. Fraser, M. R. Ashton, A. D. Baxter, M. E. C. Polywka, and E. Moses, "Preparation and use of a polymer supported BINAP hydrogenation catalyst," The Journal of Organic Chemistry, vol. 63, p. 3137, 1998.

218. D. S. Kundu, J. Schmidt, C. Bleschke, A. Thomas, and S. Blechert, "A microporous binol-derived phosphoric acid," Angewandte Chemie International Edition, vol. 51, no. 22, pp. 5456–5459, 2012.

219. Y. Huang, A. K. Unni, A. N. Thadani, and V. H. Rawal, "Single enantiomers from a chiral-alcohol catalyst," Nature, vol. 424, no. 6945, p. 146, 2003.

220. H. M. Guo, L. Cheng, L. F. Cun, L. Z. Gong, A. Q. Mi, and Y. Z. Jiang, "L-Prolinamide-catalyzed direct nitroso aldol reactions of α-branched aldehydes: a distinct regioselectivity from that with L-proline,"Chemical Communications, no. 4, pp. 429–431, 2006.

221. N. T. McDougal and S. E. Schaus, "Asymmetric Morita-Baylis-Hillman reactions catalyzed by chiral Brønsted acids," Journal of the American Chemical Society, vol. 125, no. 40, pp. 12094–12095, 2003.

222. P. Vachal and E. N. Jacobsen, "Structure-based analysis and optimization of a highly enantioselective catalyst for the strecker reaction," Journal of the American Chemical Society, vol. 124, no. 34, pp. 10012–10014, 2002.

223. G. D. Joly and E. N. Jacobsen, "Thiourea-catalyzed enantioselective hydrophosphonylation of imines: practical access to enantiomerically enriched α-amino phosphonic acids," Journal of the American Chemical

Society, vol. 126, p. 4102, 2004.

224. A. G. Wenzel and E. N. Jacobsen, "Asymmetric catalytic Mannich reactions catalyzed by urea derivatives: enantioselective synthesis of β-aryl-β-amino acids," Journal of the American Chemical Society, vol. 124, pp. 12964–12965, 2002.

225. D. E. Fuerst and E. N. Jacobsen, "Thiourea-catalyzed enantioselective cyanosilylation of ketones," Journal of the American Chemical Society, vol. 127, no. 25, pp. 8964–8965, 2005.

226. M. S. Taylor and E. N. Jacobsen, "Highly enantioselective catalytic acyl-Pictet-Spengler reactions,"Journal of the American Chemical Society, vol. 126, no. 34, pp. 10558–10559, 2004.

227. M. S. Talyor, N. Tokunaga, and E. N. Jacobsen, "Enantioselective thiourea-catalyzed acyl-mannich reactions of isoquinolines," Angewandte Chemie International Edition, vol. 44, pp. 6700–6704, 2005.

228. X. Li, C. Yang, J.-L. Jin, X.-S. Xue, and J.-P. Cheng, "Synthesis of optically enriched spirocyclic benzofuran-2-ones by bifunctional thiourea-base catalyzed double-Michael addition of benzofuran-2-ones to dienones," Chemistry—An Asian Journal, vol. 8, no. 05, pp. 997–1003, 2013.

229. S.-R. Ban, X.-X. Zhu, Z.-P. Zhang, H.-Y. Xie, and Q.-S. Li, "Benzoylthiourea–pyrrolidine as another bifunctional organocatalyst: highly enantioselective michael addition of cyclohexanone to nitroolefins,"European Journal of Organic Chemistry, vol. 15, pp. 2977–2980, 2013.

230. M. Tsakos, C. G. Kokotos, and G. Kokotos, "Primary amine-thioureas with improved catalytic properties for "difficult" Michael reactions: efficient organocatalytic syntheses of (S)-baclofen, (R)-baclofen and (S)-phenibut," Advanced Synthesis and Catalysis, vol. 354, no. 4, pp. 740–746, 2012.

231. M. J. O›Donnell, "The enantioselective synthesis of α-amino acids by phase-transfer catalysis with achiral Schiff base esters," Accounts of Chemical Research, vol. 37, pp. 506–517, 2004.

232. B. Lygo and B. J. Andrews, "The enantioselective synthesis of α-amino acids by phase-transfer catalysis with achiral Schiff base esters," Accounts of Chemical Research, vol. 37, pp. 518–525, 2004.

233. S.-K. Tian, R. Hong, and L. Deng, "Catalytic asymmetric cyanosilylation of ketones with chiral Lewis base," Journal of the American Chemical Society, vol. 125, no. 33, pp. 9900–9901, 2003.

234. T. B. Poulsen, C. Alemparte, and K. A. Jørgensen, "Enantioselective organocatalytic allylic amination,"Journal of the American Chemical Society, vol. 127, pp. 11614–11615, 2005.

235. M. Bella and K. A. Jørgensen, "Organocatalytic enantioselective conjugate addition to alkynones," Journal of the American Chemical Society, vol. 126, pp. 5672–5673, 2004.

236. P. Melchiorre, "Cinchona-based primary amine catalysis in the asymmetric functionalization of carbonyl compounds," Angewandte Chemie International Edition, vol. 51, no. 39, pp. 9748–9770, 2012.

237. E. Arceo and P. Melchiorre, "Extending the aminocatalytic HOMO-raising activation strategy: where is the limit?" Angewandte Chemie International Edition, vol. 51, no. 22, pp. 5290–5292, 2012.

238. C.-K. Pei and M. Shi, "Asymmetric cyclization reactions of allenoates with imines or α,β-unsaturated ketones catalyzed by organocatalysts derived from cinchona alkaloids," Chemistry—A European Journal, vol. 18, no. 22, pp. 6712–6716, 2012.

239. A. Russo, G. Galdi, G. Croce, and A. Lattanzi, "Highly enantioselective epoxidation catalyzed by cinchona thioureas: synthesis of functionalized terminal epoxides bearing a quaternary stereogenic center," Chemistry—A European Journal, vol. 18, no. 20, pp. 6152–6157, 2012.

240. A. Kumar and S. S. Chimni, "Organocatalytic asymmetric direct aldol reaction of pyruvic aldehyde dimethyl acetal with isatin derivatives," European Journal of Organic Chemistry, vol. 2013, no. 22, pp. 4780–4876, 2013.

241. M.-X. Zhao, H. Zhou, W.-H. Tang, W.-S. Qu, and M. Shi, "Cinchona alkaloid-derived thiourea-catalyzed diastereo- and enantioselective [3+2] cycloaddition reaction of isocyanoacetates to isatins: a facile access to optically active spirooxindole oxazolines," Advanced Synthesis & Catalysis, vol. 355, no. 7, pp. 1277–1283, 2013.

242. F. Xu, M. Zacuto, N. Yoshikawa et al., "Asymmetric synthesis of telcagepant, a CGRP receptor antagonist for the treatment of migraine," Journal of Organic Chemistry, vol. 75, no. 22, pp. 7829–7841, 2010.

243. G. Rassu, V. Zambrano, L. Pinna et al., "Direct regio-, diastereo-, and enantioselective vinylogous Michael addition of prochiral 3-alkylideneoxindoles to nitroolefins," Advanced Synthesis & Catalysis, vol. 355, no. 9, pp. 1881–1886, 2013.

244. S. Brandau, A. Landa, J. Franzén, M. Marigo, and K. A. Jørgensen, "Organocatalytic conjugate addition of malonates to alpha,beta-

unsaturated aldehydes: asymmetric formal synthesis of (-)-paroxetine, chiral lactams, and lactones," Angewandte Chemie International Edition, vol. 45, no. 26, pp. 4305–4309, 2006.·

245. L. J. Hounjet, C. Bannwarth, C. N. Garon, C. B. Caputo, S. Grimme, and D. W. Stephan, "Combinations of ethers and $B(C_6F_5)_3$ function as hydrogenation catalysts," Angewandte Chemie International Edition, vol. 52, no. 29, pp. 7492–7495, 2013.

246. X.-F. Cai, M.-W. Chen, Z.-S. Ye et al., "Asymmetric transfer hydrogenation of 3-nitroquinolines: facile access to cyclic nitro compounds with two contiguous stereocenters," Chemistry—An Asian Journal, vol. 8, no. 7, pp. 1381–1385, 2013.

247. M. Kitamura, S. Shirakawa, and K. Maruoka, "Powerful chiral phase-transfer catalysts for the asymmetric synthesis of alpha-alkyl- and alpha,alpha-dialkyl-alpha-amino acids," Angewandte Chemie International Edition, vol. 44, pp. 1549–1551, 2005.

248. T. Ooi, Y. Arimura, Y. Hiraiwa et al., "Highly enantioselective monoalkylation of p-chlorobenzaldehyde imine of glycine tert-butyl ester under mild phase-transfer conditions," Tetrahedron: Asymmetry, vol. 17, pp. 603–606, 2006.

249. T. Ooi and K. Maruoka, "Development and applications of C2-symmetric, chiral, phase-transfer catalysts," Aldrichimica Acta, vol. 40, no. 3, pp. 77–86, 2007.

250. E. J. Corey, R. K. Bakshi, and S. Shibata, "Highly enantioselective borane reduction of ketones catalyzed by chiral oxazaborolidines. Mechanism and synthetic implications," Journal of the American Chemical Society, vol. 109, pp. 5551–5553, 1987.

251. E. J. Corey, R. K. Bakshi, S. Shibata, C. P. Chen, and V. K. Singh, "A stable and easily prepared catalyst for the enantioselective reduction of ketones. Applications to multistep syntheses," Journal of the American Chemical Society, vol. 109, pp. 7925–7926, 1987.

252. E. H. M. Kirton, G. Tughan, R. E. Morris, and R. A. Field, "Rationalising the effect of reducing agent on the oxazaborolidine-mediated asymmetric reduction of N-substituted imines," Tetrahedron Letters, vol. 45, pp. 853–855, 2004.

253. B. T. Cho and Y. S. Chun, "Enantioselective synthesis of optically active metolachlor via asymmetric reduction," Tetrahedron Asymmetry, vol. 3, no. 3, pp. 337–340, 1992. ·

254. B. T. Cho and Y. S. Chun, "Asymmetric reduction of N-substituted ketimines with the reagent prepared from borane and (S)-(–)-2-amino-

3-methyl-1,1-diphenylbutan-1-ol (itsuno›s reagent): enantioselective synthesis of optically active secondary amines," Journal of the Chemical Society, Perkin Transactions, vol. 1, pp. 3200–3201, 1990.

255. R. D. Tillyer, C. Boudreau, D. Tschaen, U.-H. Dolling, and P. J. Reider, "Asymmetric reduction of keto oxime ethers using oxazaborolidine reagents. The enantioselective synthesis of cyclic amino alcohols,"Tetrahedron Letters, vol. 36, no. 25, pp. 4337–4340, 1995.

256. D. H. Ryu and E. J. Corey, "Triflimide activation of a chiral oxazaborolidine leads to a more general catalytic system for enantioselective Diels-Alder addition," Journal of the American Chemical Society, vol. 125, pp. 6388–6390, 2003.

257. Y. Tu, Z.-X. Wang, and Y. Shi, "An efficient asymmetric epoxidation method for trans-olefins mediated by a fructose-derived ketone," Journal of the American Chemical Society, vol. 118, no. 40, pp. 9806–9807, 1996.

258. Z.-X. Wang, Y. Tu, M. Frohn, J.-R. Zhang, and Y. Shi, "An efficient catalytic asymmetric epoxidation method," Journal of the American Chemical Society, vol. 119, no. 46, pp. 11224–11235, 1997.

259. Z.-X. Wang, Y. Tu, M. Frohn, J.-R. Zhang, and Y. Shi, "An efficient catalytic asymmetric epoxidation method," Journal of the American Chemical Society, vol. 119, no. 46, pp. 11224–11235, 1997.

260. M. Marigo, J. Franzén, T. B. Poulsen, W. Zhuang, and K. A. Jørgensen, "Asymmetric organocatalytic epoxidation of α,β-unsaturated aldehydes with hydrogen peroxide," Journal of the American Chemical Society, vol. 127, no. 19, pp. 6964–6965, 2005.

261. N. Marion, S. Díez-González, and S. P. Nolan, "Inside cover: a molecular solomon link (Angew. Chem. Int. Ed. 1-2/2007)," Angewandte Chemie International Edition, vol. 46, p. 2, 2007.

262. M. S. Kerr and T. Rovis, "Enantioselective synthesis of quaternary stereocenters via a catalytic asymmetric stetter reaction," Journal of the American Chemical Society, vol. 126, no. 29, pp. 8876–8877, 2004.

263. M. He, J. R. Struble, and J. W. Bode, "Highly enantioselective azadiene diels–alder reactions catalyzed by chiral N-heterocyclic carbenes," Journal of the American Chemical Society, vol. 128, pp. 8418–8420, 2006.·

264. M. He, G. J. Uc, and J. W. Bode, "Chiral N-heterocyclic carbene catalyzed, enantioselective oxodiene Diels-Alder reactions with low catalyst loadings," Journal of the American Chemical Society, vol. 128, pp. 15088–15089, 2006.

265. P.-C. Chiang, J. Kaeobamrung, and J. W. Bode, "Enantioselective,

cyclopentene-forming annulations via NHC-catalyzed benzoin-oxy-Cope reactions," Journal of the American Chemical Society, vol. 129, no. 12, pp. 3520–3521, 2007.

266. S. S. Sohn and J. W. Bode, "N-heterocyclic carbene catalyzed C–C bond cleavage in redox esterifications of chiral formylcyclopropanes," Angewandte Chemie International Edition, vol. 45, pp. 6021–6024, 2006.·

267. A. Grossmann and D. Enders, "N-heterocyclic carbene catalyzed domino reactions," Angewandte Chemie International Edition, vol. 51, no. 2, pp. 314–325, 2012.

268. D. Enders, R. Hahn, and I. Atodiresei, "Asymmetric synthesis of functionalized tetrahydronaphthalenes via an organocatalytic nitroalkane-michael/henry domino reaction," Advanced Synthesis & Catalysis, vol. 355, no. 6, pp. 1126–1136, 2013.

269. H. Pellissier, "Recent developments in asymmetric organocatalytic domino reactions," Advanced Synthesis & Catalysis, vol. 354, no. 2-3, pp. 237–294, 2012.

270. L. Wu, Y. Wang, H. Song, L. Tang, Z. Zhou, and C. Tang, "Enantioselective organocatalytic domino Michael/aldol reactions: an efficient procedure for the stereocontrolled construction of 2H-thiopyrano[2,3-b]quinoline scaffolds," Chemistry—An Asian Journal, vol. 8, pp. 2204–2210, 2013.

271. J. Zhao, C. Mück-Lichtenfeld, and A. Studer, "Cooperative N-heterocyclic carbene (NHC) and ruthenium redox catalysis: oxidative esterification of aldehydes with air as the terminal oxidant,"Advanced Synthesis & Catalysis, vol. 355, no. 6, pp. 1098–1106, 2013.

272. R. Blanc, P. Nava, M. Rajzman, L. Commeiras, and J.-L. Parrain, "N-Heterocyclic carbene mediated organocatalytic transfer of tin onto aldehydes: an easy access to syn-diols and mechanistic studies,"Advanced Synthesis & Catalysis, vol. 354, no. 10, pp. 2038–2048, 2012.

273. S. De Sarkar, A. Biswas, R. C. Samanta, and A. Studer, "Catalysis with N-heterocyclic carbenes under oxidative conditions," Chemistry—A European Journal, vol. 19, no. 15, pp. 4664–4678, 2013.

274. H. U. Vora, P. Wheeler, and T. Rovis, "Exploiting acyl and enol azolium intermediates via N-hetero-cyclic carbene-catalyzed reactions of α-reducible aldehydes," Advanced Synthesis & Catalysis, vol. 354, no. 9, pp. 1617–1639, 2012.

275. J. Mo, L. Shen, and Y. R. Chi, "Direct β-activation of saturated aldehydes to formal Michael acceptors through oxidative NHC catalysis," Angewandte Chemie International Edition, vol. 52, no. 33, pp. 8588–8591, 2013.

276. M. Hans, J. Wouters, A. Demonceau, and L. Delaude, "Mechanistic insight into the staudinger reaction catalyzed by N-heterocyclic carbenes," Chemistry—A European Journal, vol. 19, no. 29, pp. 9668–9676, 2013.

277. W. Raimondi, D. Bonne, and J. Rodriguez, "1,2-dicarbonyl compounds as pronucleophiles in organocatalytic asymmetric transformations," Angewandte Chemie International Edition, vol. 51, no. 1, pp. 40–42, 2012.

278. A. Rai and L. D. S. Yadav, "Cyclopropenone-catalyzed direct conversion of aldoximes and primary amides into nitriles," European Journal of Organic Chemistry, vol. 10, pp. 1889–1893, 2013.

279. J. Marco-Martínez, V. Marcos, S. Reboredo, S. Filippone, and N. Martín, "Asymmetric organocatalysis in fullerenes chemistry: enantioselective phosphine-catalyzed cycloaddition of allenoates onto C_{60},"Angewandte Chemie International Edition, vol. 52, no. 19, pp. 5115–5119, 2013.

280. A.-L. Lee, "Organocatalyzed carbonyl–olefin metathesis," Angewandte Chemie International Edition, vol. 52, no. 17, pp. 4524–4525, 2013.

281. F. Kniep, S. H. Jungbauer, Q. Zhang et al., "Organocatalysis by neutral multidentate halogen-bond donors," Angewandte Chemie International Edition, vol. 52, no. 27, pp. 7028–7032, 2013.

282. J. H. Kim, I. Coric, S. Vellalath, and B. List, "The catalytic asymmetric acetalization," Angewandte Chemie International Edition, vol. 52, no. 16, pp. 4474–4477, 2013.

283. P. Chauhan and S. S. Chimni, "Organocatalytic enantioselective Morita-Baylis-Hillman reaction of maleimides with isatins," Asian Journal of Organic Chemistry, vol. 2, no. 7, pp. 586–592, 2013.

284. P. Chauhan, J. Kaur, and S. S. Chimni, "Asymmetric organocatalytic addition reactions of maleimides: a promising approach towards the synthesis of chiral succinimide derivatives," Chemistry—An Asian Journal, vol. 8, no. 2, pp. 328–346, 2013.

285. C. C. J. Loh and D. Enders, "Exploiting the electrophilic properties of indole intermediates: new options in designing asymmetric reactions," Angewandte Chemie International Edition, vol. 51, no. 1, pp. 46–48, 2012.

286. A. Martínez, M. J. Webber, S. Müller, and B. List, "Versatile access to chiral indolines by catalytic asymmetric fischer indolization," Angewandte Chemie International Edition, vol. 52, no. 36, pp. 948–9490, 2013.

287. Z. Shi and T.-P. Loh, "Organocatalytic synthesis of highly functionalized

pyridines at room temperature,"Angewandte Chemie International Edition, vol. 52, no. 33, pp. 8584–8587, 2013.

288. A. V. Malkov, S. Stoncius, M. Bell et al., "Mechanistic dichotomy in the asymmetric allylation of aldehydes with allyltrichlorosilanes catalyzed by chiral pyridine N-oxides," Chemistry—A European Journal, vol. 19, no. 28, pp. 9167–9185, 2013.

289. K. Shibatomi and A. Narayama, "Catalytic enantioselective α-chlorination of carbonyl compounds,"Asian Journal of Organic Chemistry, vol. 2, no. 10, pp. 812–823, 2013.

290. C. C. J. Loh and D. Enders, "Merging organocatalysis and gold catalysis—a critical evaluation of the underlying concepts," Chemistry—A European Journal, vol. 18, no. 33, pp. 10212–10225, 2012.

291. H.-N. Yuan, S. Wang, J. Nie, W. Meng, Q. Yao, and J.-A. Ma, "Hydrogen-bond-directed enantioselective decarboxylative Mannich reaction of β-ketoacids with ketimines: application to the synthesis of anti-HIV drug DPC 083," Angewandte Chemie International Edition, vol. 52, no. 14, pp. 3869–3873, 2013.

292. Y. Hayashi, D. Sakamoto, H. Shomura, and D. Hashizume, "Asymmetric Mannich reaction of α-keto imines catalyzed by diarylprolinol silyl ether," Chemistry—A European Journal, vol. 19, no. 24, pp. 7678–7681, 2013.

293. W.-Q. Zhang, L.-F. Cheng, J. Yu, and L.-Z. Gong, "A chiral bis(betaine) catalyst for the mannich reaction of azlactones and aliphatic imines," Angewandte Chemie International Edition, vol. 51, no. 17, pp. 4085–4088, 2012.

294. L. Hong and R. Wang, "Recent advances in asymmetric organocatalytic construction of 3,3′-spirocyclic oxindoles," Advanced Synthesis & Catalysis, vol. 355, no. 6, pp. 1023–1052, 2013.

295. M.-X. Wei, C.-T. Wang, J.-Y. Du et al., "Enantioselective synthesis of Amaryllidaceae alkaloids (+)-vittatine, (+)-epi-vittatine, and (+)-buphanisine," Chemistry—An Asian Journal, vol. 8, no. 9, pp. 1966–1971, 2013.

296. H.-J. Yang, F.-J. Xiong, X.-F. Chen, and F.-E. Chen, "Highly enantioselective thiolysis of prochiral cyclic anhydrides catalyzed by amino alcohol bifunctional organocatalysts and its application to the synthesis of pregabalin," European Journal of Organic Chemistry, vol. 2013, no. 21, pp. 4495–4489, 2013.

297. R. Selke and M. Capka, "Carbohydrate phosphinites as chiral ligands for asymmetric syntheses catalyzed by complexes: part VIII: immobilization

of cationic rhodium(I) chelates of phenyl 4,6-O-(R)-benzylidene-2,3-bis(O-diphenylphosphino)-β-D-glucopyranoside on silica," Journal of Molecular Catalysis, vol. 63, pp. 319–334, 1990.

298. M. Ichikawa, "'Ship-in-Bottle' catalyst technology. Novel templating fabrication of platinum group metals nanoparticles and wires in micro/mesopores," Platinum Metals Review, vol. 44, no. 1, pp. 3–14, 2000.

299. V. Trevisan, M. Signoretto, S. Colonna, V. Pironti, and G. Strukul, "Microencapsulated chloroperoxidase as a recyclable catalyst for the enantioselective oxidation of sulfides with hydrogen peroxide,"Angewandte Chemie International Edition, vol. 43, no. 31, pp. 4097–4099, 2004.

300. Á. Zsigmond, F. Notheisz, G. Csjernyik, and J.-E. Bäckvall, "Ruthenium-catalyzed aerobic oxidation of alcohols on zeolite-encapsulated cobalt salophen catalyst," Topics in Catalysis, vol. 19, no. 1, pp. 119–124, 2002.

301. N. Legagneux, E. Jeanneau, A. Thomas et al., "Grafting reaction of platinum organometallic complexes on silica-supported or unsupported heteropolyacids," Organometallics, vol. 30, no. 7, pp. 1783–1793, 2011.

302. Y. Wan, F. Zhang, Y. Lu, and H. Li, "Immobilization of Ru(II) complex on functionalized SBA-15 and its catalytic performance in aqueous homoallylic alcohol isomerization," Journal of Molecular Catalysis A, vol. 267, pp. 165–172, 2007.

303. D. R. Burri, I. R. Shaikh, K.-M. Choi, and S.-E. Park, "Facile heterogenization of homogeneous ferrocene catalyst on SBA-15 and its hydroxylation activity," Catalysis Communications, vol. 8, no. 4, pp. 731–735, 2007.

304. M. Kuroki, T. Asefa, W. Whitnal et al., "Synthesis and properties of 1,3,5-benzene periodic mesoporous organosilica (PMO): novel aromatic PMO with three point attachments and unique thermal transformations," Journal of the American Chemical Society, vol. 124, no. 46, pp. 13886–13895, 2002.·

305. T. Maschmeyer, F. Rey, G. Sankar, and J. M. Thomas, "Heterogeneous catalysts obtained by grafting metallocene complexes onto mesoporous silica," Nature, vol. 378, pp. 159–162, 1995. ·

306. L. Hamidipour, Z. Ghasemzadeh, F. Farzaneh, and M. Ghandi, "Immobilization of Cu(II)-histidine complex on Al-MCM-41 as catalyst for epoxidation of alkenes," Journal of Sciences, vol. 23, no. 1, pp. 29–36, 2012.

307. J. G. Mesu, D. Baute, H. J. Tromp, E. E. Van Faassen, and B. M. Weckhuysen, "Synthesis and characterization of zeolite encaged enzyme-

mimetic copper histidine complexes," Studies in Surface Science and Catalysis, vol. 143, pp. 287–293, 2002.

308. B. M. Weckhuysen, A. A. Verberckmoes, L. Fu, and R. A. Schoonheydt, "Zeolite-encapsulated copper(II) amino acid complexes: synthesis, spectroscopy, and catalysis," Journal of Physical Chemistry, vol. 100, no. 22, pp. 9456–9461, 1996.

309. D. Xuereb, J. Dzierzak, and R. Raja, "Biomimetic single-site heterogeneous catalysts: design strategies and catalytic potential," in Heterogenized Homogeneous Catalysts For Fine Chemicals Production, vol. 33 of Catalysis by Metal Complexes, pp. 37–63, 2010.

310. P. Barbaro and F. Liguori, Eds., Heterogenized Homogeneous Catalysts for Fine Chemicals Production, vol. 33 of Materials and Processes Series: Catalysis by Metal Complexes, Springer, 2010.

311. Z. An, W. Zhang, H. Shi, and J. He, "An effective heterogeneous l-proline catalyst for the asymmetric aldol reaction using anionic clays as intercalated support," Journal of Catalysis, vol. 241, pp. 319–327, 2006.

312. J. W. Wiench, Y. S. Avadhut, N. Maity et al., "Characterization of covalent linkages in organically functionalized MCM-41 mesoporous materials by solid-state NMR and theoretical calculations," Journal of Physical Chemistry B, vol. 111, no. 15, pp. 3877–3885, 2007.

313. Q. Gao, W. Xu, Y. Xu et al., "Amino acid adsorption on mesoporous materials: influence of types of amino acids, modification of mesoporous materials, and solution conditions," The Journal of Physical Chemistry B, vol. 112, no. 7, pp. 2261–2267, 2008.

314. S. Luo, J. Li, L. Zhang, H. Xu, and J. P. Cheng, "Noncovalently supported heterogeneous chiral amine catalysts for asymmetric direct aldol and Michael addition reactions," Chemistry—A European Journal, vol. 14, no. 4, pp. 1273–1281, 2008.

315. C. Aprile, F. Giacalone, M. Gruttadauria et al., "New ionic liquid-modified silica gels as recyclable materials for l-proline- or H-Pro-Pro-Asp-NH2-catalyzed aldol reaction," Green Chemistry, vol. 9, no. 12, pp. 1328–1334, 2007.

316. I. Hermans, J. Van Deun, K. Houthoofd, J. Peeters, and P. A. Jacobs, "Silica-immobilized N-hydroxyphthalimide: an efficient heterogeneous autoxidation catalyst," Journal of Catalysis, vol. 251, no. 1, pp. 204–212, 2007.

317. K. Yamaguchi, T. Imago, Y. Ogasawara, J. Kasai, M. Kotani, and N. Mizuno, "An immobilized organocatalyst for cyanosilylation and epoxidation," Advanced Synthesis and Catalysis, vol. 348, no. 12-13, pp.

1516–1520, 2006.

318. H. Hagiwara, S. Inotsume, M. Fukushima, T. Hoshi, and T. Suzuki, "Heterogeneous amine catalyst grafted on amorphous silica: an effective organocatalyst for microwave-promoted Michael reaction of 1,3-dicarbonyl compounds in water," Chemistry Letters, vol. 35, no. 8, pp. 926–927, 2006.

319. A. Corma and H. Garcia, "Silica-bound homogenous catalysts as recoverable and reusable catalysts in organic synthesis," Advanced Synthesis Catalysis, vol. 348, pp. 1391–1412, 2006.

320. C. Ayats, A. H. Henseler, and M. A. Pericás, "A solid-supported organocatalyst for continuous-flow enantioselective aldol reactions," ChemSusChem, vol. 5, no. 2, pp. 320–325, 2012.

321. D. Font, C. Jimeno, and M. A. Pericàs, "Polystyrene-supported hydroxyproline: an insoluble, recyclable organocatalyst for the asymmetric aldol reaction in water," Organic Letters, vol. 8, pp. 4653–4655, 2006.·

322. D. Font, S. Sayalero, C. Jimeno, and M. A. Pericàs, "Toward an artificial aldolase," Organic Letters, vol. 10, pp. 337–340, 2008.

323. D. Font, S. Sayalero, A. Bastero, C. Jimeno, and M. A. Pericàs, "Toward an artificial aldolase," Organic Letters, vol. 12, p. 2678, 2010.

324. M. Benaglia, M. Cinquini, F. Cozzi, A. Puglisi, and G. Celentano, "Poly(ethylene-glycol)-supported proline: a recyclable aminocatalyst for the enantioselective synthesis of γ-nitroketones by conjugate addition," Journal of Molecular Catalysis A, vol. 204-205, pp. 157–163, 2003.

325. D. Q. Xu, S. P. Luo, Y. F. Wang et al., "Organocatalysts wrapped around by poly(ethylene glycol)s (PEGs): a unique host-guest system for asymmetric Michael addition reactions," Chemical Communications, no. 42, pp. 4393–4395, 2007.

326. A. M. Caminade, A. Ouali, M. Keller, and J. P. Majoral, "Organocatalysis with dendrimers," Chemical Society Reviews, vol. 41, pp. 4113–4125, 2012.

327. L. Zhang, S. Luo, and J.-P. Cheng, "Non-covalent immobilization of asymmetric organocatalysts,"Catalysis Science and Technology, vol. 1, no. 4, pp. 507–516, 2011.

328. B.-C. Hong, H.-C. Tseng, and S.-H. Chen, "Synthesis of aromatic aldehydes by organocatalytic [4+2] and [3+3] cycloaddition of α,β-unsaturated aldehydes," Tetrahedron, vol. 63, no. 13, pp. 2840–2850,

2007.·

329. M. J. Gaunt, C. C. C. Johansson, A. McNally, and N. T. Vo, "Enantioselective organocatalysis," Drug Discovery Today, vol. 12, no. 1-2, pp. 8–27, 2007.

330. D. J. Xuereb and R. Raja, "Design strategies for engineering selectivity in bio-inspired heterogeneous catalysts," Catalysis Science and Technology, vol. 1, no. 4, pp. 517–534, 2011.· ·

331. D. J. Xuereb, Strategies for organocatalyst heterogenisation and performance in selective transformations [Ph.D. thesis], University of Southampton, 2012.

332. D. E. De Vos, I. F. J. Vankelecom, and P. A. Jacobs, Eds., Chiral Catalyst Immobilization and Recycling, Wiley-VCH, Weinheim, Germany, 2000.

333. J. Zhou, J. Wan, X. Ma, and W. Wang, "Copolymer-supported heterogeneous organocatalyst for asymmetric aldol addition in aqueous medium," Organic & Biomolecular Chemistry, vol. 10, pp. 4179–4185, 2012.

334. P. C. Bulman Page, A. Mace, D. Arquier et al., "Towards heterogeneous organocatalysis: chiral iminium cations supported on porous materials for enantioselective alkene epoxidation," Catalysis Science & Technology, vol. 3, pp. 2330–2339, 2013.

335. J. Franzén, M. Marigo, D. Fielenbach, T. C. Wabnitz, A. Kjaersgaard, and K. A. Jørgensen, "A general organocatalyst for direct α-functionalization of aldehydes: stereoselective C–C, C–N, C–F, C–Br, and C–S bond-forming reactions. Scope and mechanistic insights," Journal of the American Chemical Society, vol. 127, no. 51, pp. 18296–18304, 2005.

336. M. Fochi, L. Gramigna, A. Mazzanti et al., "Solvent-free non-covalent organocatalysis: enantioselective addition of nitroalkanes to alkylideneindolenines as a flexible gateway to optically active tryptamine derivatives," Advanced Synthesis & Catalysis, vol. 354, pp. 71373–71380, 2012.

337. Y. Zhang, Y. Zhang, Y. L. Sun et al., "4-(N,N-dimethylamino) pyridine-embedded nanoporous conjugated polymer as a highly active heterogeneous organocatalyst," Chemistry—A European Journal, vol. 18, no. 20, pp. 6328–6334, 2012.

338. C. A. Wang, Y. Zhang, J. Y. Shi, and W. Wang, "A self-supported polymeric MacMillan catalyst for homogeneous organocatalysis and heterogeneous recycling," Chemistry—An Asian Journal, vol. 8, no. 6, pp. 1110–1114, 2013.

339. V. Lucchini, M. Noè, M. Selva, M. Fabris, and A. Perosa, "Cooperative nucleophilic-electrophilic organocatalysis by ionic liquids," Chemical Communications, vol. 48, no. 42, pp. 5178–5180, 2012.

340. D. Kühbeck, G. Saidulu, K. R. Reddy, and D. D. Díaz, "Critical assessment of the efficiency of chitosan biohydrogel beads as recyclable and heterogeneous organocatalyst for C–C bond formation," Green Chemistry, vol. 14, pp. 378–392, 2012.

341. S. Verma, S. L. Jain, and B. Sain, "An efficient biomaterial supported bifunctional organocatalyst (ES–SO3- $C_5H_5NH^+$) for the synthesis of β-amino carbonyls," Organic & Biomolecular Chemistry, vol. 9, pp. 2314–2318, 2011.

342. M. S. DeClue and J. S. Siegel, "Polysiloxane-bound ligand accelerated catalysis: a modular approach to heterogeneous and homogeneous macromolecular asymmetric dihydroxylation ligands," Organic & Biomolecular Chemistry, vol. 2, pp. 2287–2298, 2004.

343. G. Kardos and T. Soós, "Tether-free immobilized bifunctional squaramide organocatalysts for batch and flow reactions," European Journal of Organic Chemistry, vol. 2013, no. 21, pp. 4490–4494, 2013.

344. P. Kasaplar, P. Riente, C. Hartmann, and M. A. Pericàs, "A polystyrene-supported, highly recyclable squaramide organocatalyst for the enantioselective Michael addition of 1,3-dicarbonyl compounds to β-nitrostyrenes," Advanced Synthesis & Catalysis, vol. 354, no. 16, pp. 2905–2910, 2013.

345. K. E. Alza, C. Rodríguez-Escrich, S. Sayalero, A. Bastero, and M. A. Pericàs, "A solid-supported organocatalyst for highly stereoselective, batch, and continuous-flow mannich reactions," Chemistry—A European Journal, vol. 15, no. 39, pp. 10167–10172, 2009.

346. P. Riente, J. Yadav, and M. A. Pericàs, "A solid-supported organocatalyst for continuous-flow enantioselective aldol reactions," ChemSusChem, vol. 5, no. 2, pp. 320–325, 2012.

347. M. Kotke and P. Schreiner, "(Thio)Urea Organocatalysts," in Hydrogen Bonding in Organic Synthesis, P. M. Pihko, Ed., pp. 141–352, Wiley-VCH, 2009.

348. Y. Takemoto, "Development of chiral thiourea catalysts and its application to asymmetric catalytic reactions," Chemical and Pharmaceutical Bulletin, vol. 58, pp. 593–601, 2010.

349. M. Tsakos, C. G. Kokotos, and G. Kokotos, "Primary amine-thioureas with improved catalytic properties for "difficult" Michael reactions: efficient organocatalytic syntheses of (S)-baclofen, (R)-baclofen and

(S)-phenibut," Advanced Synthesis and Catalysis, vol. 354, no. 4, pp. 740–746, 2012.

350. M. Tsakos and C. G. Kokotos, "Organocatalytic "Difficult" Michael reaction of ketones with nitrodienes utilizing a primary amine—thiourea based on di-tert-butyl aspartate," European Journal of Organic Chemistry, vol. 2012, no. 3, pp. 576–580, 2012.

351. J. Hine, S.-M. Linden, and V. M. Kanagasabapathy, "1,8-Biphenylenediol is a double-hydrogen-bonding catalyst for reaction of an epoxide with a nucleophile," Journal of the American Chemical Society, vol. 107, no. 4, pp. 1082–1083, 1985.

352. J. Hine, S. M. Linden, and V. M. Kanagasabapathy, "Double-hydrogen-bonding catalysis of the reaction of phenyl glycidyl ether with diethylamine by 1,8-biphenylenediol," The Journal of Organic Chemistry, vol. 50, pp. 5096–5099, 1985.

353. C. K. De, E. G. Klauber, and D. Seidel, "Merging nucleophilic and hydrogen bonding catalysis: an anion binding approach to the kinetic resolution of amines," Journal of the American Chemical Society, vol. 131, pp. 17060–17061, 2009.

354. R. P. Herrera, V. Sgarzani, L. Bernardi, and A. Ricci, "Catalytic enantioselective friedel-crafts alkylation of indoles with nitroalkenes by using a simple thiourea organocatalyst," Angewandte Chemie International Edition, vol. 44, pp. 6576–6579, 2005.

355. "Special issue on "Organocatalysis"," Advanced Synthesis & Catalysis, vol. 346, no. 9-10, 2004.

356. F. Cozzi, "Immobilization of organic catalysts: when, why, and how," Advanced Synthesis & Catalysis, vol. 348, pp. 1367–1390, 2006.

357. T. E. Kristensen and T. Hansen, "Polymer-supported chiral organocatalysts: synthetic strategies for the road towards affordable polymeric immobilization," European Journal of Organic Chemistry, no. 17, pp. 3179–3204, 2010.

358. M. Benaglia, F. Cozzi, and A. Puglisi, "Polymer-supported organic catalysts," Chemical Reviews, vol. 103, no. 9, pp. 3401–3429, 2003.

359. M. Benaglia, Ed., Recoverable and Recyclable Catalysts, John Wiley & Sons, Chichester, UK, 2009.

360. J. Lu and P. H. Toy, "Organic polymer supports for synthesis and for reagent and catalyst immobilization," Chemical Reviews, vol. 109, no. 2, pp. 815–838, 2009.

361. M. Gruttadauria, F. Giacalone, and R. Noto, "Supported proline and

proline-derivatives as recyclable organocatalysts," Chemical Society Reviews, vol. 37, no. 8, pp. 1666–1688, 2008.

362. F. Calderón, R. Fernández, F. Sánchez, and A. Fernández-Mayoralas, "Asymmetric aldol reaction using immobilized proline on mesoporous support," Advanced Synthesis & Catalysis, vol. 347, no. 10, pp. 1395–1403, 2005.

363. L. Qin, L. Zhang, Q. Jin, J. Zhang, B. Han, and M. Liu, "Supramolecular assemblies of amphiphilic L-proline regulated by compressed CO_2 as a recyclable organocatalyst for the asymmetric aldol reaction,"Angewandte Chemie International Edition, vol. 52, no. 30, pp. 7761–7765.

364. S. Luo, J. Li, L. Zhang, H. Xu, and J.-P. Cheng, "Noncovalently supported heterogeneous chiral amine catalysts for asymmetric direct aldol and Michael addition reactions," Chemistry—A European Journal, vol. 14, no. 4, pp. 1273–1281, 2008.

365. H. Hagiwara, S. Inotsume, M. Fukushima, T. Hoshi, and T. Suzuki, "Heterogeneous amine catalyst grafted on amorphous silica: an effective organocatalyst for microwave-promoted Michael reaction of 1,3-dicarbonyl compounds in water," Chemistry Letters, vol. 35, no. 8, pp. 926–927, 2006.

366. I. Hermans, J. Van Deun, K. Houthoofd, J. Peeters, and P. A. Jacobs, "Silica-immobilized N-hydroxyphthalimide: an efficient heterogeneous autoxidation catalyst," Journal of Catalysis, vol. 251, no. 1, pp. 204–212, 2007.

367. A. Corma and H. Garcia, "Silica-bound homogenous catalysts as recoverable and reusable catalysts in organic synthesis," Advanced Synthesis & Catalysis, vol. 348, pp. 1391–1412, 2006. ·

368. K. Yamaguchi, T. Imago, Y. Ogasawara, J. Kasai, M. Kotani, and N. Mizuno, "An immobilized organocatalyst for cyanosilylation and epoxidation," Advanced Synthesis and Catalysis, vol. 348, no. 12-13, pp. 1516–1520, 2006.

369. A. Córdova and J. Hafrén, "Direct Homogeneous and Heterogeneous Organic Acid and Amino Acid-Catalyzed Modification of Amines and Alcohols," International Patent WO 2006068611 A1 20060629.

370. R. S. Verma and V. Polshettiwar, "Magnetic nanoparticle-supported glutathione as a sustainable organocatalyst," Patent no.: US, 8324125 B2, December 2012.

371. R. C. Garrett and H. Yang, "Sulfonamide-Based Organocatalysis and Method for their Use," Patent no.: US 8, 399, 684 B2, March 2013.

372. A. M. Caminade, C. O. Turrin, R. Laurent, A. Ouali, and B. Delavaux-Nicot, Eds., Dendrimers Towards Catalytic Material and Biomedical Uses, John Wiley & Sons, Chichester, UK, 2011.

373. P. Servin, R. Laurent, L. Gonsalvi et al., "Grafting of water-soluble phosphines to dendrimers and their use in catalysis: positive dendritic effects in aqueous media," Dalton Transactions, no. 23, pp. 4432–4434, 2009.

374. M. Keller, A. Perrier, R. Linhardt, et al., "Dendrimers or nanoparticles as supports for the design of efficient and recoverable organocatalysts?" Advanced Synthesis & Catalysis, vol. 355, no. 9, pp. 1748–1754, 2013.

375. Thematic Series "Organocatalysis" (Guest Editor: Benjamin List) Beilstein Journal of Organic Chemistry, 2012, http://www.beilstein-journals.org/bjoc/browse/singleSeries.htm?sn=27.

376. Y. S. Lee, M. M. Alam, and R. S. Keri, "Enantioselective reactions of N-acyliminium ions using chiral organocatalysts," Chemistry—An Asian Journal, vol. 8, no. 12, pp. 2906–2919, 2013.

377. F. Lv, S. Liu, and W. Hu, "Recent advances in the use of chiral brønsted acids as cooperative catalysts in cascade and multicomponent reactions," Asian Journal of Organic Chemistry, vol. 2, no. 10, pp. 824–836, 2013.

378. W. Yan, X. Shi, and C. Zhong, "Secondary amines as lewis bases in nitroalkene activation," Asian Journal of Organic Chemistry, vol. 2, no. 11, pp. 904–914, 2013.

379. S. Ulf and M. Rainer, "Recent advances in organocatalytic methods for asymmetric C–C bond formation," Chemistry—A European Journal, vol. 19, no. 43, pp. 14346–14396, 2013.

380. S. Narayanaperumal, D. G. Rivera, R. C. Silva, and M. W. Paixão, "Terpene-derived bifunctional thioureas in asymmetric organocatalysis," ChemCatChem, vol. 5, no. 10, pp. 2756–2773, 2013.

381. L.-W. Xu, "Powerful amino acid derived bifunctional phosphine catalysts bearing a hydrogen bond donor in asymmetric synthesis," ChemCatChem, vol. 5, no. 10, pp. 2775–2784, 2013.

382. N. Kielland, C. J. Whiteoak, and A. W. Kleij, "Stereoselective synthesis with carbon dioxide," Advanced Synthesis & Catalysis, vol. 355, no. 11-12, pp. 2115–2138, 2013.

383. S. Mohammadi, R. Heiran, R. P. Herrera, and E. Marqués-López, "Isatin as a strategic motif for asymmetric catalysis," ChemCatChem, vol. 5, no. 8, pp. 2131–2148, 2013.

Chapter 8

ENVIRONMENTALLY BENIGN MORTAR-PESTLE-INDUCED ACYLATION AND O-ALKYLATION OF AROMATIC AND HETEROAROMATIC COMPOUNDS UNDER SOLVENT-FREE MICELLAR CONDITIONS AND COMPUTATION OF THEIR DRUG LIKELINESS PROPERTIES

Kancharla Rajendar Reddy, Kamatala Chinna Rajanna, Kusampally Uppalaiah, Mukka Satish Kumar, and Marri Venkateshwarlu

Department of Chemistry, Osmania University, Hyderabad 500 007, India

ABSTRACT

Environmentally benign mortar-pestle-induced practical methods have been developed for the acylation and O-alkylation of aromatic and hetero aromatic compounds under solvent free micellar conditions, which were found to efficiently afford moderate to excellent yields of products.

INTRODUCTION

The diverse nature of chemical universe requires various green strategic pathways in our quest towards attaining sustainability. The emerging area of green chemistry envisages minimum hazard as the performance criteria while designing new chemical processes. One of the thrust areas for achieving this target is to explore alternative reaction conditions and reaction media to accomplish the desired chemical transformations with minimized byproducts or waste as well as eliminating the use of conventional organic solvents, wherever possible. Consequently, several newer strategies have appeared such as solvent-free reactions (grinding), multicomponent reactions under solvent-free conditions could enhance their efficiency from an economic as well as an ecological point of view the solid-state organic reactions is gaining significance both from the mechanistic and synthetic point of view [1–3]. Number of articles are available reporting solid-state reactions by grinding such as, Grignard reaction [4], Reformatsky reaction [5], Aldol condensation [6], Dickhmann condensation [7], phenol coupling reaction [8], reduction reaction [9], Wittig reaction [10], Grignard and McMurry reaction [11], and Synthesis of Polyhydroquinolines [12]. In recent past, our research group is actively involved in the exploration of in the exploration of nonconventional methods for nitration, bromination, acetylation, and benzylation reactions in the direction of achieving greenery conditions [13–16]. Most of these griding reactions are carried out at room temperature, absolutely solvent-free and use only a mortar and pestle and also economical and green procedures. Therefore, we focus on developing the novel procedure involving a solid-state reaction performed by grinding.

Alkylated and acylated aromatic and heteroaromatic compounds are significant precursors in drug intermediates and industrial uses [17]. Specifically naphthalene derivatives such as naphthol ethers and acylated naphthol ethers have been identified as one of the best ranges of potent antimicrobials effective against wide range of human pathogens. A perusal of literature indicated that Williamson synthesis is probably one of the most common classical methods being used for the preparation of symmetrical and unsymmetrical ethers [18–24]. Recent publications of Swamy et al. [25], Debabrata and coworkers [26, 27], provide excellent bibliography and information on the synthesis of aryl, heteroaryl ethers derived from various roots.

The Friedel-Crafts reaction has a long history in organic synthesis for electrophilic aromatic substitution reactions such as alkylations and acylations. To date these reactions are of great importance for the synthesis of aromatic carbonyl compounds [28]. Aromatic ketones can also be prepared by the reaction of carboxylic acids with aromatic hydrocarbons catalyzed by Nafion-H [29]. A significant number of Lewis acid catalysts ($AlCl_3$, $FeCl_3$, $SnCl_4$, and $BF_3 \cdot OEt_2$) have been shown to be very successful for the acylation of aromatic substrates with acid chlorides or anhydrides, while the reaction is usually carried out using a stoichiometric amount of $AlCl_3$, reactive aromatic compounds are known to undergo acylation in the presence of a catalytic amount of Lewis acid. Recently, several effective catalyst systems for the acylation of anisole and its derivatives have also been developed [30–34]. A perusal of literature indicated that surfactants have been used to promote a variety of synthetic organic reactions [35–37]. Our preliminary studies in this direction ended up with fruitful results when we have performed the title reactions in presence of micelle forming anionic (SDS), cationic (CTAB, CTAC), and non-ionic (Tx-100) surfactants. This study is aimed at developing "greener synthetic protocols" for etherification and acylation reactions under solvent-free conditions, as encouraged by the "Green Chemistry strategies" of Paul Anastas and Warner [38]. We have used mortar-pestle grinding technique in this study because it is economically cheap and safer method and yet is known to good yields of products in the synthesis of several organic compounds [1–3]. These reactions underwent dramatic rate accelerations under these conditions. We were also successful to develop solvent-free synthetic methods in the above protocols by replacing acids with a variety of micelle forming surfactants such as sodium dodecyl sulphate (SDS), cetyl trimethyl ammonium bromide (CTAB), cetyl trimethyl ammonium chloride (CTAC), and Triton-X 100.

RESULTS AND DISCUSSION

Synthesis of naphthol ethers under Williamson's conditions (refluxing in sulfuric acid and organic solvent) required several hours ($\geq$20 h) at relatively high temperature. Even though there are some reports to modify the drastic conditions of these reactions, many of them exhibited long reaction times, accumulation of unwanted byproducts, which ultimately involved tedious work-up procedures. Encouraged by this aspect, we have conducted O-alkylation of aromatic and heteroaromatic alcohol reaction (Scheme 1) in micellar media under acid-free conditions.

$$Ar\!-\!OH \xrightarrow[\text{Grinding conditions}]{\text{ROH / micelles}} Ar\!-\!O\!-\!R$$

Ar = aromatic/heteroaromatic

R = CH$_3$, C$_2$H$_5$, C$_3$H$_7$;

Micelles = CTAB, SDS, TX-100

Scheme 1: O-alkylation of Aromatic/heteroaromatic compounds in micellar media.

However, for comparison, we have also conducted etherification of β-naphthol reactions under Williamson's classical conditions in acidic media. Results obtained under acidic and acid-free micellar conditions are compiled in Table 1 and O-alkylation of hydroxy pyridines in Table 2, and the reaction time versus conversion of aromatic/heteroaromatic alcohols using micelles as a catalyst were reported in Table 3, which clearly indicate highly significant rate accelerations followed by very good yield of end products. Catalytic activity of different micelles is in the order: CTAB > SDS > Tx-100. It is believed that micelles themselves act as micro-reactors. By introducing a surfactant (CTAB), dehydration was successfully achieved. The catalytic effect of the micellar solution of CTAB may be attributed to the hydrophobic nature of organic substrates. Formation of emulsion droplets takes place in water in the presence of surfactant and substrate molecules. It is suggested that most of the organic substrates are concentrated in these spherical droplets, which act as a hydrophobic reaction sites and results in an increase in the effective concentration of the organic reactants, which might increase the reaction rate via a concentration effect. In micellar solution, organic substrates are pushed away from water molecules towards the hydrophobic core of micelle droplets thus inducing efficient collisions between organic substrates which eventually enhance the reaction rates. The hydrophobic interior of the micelles swiftly excludes the water molecules generated during the reaction, thus shifting the equilibrium towards the desired product that ultimately leads to an increase in the reaction yield [39, 40]. This explanation is schematically represented in Figures 1 and 2.

Table 1: Mortar-pestle-induced O-alkylation of β-naphthols under solvent-free conditions

Reactants		H_2SO_4 (5 hr) % yield	CTAB (2 hr) % yield	SDS (2.5 hr) % yield	TX-100 (3.0 hr) % yield	M.P/B.P (°C)
1	2					
β-Naphthol	MeOH	89	93	90	90	70-71
	EtOH	90	93	90	90	35-37
	1-PrOH	88	91	91	91	37-39
	2-PrOH	91	94	91	91	—
1-Bromonaphthalen-2-ol	MeOH	90	91	89	90	80-81
	EtOH	88	92	90	90	64-66
	1-PrOH	88	96	90	89	—
	2-PrOH	91	94	89	89	37-39
1-Chloronaphthalen-2-ol	MeOH	90	90	93	89	64-65
	EtOH	90	90	92	90	56-58
	1-PrOH	91	91	91	88	—
	2-PrOH	91	91	94	91	—
1-Iodonaphthalen-2-ol	MeOH	90	89	91	90	87-89
	EtOH	90	90	92	88	35-37
	1-PrOH	89	90	96	88	37-39
	2-PrOH	89	89	94	91	—

Table 2: Mortar-pestle-induced micellar mediated O-alkylation of hydroxy pyridines under solvent-free conditions

Reactants		CTAB (1.5 hr)	SDS (2.0 hr)	TX-100 (2.5 hr)
1	2	% yield	% yield	% yield
2-Hydroxy pyridine	MeOH	89	85	86
	EtOH	90	87	87
	1-PrOH	90	86	82
	2-PrOH	90	87	89
3-Hydrox pyridine	MeOH	91	85	82
	EtOH	89	84	81
	1-PrOH	85	83	80
	2-PrOH	89	85	79
4-Hydroxy pyridine	MeOH	91	88	81
	EtOH	88	85	83
	1-PrOH	88	83	82
	2-PrOH	89	84	81
5-Bromopyrimidin-2-ol	MeOH	91	87	89
	EtOH	89	87	87
	1-PrOH	89	86	86
	2-PrOH	91	84	86
6-Hydroxypicolinaldehyde	MeOH	92	88	89
	EtOH	88	86	89
	1-PrOH	87	85	84
	2-PrOH	88	87	86

Table 3: Reaction of aromatic/heteroaromatic alcohol (10 mmol) with methanol (20 mmol) using micelles as catalysts

Aromatic/heteroaromatic alcohol (1)	CTAB		SDS		TX-100	
	Time (min)	Conversion of 1 (%)	Time (min)	Conversion of 1 (%)	Time (min)	Conversion of 1 (%)
β-Naphthol	60	50	60	50	60	45
	80	63	90	65	100	60
	100	75	120	75	140	78
	120	93	150	90	180	90
2-Hydroxy pyridine	60	60	60	55	60	50
	70	73	90	68	90	65
	80	81	120	77	120	77
	90	89	120	85	150	86

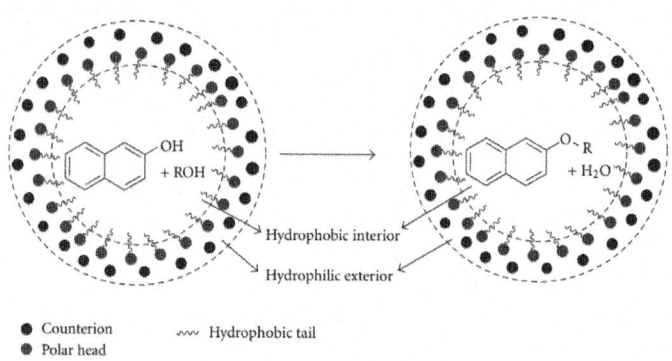

Figure 1: Schematic diagram representing the role of micelles in the reaction.

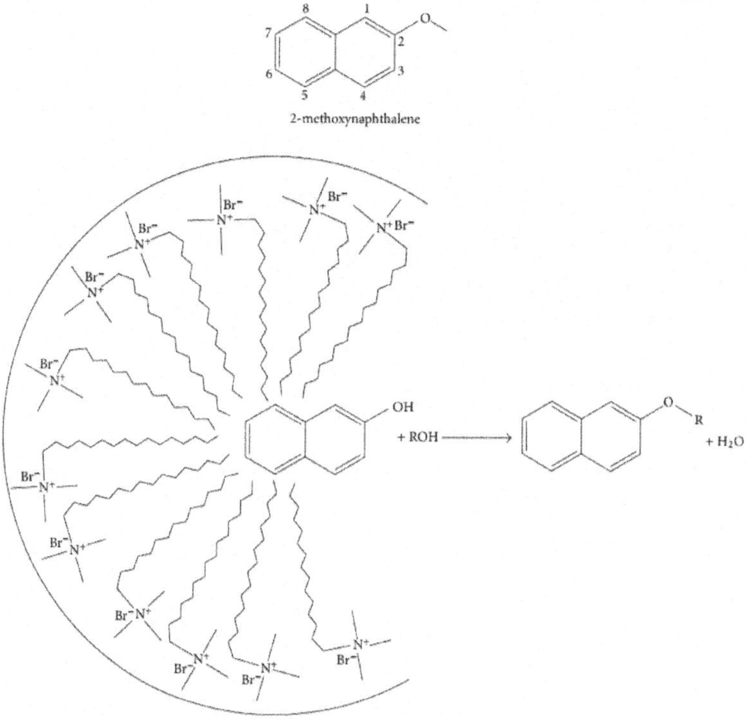

2-methoxynaphthalene

Figure 2: Schematic diagram representing the role of CTAB in the reaction.

The Friedel-Crafts acylation of 2-methoxy naphthalene in the presence of the electron-donating group (OMe) must be taken into account in order to figure out the reaction. Presence of the electron-donating group activates the 1, 6, and 8 positions of the naphthalene ring.

The 1-position is more active than the other two positions and the 6-position is more stable than the other two positions, so acylation of 2-methoxynaphthalene generally occurs at this kinetically favoured 1-position at low temperatures and at the thermodynamically favoured 6-position at high temperatures. At high temperatures, in the reaction mixture, migration of the acyl group from 1- to 6-position which is named as transacylation and protiodeacylation of the acyl group at the 1-position of 1-acyl-2-methoxy naphthalene are possible. These two reactions may result in the formation of the thermodynamically favoured product which is 6-acyl-2-methoxynaphthalene. Steric hindrance to acylation is reported in the order 1 > 8 > 6-position (Das, 2000). Therefore, the isomerisation of the sterically hindered ketone 1-acyl-2-methoxynaphthalene to sterically less hindered isomers like 6-acyl-2-methoxy naphthalene and 8-acyl-2-methoxynaphthalene will be favoured.

Table 4: Reaction of 1-halo-2-methoxy naphthalene (2 mmol) with acylchlorides (2 mmol) using CTAB/CTAC as a catalyst

Naphthalene	Acylchloride	Total (%) Yield	Products ratio (3:4)
1a	2a	93	98 : 02
1a	2b	90	98 : 02
1a	2c	89	98 : 02
1a	2d	92	98 : 02
1a	2e	90	98 : 02
1a	2f	93	98 : 02
1a	2g	94	98 : 02
1b	2a	93	97 : 03
1b	2b	90	98 : 02
1b	2c	89	98 : 02
1b	2d	92	98 : 02
1b	2e	90	98 : 02
1b	2f	93	98 : 02
1b	2g	94	98 : 02

Catalytic reaction of 1-halo-2-methoxy naphthalene (2 mmol) was treated with acylchloride (2 mmol) in the presence of CTAB or CTAC, (0.1 mmol, 5 mol%) were grounded with pestle for 2 hr it gave 6-acyl-1-halo-2-methoxynaphthalene(3) as the major product along with 8-acyl-1-halo-2-methoxy naphthalene(4) selectivity (98 : 02) represented as in Scheme 2, Table 4 and the reaction time versus conversion of 1-halo-2-methoxy naphthalenes using CTAB/CTAC (various amounts) as a catalyst were reported in Table 5.

Table 5: Reaction of 1-halo-2-methoxynaphthalene (2 mmol) With 2g (2 mmol) using CTAB/CTAC (various amounts) as a catalyst

Naphthalene	CTAB/CTAC(mmol)	Time (min)	Conversion of 1 (%)
1a	0.02	60	70
		80	73
		100	75
		120	80
1a	0.04	60	72
		80	78
		100	82
		120	85
1a	0.06	60	74
		80	80
		100	84
		120	86
1a	0.08	60	76
		80	83
		100	86
		120	90
1a	0.10	60	80
		80	85
		100	89
		120	93
1b	0.02	60	70
		80	75
		100	78
		120	83
1b	0.04	60	72
		80	75
		100	79
		120	85
1b	0.06	60	74
		80	79
		100	80
		120	87
1b	0.08	60	76
		80	81
		100	83
		120	89
1b	0.10	60	80
		80	85
		100	88
		120	93

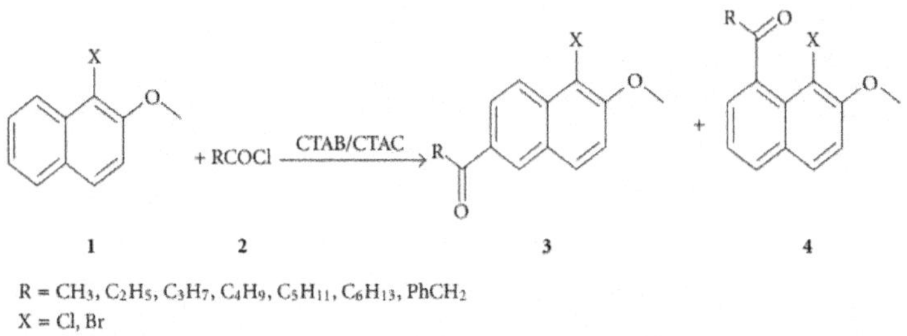

$R = CH_3, C_2H_5, C_3H_7, C_4H_9, C_5H_{11}, C_6H_{13}, PhCH_2$
$X = Cl, Br$

Scheme 2: Acylation of 1-halo-2-methoxy naphthalenes in micellar media.

Mortar-pestle grinding synthetic methods were to avoid solvent. The reaction times fairly reduced from 24 h to about 2-3 h (under Williamson's method to Mortar-pestle method) without many changes in the yield of products (Tables 1 and 2). Rate accelerations under grinding method could be attributed to bulk activation of molecules due to the conversion of mechanical energy (exerted due to grinding) into heat energy due to frictional forces operating between solid state reagents [1–3].

Calculation of Molecular Physicochemical Properties by Molinspiration

Drug likeness may be defined as a complex balance of various molecular properties and structure features which determine whether particular molecule is similar to the known drugs. These properties, mainly hydrophobicity, electronic distribution, hydrogen bonding characteristics, molecule size and flexibility, and various pharmacophoric features influence the behavior of molecule in a living organism, including bioavailability, transport properties, affinity to proteins, reactivity, toxicity, metabolic stability, and many others. Mol inspiration calculations are a set of simple molecular properties obtained by Lipinski Rule-5 of 5" [41, 42], which are useful to explore drug likeliness nature of a molecule. Lipinski's fifth rule states that a molecule possess "drug-likeliness" if the molecule has the properties: log P ≤ 5, molecular weight ≤500, number of hydrogen bond acceptors ≤10, and number of hydrogen bond donor's ≤5. Molecules violating more than one of these rules may have problems with bioavailability. For the prediction of oral bioavailability of drug molecules [43], lipophilicity log P values, and polar surface area (PSA) [44] values are the most important properties. The Calculated parameters from Mol inspiration methodology for each newly synthesized 6-acyl-1-halo-2-methoxynaphthalene derivative (entries 1 to 14) are compiled in Tables 6 and 7 and tested for specific activity with standard drugs Naproxen. For all the compounds, the calculated log P values were ≤5 (in comparison with the accepted the upper limit) indicating the ability of the drug to penetrate through biomembranes and also good water solubility (according to Lipinski's rules). Further it is also interesting to note that all the synthesized molecules exhibit less than 140 PSA values indicating good intestinal absorption. Data presented in Table 7 are GPCR ligand activity, ion channel modulation, kinase inhibition activity, nuclear receptor ligand activity, Protease inhibitor, and Enzyme inhibitor are compared with Naproxen as standard drug indicate that most of the synthesized compounds depict by and large consistent negative values, which are in consonance with Lipinski's Rule -5 and are also comparable with Naproxen drug.

Table 6: Molinspiration calculations of compounds

S. number	6-Acyl-1-halo-2-methoxynaphthalene	M.Wt	miLog P	TPSA	OH–NH	n-Vio	Volume
1	1-(1-Bromo-2-methoxynaphthalen-6-yl)ethanone	279.13	3.789	26.305	0	2	0
2	1-(1-Bromo-2-methoxynaphthalen-6-yl)pentan-1-one	321.21	5.356	26.305	0	2	1
3	1-(1-Bromo-2-methoxynaphthalen-6-yl)hexan-1-one	335.24	5.861	26.305	0	2	1
4	1-(1-Bromo-2-methoxynaphthalen-6-yl)-4-methylpentan-1-one	335.24	5.337	26.305	0	2	1
5	1-(1-Bromo-2-methoxynaphthalen-6-yl)-5-methylhexan-1-one	349.26	5.842	26.305	0	2	1
6	1-(1-Bromo-2-methoxynaphthalen-6-yl)-3,3-dimethylbutan-1-one	335.24	5.386	26.305	0	2	1
7	1-(1-Bromo-2-methoxynaphthalen-6-yl)-2-phenylethanone	355.23	5.217	26.305	0	2	1
8	1-(1-Chloro-2-methoxynaphthalen-6-yl)ethanone	234.68	3.658	26.305	0	2	0
9	1-(1-Chloro-2-methoxynaphthalen-6-yl)pentan-1-one	276.76	5.225	26.305	0	2	1
10	1-(1-Chloro-2-methoxynaphthalen-6-yl)hexan-1-one	290.79	5.73	26.305	0	2	1
11	1-(1-Chloro-2-methoxynaphthalen-6-yl)-4-methylpentan-1-one	290.79	5.206	26.305	0	2	1
12	1-(1-Chloro-2-methoxynaphthalen-6-yl)-5-methylhexan-1-one	304.81	5.711	26.305	0	2	1
13	1-(1-Chloro-2-methoxynaphthalen-6-yl)-3,3-dimethylbutan-1-one	290.79	5.255	26.305	0	2	1
14	1-(1-Chloro-2-methoxynaphthalen-6-yl)-2-phenylethanone	310.78	5.086	26.305	0	2	1
STD	Naproxen	264.70	3.983	46.533	1	3	0

Table 7: Drug likeness of compounds

S. number	6-Acyl-1-halo-2-methoxynaphthalene	GPCRL	ICM	KI	NRL
1	1-(1-Bromo-2-methoxynaphthalen-6-yl)ethanone	−0.56	−0.51	−0.71	−0.57
2	1-(1-Bromo-2-methoxynaphthalen-6-yl)pentan-1-one	−0.18	−0.36	−0.42	−0.20
3	1-(1-Bromo-2-methoxynaphthalen-6-yl)hexan-1-one	−0.13	−0.34	−0.35	−0.14
4	1-(1-Bromo-2-methoxynaphthalen-6-yl)-4-methylpentan-1-one	−0.14	−0.33	−0.39	−0.14
5	1-(1-Bromo-2-methoxynaphthalen-6-yl)-5-methylhexan-1-one	−0.09	−0.31	−0.34	−0.08
6	1-(1-Bromo-2-methoxynaphthalen-6-yl)-3,3-dimethylbutan-1-one	−0.19	−0.27	−0.46	−0.32
7	1-(1-Bromo-2-methoxynaphthalen-6-yl)-2-phenylethanone	−0.06	−0.30	−0.24	−0.05
8	1-(1-Chloro-2-methoxynaphthalen-6-yl)ethanone	−0.50	−0.22	−0.67	−0.49
9	1-(1-Chloro-2-methoxynaphthalen-6-yl)pentan-1-one	−0.13	−0.12	−0.39	−0.14
10	1-(1-Chloro-2-methoxynaphthalen-6-yl)hexan-1-one	−0.08	−0.11	−0.32	−0.08
11	1-(1-Chloro-2-methoxynaphthalen-6-yl)-4-methylpentan-1-one	−0.10	−0.10	−0.37	−0.08
12	1-(1-Chloro-2-methoxynaphthalen-6-yl)-5-methylhexan-1-one	−0.04	−0.09	−0.31	−0.02
13	1-(1-Chloro-2-methoxynaphthalen-6-yl)-3,3-dimethylbutan-1-one	−0.14	−0.04	−0.43	−0.26
14	1-(1-Chloro-2-methoxynaphthalen-6-yl)-2-phenylethanone	−0.02	−0.10	−0.21	0.00
STD	Naproxen	−0.06	0.00	−0.36	0.18

GPCR: ligand, ICM: Ion channel modulator, KI: Kinase inhibitor, NRL: Nuclear receptor ligand.

CONCLUSIONS

There are different shades of greener processes as we continue exploring several alternatives to conventional chemical transformations. That approach will require new environmentally benign syntheses of O-alkylation and acylation of naphthalene derivatives in aqueous micellar media. Thus, the present protocols show rate accelerations associated with high products yields, when compared with the similar reactions performed under classical conditions. Surfactants catalyze the reaction efficiently with short reaction times without using any harmful organic reagents and solvents. Molecular properties computed from Molinspiration methodology for all the synthesized 6-acyl-1-halo-2-methoxynaphthalene derivatives by and large indicate Drug likeliness behavior that is comparable to standard drugs. Thus, it is believed that the present work is a major breakthrough in the area of synthesis of 6-acyl, 1-halo-2-methoxy naphthalenes with potential biological activity.

EXPERIMENTAL DETAILS

All reactions were followed by TLC with detection by UV light. Melting points were recorded on BUCHI B-545 capillary melting point apparatus and are uncorrected. Infrared (IR) spectra were recorded on a Perkin Elmer FT-IR spectrometer. ^1H and ^{13}C NMR spectra were recorded at a Varian VNMRS-400 and 100 MHz spectrometer. The Samples were analyzed in $CDCl_3$, DMSO-d_6 chemical shift values are reported in ppm relative to TMS as the internal reference. Mass spectra were recorded on a ZAB-HS mass spectrometer using ESI ionization. The isolation of pure products was carried out via preparative thin layer chromatography (silica gel 60 GF_{254}, Merck). Excess of solvent was evaporated under reduced pressure at a bath temperature of 50°C. All solvents, organic and inorganic compounds, were purchased from Aldrich, Merck, Fluka, and Sdfine and used without further purification.

General Procedure for Etherification of -Naphthols under Mortar Pestle Conditions

A mixture of substrate (10 mmol), alcohol (20 mmol), and micelles (1.0 mmol, 25 mol %) were added to a mortar. The mixture was grounded by mortar and pestle at room temperature, after the indicated reaction time, the reaction mixture was purified by thin layer chromatography (silicagel, EtOAc-petroleum ether, 1 : 9) to obtain the desired product.

General Procedure for Acylation of -Naphthols under Mortar Pestle Conditions

1-halo-2-methoxynaphthalene (2 mmol) and acylchloride (2 mmol) and CTAB (0.1 mmol, 5 mol %) were added to a mortar and grinded with pestle till the reaction is completed as ascertained by TLC. The final products were isolated by absorbing the reaction mixture into silica gel and purifying it by column chromatography using ethyl acetate/hexane gradient.

Spectroscopic Analysis of Representative Compounds

2-Methoxynaphthalene

^{1}H NMR (400 MHz, CDCl$_3$) δ 3.91 (s, 3H), 7.15–7.17 (d, J = 8.8 Hz, 1 H), 7.29–7.36 (m, 2H), 7.43–7.47 (m, 1H), 7.72–7.77 (m, 3H); ^{13}C NMR (50 MHz, CDCl$_3$) δ 55.6, 105.9, 113.7, 124.4, 126.2, 127.8, 128.2, 129.1, 129.9, 133.2, 153.8; IR (KBr, cm^{-1}) 3067, 3008, 2963, 1632, 1599, 1477, 1462, 1452, 1440, 1398, 1391, 1368, 1262, 1218, 1197, 1173, 1152, 1141, 1118, 1031, 1017, 963, 948, 874, 838, 818, 753, 743, 622, 481, 470; MS (EI)m/z 159.2 (M)$^+$.

2-Ethoxynaphthalene

^{1}H NMR (400 MHz, CDCl$_3$) δ 1.46 (t, J = 7.2 Hz, 3H), 4.127 (d, J = 6.8 Hz, 2H), 7.14–7.15 (d, J = 8.8 Hz, 1H), 7.28–7.39 (m, 2H), 7.39–7.43 (m, 1H), 7.69–7.75 (m, 3H); ^{13}C NMR (50 MHz, CDCl$_3$) δ 14.8, 64.1, 106.5, 118.9, 123.4, 126.2, 126.7, 128.2, 129.1, 129.2, 133.6, 156.8; IR (KBr, cm^{-1}) 3067, 2984, 2940, 2876, 1829, 1601, 1579, 1511, 1457, 1440, 1395, 1390, 1367, 1358, 1350, 1269, 1259, 1185, 1166, 1144, 1122, 1046, 1020, 958, 928, 873, 823, 762, 719, 623, 477; MS (EI) m/z 173.2 (M)$^+$.

2-Propoxynaphthalene

^{1}H NMR (400 MHz, CDCl$_3$) δ 1.06 (t, J = 6.0 Hz, 3H), 1.83–1.86 (m, 2H), 4.01 (t, 2H), 7.10–7.14 (m, 2H), 7.30–7.41 (m, 2H), 7.69–7.74 (m, 3H); ^{13}C NMR (50 MHz, CDCl$_3$) δ 10.5, 22.6, 69.4, 106.6, 119.1, 123.4, 126.2, 126.6, 127.6, 128.8, 129.3, 134.6, 156.7; IR (KBr, cm^{-1}) 3060, 2976, 2934, 2908, 2878, 1948, 1904, 1832, 1628, 1600, 1513, 1479, 1467, 1448, 1441, 1392, 1371, 1357, 1270, 1260, 1216, 1146, 1121, 1046, 1024, 1017, 1017, 986, 958, 916, 842, 819, 770, 746, 727, 727, 645, 623, 474; MS (EI) m/z 187.10 (M)$^+$.

2-Isopropoxynaphthalene

^{1}H NMR (400 MHz, CDCl$_3$) δ 1.45 (d, J = 6.0 Hz, 6H), 4.66–4.82 (m, 1H), 7.13–7.18 (m, 2H), 7.28–7.48 (m, 2H), 7.68–7.82 (m, 3H); ^{13}C NMR (50 MHz, CDCl$_3$) δ 22.0, 69.8, 108.4, 119.7, 123.4, 126.2, 126.6, 127.6, 128.8, 129.3, 134.6, 155.7; IR (KBr, cm^{-1}) 3580, 2981, 2936, 2876, 1948, 1904, 1832, 1628, 1600, 1581, 1510, 1468, 1440, 1388, 1373, 1356, 1334, 1216, 1188, 1171, 1137, 1118, 1019, 974, 941, 900, 871, 842, 814, 675, 645, 623, 532; MS (EI) m/z 187.10 (M)$^+$.

1-Chloro-2-methoxynaphthalene

^{1}H NMR (400 MHz, CDCl$_3$) δ 4.03 (s, 3H), 7.25 (d, J = 8.6 Hz, 1H), 7.37–7.52 (m, 1H), 7.53–7.61 (m, 1H), 7.79–7.69 (m, 2H), 8.25 (d, J = 8.6 Hz, 1H); ^{13}C NMR (50 MHz, CDCl$_3$) δ 57, 105.9, 113.7, 124.4, 126.2, 127.8, 128.2, 129.1, 129.9, 133.2, 153.8; IR (KBr, cm^{-1}) 3430, 3049, 2972, 2948, 2846, 2541, 1948, 1763, 1625, 1590, 1505, 1469, 1355, 1337, 1273, 1246, 1187, 1148, 1068, 1018, 985, 894, 865, 804, 764, 740, 657, 588, 532; MS (EI)m/z 192 (M)$^+$.

1-Bromo-2-methoxynaphthalene

^{1}H NMR (400 MHz, CDCl$_3$) δ 4.04 (s, 3H), 7.28 (d, J = 8.6 Hz, 1H), 7.39–7.53 (m, 1H), 7.53–7.61 (m, 1H), 7.76–7.89 (m, 2H), 8.23 (d, J = 8.6 Hz, 1H); ^{13}C NMR (50 MHz, CDCl$_3$) δ 57.1, 105.9, 113.7, 124.4, 126.2, 127.8, 128.2, 129.1, 129.9, 133.2, 153.8; IR (KBr, cm^{-1}) 3430, 3045, 2970, 2941, 2841, 1620, 1594, 1500, 1466, 1454, 1351, 1334, 1270, 1245, 1185, 1153, 1134, 1061, 1021, 968, 890, 855, 803, 761, 743, 708, 644, 579, 516 cm^{-1}; MS (EI) m/z (M-H)$^+$ 236.10.

1-Iodo-2-methoxynaphthalene

^{1}H NMR (400 MHz, CDCl$_3$) δ 4.03 (s, 3H), 7.22 (d, J = 9.0 Hz, 1H), 7.34–7.42 (m, 1H), 7.50–7.56 (m, 1H), 7.75 (d, J = 8.6 Hz, 1H), 7.84 (d, J = 9.0 Hz, 1H), 8.15 (d, J = 8.6 Hz, 1H); ^{13}C NMR (50 MHz, CDCl$_3$) δ 57.3, 87.8, 113.0, 124.4, 128.2, 128.3, 130.4, 131.3, 130.0, 135.7, 156.7; IR (KBr, cm^{-1}) 3042, 3006, 2969, 2937, 2838, 1617, 1587, 1551, 1497, 1451, 1423, 1346, 1328, 1263, 1242, 1181, 1153, 1132, 1058, 1021, 959, 887, 801, 761, 743; MS (EI) m/z 285.06 (M)$^+$.

1-Bromo-2-ethoxynaphthalene

^{1}H NMR (400 MHz, CDCl$_3$) δ 1.38–1.41 (t, J = 6.8 Hz, 3H), 4.13–4.14 (q, J = 6.8 Hz, 2H), 7.15–7.17 (d, J = 8.8 Hz, 1 H), 7.29–7.36 (m, 2H), 7.43–7.47 (m, 1H), 7.78–7.83 (m, 3H); ^{13}C NMR (50 MHz, CDCl$_3$) δ 22.8, 64.6, 105.9, 113.7, 124.4, 126.2, 127.8, 128.2, 129.1, 129.9, 133.2, 153.8.

1-Bromo-2-propoxynaphthalene

^{1}H NMR (400 MHz, CDCl$_3$) δ 1.03 (t, J = 7.5 Hz, 3H), 1.80–1.86 (m, 2H), 4.51 (t, 7.5 Hz, 2H), 7.10–7.14 (m, 2H), 7.30–7.41 (m, 2H), 7.69–7.74 (m, 3H); ^{13}C NMR (50 MHz, CDCl$_3$) δ 10.5, 22.6, 78.4, 113.6, 123.4, 126.2, 126.6, 127.6, 128.8, 129.3, 131.3, 132.6, 154.7; MS (EI) m/z 266.04 (M)$^+$.

1-Bromo-2-isopropoxy Naphthalene

^{1}H NMR (400 MHz, CDCl$_3$) δ 1.45 (d, J = 6.0 Hz, 6H), 4.61–4.87 (m, 1H), 7.28 (d, J = 8.6 Hz, 1H), 7.39–7.53 (m, 1H), 7.53–7.61 (m, 1H), 7.76–7.85 (m, 2H), 8.23 (d, J = 8.6 Hz, 1H); ^{13}C NMR (50 MHz, CDCl$_3$) δ 22.0, 69.8, 108.4, 119.7, 123.4, 126.2, 126.6, 127.6, 128.8, 129.3, 134.6, 155.7; IR (KBr, cm^{-1}) 3580, 2981, 2936, 2876, 1948, 1904, 1832, 1628, 1600, 1581, 1510, 1468, 1440, 1388, 1373, 1356, 1334, 1216, 1188, 1171, 1137, 1118, 1019, 974, 941, 900, 871, 842, 814, 675, 645, 623, 532; MS (EI) m/z 266.10 (M)$^+$.

4-Methoxy Pyridine

^{1}H NMR (400 MHz, CDCl$_3$) δ 3.84 (s, 3H), 6.81 (d, J = 8.6 Hz, 2H), 8.43 (d, J = 8.6 Hz, 1H); ^{13}C NMR (50 MHz, CDCl$_3$) δ 55.1, 109.9, 151.4, 165.4, MS (EI) m/z 110.3(M)$^+$, (B.P.168.2°C).

3-Methoxy Pyridine

^{1}H NMR (400 MHz, CDCl$_3$) δ 3.84 (s, 3H), 7.34 (t, J = 8.65, J = 4.75 Hz, 1H), 7.38 (d, J = 8.65 Hz, 1H); 8.19 (d,J = 4.75 Hz, 1H); 8.32 (s, 1H); MS (EI) m/z 110.3(M)$^+$, (B.P.168.4°C).

2-Methoxy Pyridine

^{1}H NMR (400 MHz, CDCl$_3$) δ 3.94 (s, 3H), 6.72 (d, J = 8.65 Hz, 1H); 6.82 (t, J = 8.65, J = 4.75 Hz, 1H), 7.59 (t, J= 8.65, J = 4.75 Hz 1H); 8.18 (d, J = 4.75 Hz, 1H); ^{13}C NMR (50 MHz, CDCl$_3$) δ 53.1, 110.9, 116.4, 138, 147, 165.4, MS (EI) m/z 110.3 (M)$^+$, (B.P.142.5°C).

4-Ethoxy Pyridine

^{1}H NMR (400 MHz, CDCl$_3$) δ 1.34 (q, 2H), 3.94 (t, 3H), 6.81 (d, J = 8.4 Hz, 2H), 8.43 (d, J = 8.4 Hz, 2H); MS (EI) m/z 123.8(M)$^+$, (B.P.165°C).

6-METHOXYPICOLINALDEHYDE

^{1}H NMR (400 MHz, DMSO) δ 3.971 (s, 3H), 6.986 (d, J = 8.3 Hz, 1H), 8.108–8.136 (m, 1H), 8.770 (d, J = 2.5 Hz, 1H); 9.968 (s, 1H); MS (EI) m/z 137.9(M)$^+$.

5-Bromo-2-methoxypyrimidine

^{1}H NMR (400 MHz, DMSO) δ 3.915 (s, 3H), 8.765 (s, 2H); MS (EI) m/z 188.9(M)$^+$.

1-(5-Bromo-6-methoxynaphthalen-2-yl)hexan-1-one

^{1}H NMR (400 MHz, CDCl$_3$) δ 0.95 (t, J = 6.6 Hz, 3H), 1.38−1.42 (m, 4H), 1.93 (d, J = 7.7 Hz, 2H), 3.03 (t, J = 7.7 Hz, 2H), 4.02 (s, 3H), 7.28 (d, J = 8.7 Hz, 1H), 7.98 (d, J = 8.7 Hz, 1H), 8.04 (s, 1H), 8.23 (d, J = 8.7 Hz, 1H), 8.41 (d, J = 8.7 Hz, 1H); IR (KBr) 3430, 3057, 3016, 2957, 2922, 2882, 2126, 1943, 1716, 1672, 1621, 1559, 1492, 1472, 1440, 1408, 1357, 1328, 1272, 1252, 1219, 1177, 1065, 965, 908, 863, 822, 801, 768, 732, 663, 596, 520, 467 cm^{-1}; MS (ES) m/z (337.0 M)$^{+2}$.

ACKNOWLEDGMENTS

The authors are thankful to Professor P. K. Saiprakash for constant encouragement and Heads of the Chemistry Department at Osmania University and Nizam College, Hyderabad for providing facilities.

REFERENCES

1. N. B. Singh, R. J. Singh, and N. P. Singh, "Organic solid state reactivity," Tetrahedron, vol. 50, no. 22, pp. 6441–6493, 1994. View at Publisher

2. G. R. Desiraju, "Reactivity of organic solids," Solid State Ionics, vol. 101–103, no. 1, pp. 839–842, 1997.

3. K. Tanaka and F. Toda, "Solvent-free organic synthesis," Chemical Reviews, vol. 100, no. 3, pp. 1025–1074, 2000.

4. R. A. Sheldon, "Green solvents for sustainable organic synthesis: state of the art," Green Chemistry, vol. 7, no. 5, pp. 267–278, 2005.

5. A. L. Garay, A. Pichon, and S. L. James, "Solvent-free synthesis of metal complexes," Chemical Society Reviews, vol. 36, no. 6, pp. 846–855, 2007.

6. A. Orita, L. Jiang, T. Nakano, N. Ma, and J. Otera, "Solventless reaction dramatically accelerates supramolecular self-assembly," Chemical Communications, no. 13, pp. 1362–1363, 2002.

7. F. Toda, K. Tanaka, and S. Iwata, "Oxidative coupling reactions of phenols with FeCl3 in the solid state,"Journal of Organic Chemistry, vol. 54, no. 13, pp. 3007–3009, 1989.

8. F. Toda, K. Kiyoshige, and M. Yagi, "NaBH$_4$ reduction of ketones in the solid state," Angewandte Chemie International Edition in English, vol. 28, pp. 320–321, 1989.

9. V. P. Balema, J. W. Wiench, M. Pruski, and V. K. Pecharsky, "Mechanically induced solid-state generation of phosphorus ylides and the solvent-free

wittig reaction," Journal of the American Chemical Society, vol. 124, no. 22, pp. 6244–6245, 2002.

10. J. M. Harrowfield, R. J. Hart, and C. R. Whitaker, "Magnesium and aromatics: mechanically-induced grignard and McMurry reactions," Australian Journal of Chemistry, vol. 54, no. 7, pp. 423–425, 2001.

11. Z. X. Wang and L. Q. Hue, "Solventless syntheses of pyrazole derivatives," Green Chemistry, vol. 6, no. 2, pp. 90–92, 2004.

12. V. M. Markhele, S. A. Sadaphal, and M. S. Shingare, "An efficient one-pot synthesis of polyhydroquinolines at room temperature using MCM-41 catalyst under solvent-free conditions,"Bulletin of the Catalysis Society of India, vol. 6, pp. 125–131, 2007.

13. K. C. Rajanna, K. Ramesh, S. Ramgopal, S. Shylaja, P. G. Reddy, and P. K. Saiprakash, "Poly ethylene glycols as efficient media for the synthesis of β-nitro styrenes from α, β-unsaturated carboxylic acids and metal nitrates under conventional and non-conventional conditions," Green and Sustainable Chemistry, vol. 1, pp. 132–148, 2011.

14. S. Ramgopal, K. Ramesh, A. Chakradhar, N. M. Reddy, and K. C. Rajanna, "Metal nitrate driven nitro Hunsdiecker reaction with α,β-unsaturated carboxylic acids under solvent-free conditions," Tetrahedron Letters, vol. 48, no. 23, pp. 4043–4045, 2007.

15. A. Chakradhar, R. Roopa, K. C. Rajanna, and P. K. Saiprakash, "Vilsmeier-haack bromination of aromatic compounds with KBr and N-bromosuccinimide under solvent-free conditions," Synthetic Communications, vol. 39, no. 10, pp. 1817–1824, 2009.

16. M. Venkateswarlu, K. C. Rajanna, M. S. Kumar, U. U. Kumar, S. Ramgopal, and P. K. Saiprakash, "Rate enhancements in the acetlation and benzoylation of certain aromatic compounds with vilsmeier-haack reagents using acetamide, benzamide and oxychlorides under non-conventional conditions,"International Journal of Organic Chemistry, vol. 1, pp. 233–241, 2011.

17. M. Pagliaro, R. Ciriminna, H. Kimura, M. Rossi, and C. D. Pina, "From glycerol to value-added products," Angewandte Chemie International Edition, vol. 46, no. 24, pp. 4434–4440, 2007. · ·

18. N. Baggett, Comprehensive Organic Chemistry, vol. 1 of Edited by D. Barton, W. D. Ollis and J. F. Stoddart, Pergaman, Oxford, UK, 1979.

19. J. March, Advanced Organic Chemistry, Reactions, Mechanism and Structure, John Wiley & Sons, New York, NY, USA, 4th edition, 1992.

20. S. Kim, K. N. Chung, and S. Yang, "Direct synthesis of ethers via zinc chloride mediated etherification of alcohols in dichloroethane," Journal of Organic Chemistry, vol. 52, no. 17, pp. 3917–3919, 1987.

21. L. Karas and W. J. Piel, Kirk-Othmer Encyclopedia of Chemical Technology, vol. 9, John Wiley & Sons, New York, NY, USA, 4th edn edition, 1992.

22. P. Salehi, N. Iranpoor, and F. K. Behbahani, "Selective and efficient alcoholyses of allylic, secondary- and tertiary benzylic alcohols in the presence of iron(III)," Tetrahedron, vol. 54, no. 5-6, pp. 943–948, 1998.

23. J. Pozniczek, A. Micek-llnicka, A. Lubanska, and A. Bielanski, "Catalytic synthesis of ethyl-tert-butyl ether on Dawson type heteropolyacid," Applied Catalysis A, vol. 286, pp. 52–60, 2005.

24. T. Ollevier and T. M. Mwene-Mbeja, "Bismuth triflate catalyzed Claisen rearrangement of allyl naphthyl ethers," Tetrahedron Letters, vol. 47, no. 24, pp. 4051–4055, 2006.

25. K. C. K. Swamy, N. N. B. Kumar, E. Balaraman, and K. V. P. P. Kumar, "Mitsunobu and related eactions: advances and applications," Chemical Reviews, vol. 109, pp. 551–2651, 2009.

26. M. Debabrata and S. L. Buchwald, "Cu-catalyzed arylation of phenols: synthesis of sterically hindered and heteroaryl diaryl ethers," Journal of Organic Chemistry, vol. 75, no. 5, pp. 1791–1794, 2010.

27. M. D. Maiti, "Chemoselectivity in the Cu-catalyzed O-arylation of phenols and aliphatic alcohols,"Chemical Communications, vol. 47, no. 29, pp. 8340–8342, 2011.

28. G. A. Olah, Friedel-Crafts Chemistry, Wiley-Interscience, New York, NY, USA, 1973.

29. T. Yamato, C. Hideshima, G. K. S. Prakash, and G. A. Olah, "Organic reactions catalyzed by solid superacids. 5. Perfluorinated sulfonic acid resin (Nafion-H) catalyzed intramolecular Friedel-Crafts acylation," Journal of Organic Chemistry, vol. 56, no. 12, pp. 3955–3957, 1991.

30. T. Mukaiyama, T. Ohno, T. Nishimura, and S. Suda, "The catalytic friedel-crafts acylation reaction using a catalyst generated from GaCl3 and a silver salt," Chemistry Letters, pp. 1059–1062, 1991.

31. T. Mukaiyama, K. Suzuki, J. S. Han, and S. Kobayashi, "A novel catalyst system, antimony(V) chloride-lithium perchlorate (SbCl5-LiClO4), in the friedel-crafts acylation reaction," Chemistry Letters, pp. 435–438, 1992.

32. A. Kawada, S. Mitamura, and S. Kobayashi, "Lanthanide

trifluoromethanesulfonates as reusable catalysts: catalytic Friedel-Crafts acylation," Journal of the Chemical Society, Chemical Communications, no. 14, pp. 1157–1158, 1993.

33.   P. A. Evans, J. D. Nelson, and A. L. Stanley, "Directed lithiation/ transmetalation approach to palladium-catalyzed cross-coupling acylation reactions," Journal of Organic Chemistry, vol. 60, no. 7, pp. 2298–2301, 1995.

34.   S. H. Kim, M. V. Hanson, and R. D. Rieke, "Direct formation of organomanganese bromides using Rieke manganese," Tetrahedron Letters, vol. 37, no. 13, pp. 2197–2200, 1996.

35.   J. H. Fendler and E. J. Fendler, Catalysis in Micellar and Macromolecular Systems, Academic Press, London, UK, 1975.

36.   F. M. Menger, J. U. Rhee, and H. K. Rhee, "Applications of surfactants to synthetic organic chemistry,"Journal of Organic Chemistry, vol. 40, no. 25, pp. 3803–3805, 1975.

37.   P. T. Anastas and J. C. Warner, Green Chemistry: Theory and Practice, Oxford University Press, New York, NY, USA, 1998.

38.   L. M. Wang, N. Jiao, J. Qiu et al., "Sodium stearate-catalyzed multicomponent reactions for efficient synthesis of spirooxindoles in aqueous micellar media," Tetrahedron, vol. 66, no. 1, pp. 339–343, 2010.

39.   Y. Watanabe, K. Sawada, and M. Hayashi, "A green method for the self-aldol condensation of aldehydes using lysine," Green Chemistry, vol. 12, no. 3, pp. 384–386, 2010.

40.   H. Firouzabadi, N. Iranpoor, S. Kazemi, A. Ghaderi, and A. Garzan, "Highly efficient halogenation of organic compounds with halides catalyzed by cerium(III) chloride heptahydrate using hydrogen peroxide as the terminal oxidant in water," Advanced Synthesis and Catalysis, vol. 351, no. 11-12, pp. 1925–1932, 2009.

41.   http://www.molinspiration.com.

42.   C. A. Lipinski, F. Lombardo, B. W. Dominy, and P. J. Feeney, "Experimental and computational approaches to estimate solubility and permeability in drug discovery and development settings,"Advanced Drug Delivery Reviews, vol. 23, no. 1–3, pp. 3–25, 1997.

43.   D. F. Veber, S. R. Johnson, H. Y. Cheng, B. R. Smith, K. W. Ward, and K. D. Kopple, "Molecular properties that influence the oral bioavailability of drug candidates," Journal of Medicinal Chemistry, vol. 45, no. 12, pp. 2615–2623, 2002.

44.   P. Ertl, B. Rohde, and P. Selzer, "Fast calculation of molecular polar

surface area as a sum of fragment-based contributions and its application to the prediction of drug transport properties," Journal of Medicinal Chemistry, vol. 43, no. 20, pp. 3714–3717, 2000.

CITATION

CHAPTER 1

Wagner, M. , Contie, Y. , Ferroud, C. and Revial, G. (2014) Enantioselective Aldol Reactions and Michael Additions Using Proline Derivatives as Organocatalysts. International Journal of Organic Chemistry, 4, 55-67. doi:10.4236/ijoc.2014.41008.

CHAPTER 2

Claudia Piutti, and Francesca Quartieri, The Piancatelli Rearrangement: New Applications for an Intriguing Reaction, doi:10.3390/molecules181012290

CHAPTER 3

Maurizio D'Auria, and Rocco Racioppi, Oxetane Synthesis through the Paternò-Büchi Reaction, doi:10.3390/molecules180911384

CHAPTER 4

K. Ishimaru, D. Maeda, K. Ono and Y. Tanimura, "Mannich-Type Reactions of Aldimines and Hetero Diels-Alder Reactions of Aldehydes Catalyzed by Anion-Type Lewis Bases Derived from a Single Molecule," International Journal of Organic Chemistry, Vol. 2 No. 3, 2012, pp. 188-193. doi: 10.4236/ijoc.2012.23028.

CHAPTER 5

Gianluca Papeo, and Maurizio Pulici, Italian Chemists' Contributions to Named Reactions in Organic Synthesis: An Historical Perspective, doi:10.3390/molecules180910870

CHAPTER 6

El Sayed H. El Ashry, Hamida Abdel Hamid, Ahmed A. Kassem and Mahmoud Shoukry, Synthesis and Reactions of Acenaphthenequinones-Part-2. The Reactions of Acenaphthenequinones, doi:10.3390/70200155

CHAPTER 7

Isak Rajjak Shaikh, "Organocatalysis: Key Trends in Green Synthetic Chemistry, Challenges, Scope towards Heterogenization, and Importance from Research and Industrial Point of View," Journal of Catalysts, vol. 2014, Article ID 402860, 35 pages, 2014. doi:10.1155/2014/402860.

CHAPTER 8

Kancharla Rajendar Reddy, Kamatala Chinna Rajanna, Kusampally Uppalaiah, Mukka Satish Kumar, and Marri Venkateshwarlu, "Environmentally Benign Mortar-Pestle-Induced Acylation and O-Alkylation of Aromatic and Heteroaromatic Compounds under Solvent-Free Micellar Conditions and Computation of Their Drug Likeliness Properties," Organic Chemistry International, vol. 2012, Article ID 647386, 10 pages, 2012. doi:10.1155/2012/647386.

INDEX

C

C4-O 79
Catalysed epimerisation 70
Cetyl trimethyl ammonium bromide
(CTAB) 281
Cetyl trimethyl ammonium chloride
(CTAC) 281
Corresponding oxetane 49, 53, 57, 63,
64, 65, 66, 86

D

Dichloroethane (DCE) 41
Diethyl tartrate (DET) 149
Donor-acceptor (D-A) 33

E

Electron donating 62
Epimeric mixture 82

F

Fluidized catalytic cracking (FCC) 208

I

Iodonium di-sym-collidine perchlorate
(IDCP) 71

L

Layered double hydroxides (LDH) 238

N

N-heterocyclic carbene (NHC) 227, 267
Nuclear magnetic resonance (NMR) 240

O

Organization of the Arab Petroleum Exporting Countries (OAPEC) 146

P

Phase transfer catalysis (PTC) 224
Phosphomolybdic acid (PMA) 33
Photochemical coupling 52, 70, 86
Polar surface area (PSA) 287
Polycyclic compounds 181

S

Scanning electron microscopy (SEM)
240
Sodium dodecyl sulphate (SDS) 281
Spin-orbit coupling (SOC) 54

T

Thermal cyclization 181
Transmission electron microscopy
      (TEM) 240

U

UV radiation 79

X

X-ray diffraction (XRD) 240